Quantitative Genetic Studies of Behavioral Evolution

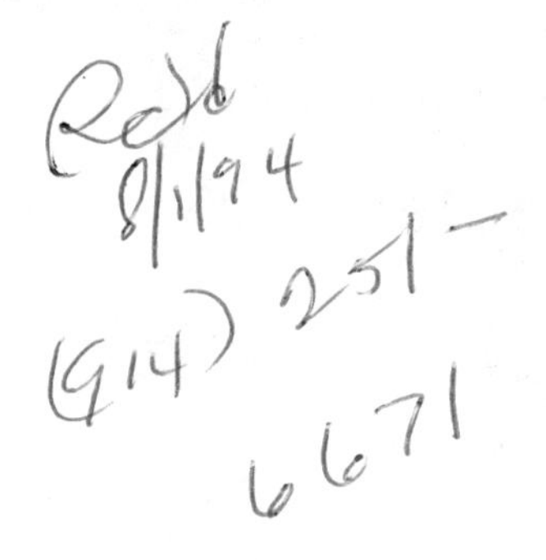

Quantitative Genetic Studies of Behavioral Evolution

Edited by Christine R. B. Boake

The University of Chicago Press *Chicago and London*

Christine R. B. Boake is an associate professor in the Department of Zoology and in the Programs in Ecology and Ethology at the University of Tennessee, Knoxville.

The University of Chicago Press, Chicago 60637
The University of Chicago Press, Ltd., London

Printed in the United States of America

03 02 01 00 99 98 97 96 95 94   1 2 3 4 5
ISBN: 0-226-06215-5 (cloth)
0-226-06216-3 (paper)

Library of Congress Cataloging-in-Publication Data

Quantitative genetic studies of behavioral evolution / Christine R. B. Boake, editor.
p. cm.
Includes bibliographical references and index.
1. Behavior genetics. 2. Behavior evolution. 3. Quantitative genetics. I. Boake, Christine R. B.
QH457.Q36 1994
591.5′1—dc20   93-33824
CIP

♾ The paper used in this publication meets the minimum requirements of the American National Standard for Information Sciences—Permanence of Paper for Printed Library Materials, ANSI Z39.48-1984.

For my parents

Contents

Applications of Quantitative Genetics to Studies of Behavioral Evolution in Natural Populations

Conclusions

Preface

This book is designed for the use of behavioral ecologists who wish to learn how quantitative genetics might be applied to studies of behavioral evolution. The idea to publish it began with a symposium that Ann Hedrick and I organized for the 1990 Animal Behavior Society meetings at the State University of New York at Binghamton. Because I wanted to edit a book that would have greater breadth and depth than the symposium, I invited a number of persons who had not spoken at the symposium to write chapters.

I envision the book as being used for graduate seminars, or as a reference for someone wanting an entrance to the literature. I did not want the book to be intensely statistical, but I wanted readers to have access to the relevant statistical works. Accordingly, I asked the authors to outline their reasons for using particular breeding designs, and to provide extensive references to the primary literature. The chapters are written with the assumption that readers have some familiarity with analysis of variance and with regression analyses, and that they have access to a copy of Falconer's (1989) text, in which the details of obtaining genetic information from various statistical analyses are laid out. This book provides concrete examples of how quantitative genetic procedures and theory are being applied to a variety of behavioral problems.

Each chapter of the book was reviewed by two scientists: a quantitative geneticist who was familiar with the particular behavioral issues, and a behavioral ecologist who took the role of a prospective user of the book. All reviewers had the option of being anonymous. Some of the authors served as reviewers for each others' chapters, and in addition to them, I wish to thank the following persons for their careful reviews: Felix Breden, Buck Buchanan, John Damuth, Susan Foster, John Gittleman, Charles Goodnight,

Patty Gowaty, Zuleyma Tang Martinez, Rick Howard, Lyle Konigsberg, Russ Lande, Dave McCauley, Frank Messina, Tim Mousseau, Don Price, Susan Riechert, Jim Schwartz, Steve Shuster, Mike Wade, Bruce Waldman, Jerry Waldvogel, Marta Wells, Kent Wells, and Jerry Wilkinson. The book was the topic of a seminar that I offered in the spring of 1992, and comments from the seminar's participants were valuable in guiding my suggestions for the final revisions and in developing the index; I thank Andrea Bixler, Stan Guffey, Luke Hasty, Don Price, and Teresa Roberts for their input. The University of Chicago Press sent the entire manuscript to two reviewers who were anonymous to us. One reviewer, Felix Breden, has agreed to forgo anonymity. Both reviewers' comments were exceedingly helpful in guiding our revisions.

Some other individuals deserve special thanks. Stim Wilcox's suggestion that we organize the symposium was the impetus that resulted in the book. Mark Courtney provided additional encouragement. Susan Abrams's enthusiasm and encouragement throughout the editing process was essential to the book's completion. Finally, my interest in using quantitative genetics to study behavioral evolution began while I was a postdoc at the University of Chicago, where I was able to have extensive interactions with Steve Arnold, Russ Lande, and Mike Wade. In particular, I was based in Mike Wade's lab, and the education that he gave me, directly and by example, has formed the foundation for all my forays into evolutionary genetics, including this one.

Chris R. B. Boake

Introduction

1

Outlining the Issues

Christine R. B. Boake

The study of processes by which populations change from one average phenotype to another over time is one of the major topics in evolutionary biology. Research into evolutionary processes involves identifying and measuring selective factors and evaluating genetic variation. The study of behavioral evolution has largely involved analyses of selection on behavior, with relatively little emphasis on genetics. Evolutionary theory and empirical studies of many kinds of characters have demonstrated that both the rate and direction of evolution can be strongly influenced by genetic factors. In extreme cases, genetic influences can move a trait in a direction opposite to that predicted by the results of analyses of selection. Genetic studies of behavior could be based on single-locus models and analyses of mutants at single loci, but behavioral traits are exceedingly well suited for analyses with the methods of quantitative genetics.

Quantitative genetics is the study of the inheritance of continuously varying traits, as opposed to traits that vary in a qualitative fashion within a population. These traits are usually polygenically controlled; that is, their expression is influenced by many genes. Consequently, many of the methods of population and molecular genetics, which focus on identifying a particular locus that has a major effect on a trait, are inappropriate for quantitatively varying traits. The methods that form the basis of quantitative genetics rely on measuring the phenotypes of related and unrelated groups of individuals in a population and determining the degree to which relatives resemble one another. These methods were developed as part of the Modern Synthesis by Fisher, Wright, and Haldane, and formed the substance of the reconciliation between Mendelian and Darwinian approaches to evolution (Provine 1971). However, as Darwin (1859) pointed out, artificial selection had been used for millenia to change both the morphology and behavior of domestic species; for most of that time, the methods were qualitative in nature, and perhaps unconscious. The quantitative methods were further developed in this century to assist in improving livestock and crops, and because of their major economic importance, have received extensive attention from both theoreticians and empiricists. The methods were not applied extensively to evolutionary problems for many decades, despite their original development in an evolutionary context. Quantitative genetics began to be applied to evolutionary problems largely as a result of Lande's efforts, beginning in 1976, which stimulated many behavioral ecologists to apply quantitative genetics to their research.

Many of those who use quantitative genetics in behavioral studies began to do so well before Lande began publishing. The procedures have been used by psychologists in studies of behavior for decades (Fuller and Thompson 1960), and were being used for analyses of the evolution of behavior long before the recent spate of models (Parsons 1973; Ehrman and Parsons 1976, 1981). The early evolutionary applications of behavioral genetics tended to use small mammals and *Drosophila.* These studies took place before behavioral ecology was developed into a major component of evolutionary biology, and many had their origins in comparative psychology. Tolman (1924) selected rats for maze-learning ability, and Tryon (1934) explained such research within the context of evolutionary theory. Scott conducted an extremely large study of the inheritance of behavioral differences between different breeds of dogs, with the goal of modeling the evolution of differences in social interactions (Scott and Fuller 1965). Studies of the behavior of ducks and their hybrids were used to develop a tree that approximated their phylogeny and to demonstrate that Mendelian principles could be applied to behavioral differences between species, thus validating the use of behavior in phylogenies (Lorenz 1958; Sharpe and Johnsgard 1966).

Studies of *Drosophila* behavior have incorporated genetic analyses for decades. Manning (1961, 1963) applied artificial selection to studies of mating speed in flies and showed that a natural behavior would respond to selection. Ewing (1961) used lines that had been selected for body size to understand the role of acoustic signals in male mating success, a question that is important in many modern studies of sexual selection. Ehrman (1960, 1961) demonstrated that behavioral reproductive isolation between species of *Drosophila* was under polygenic control; her focus was on the failure of pairs to mate rather than on detailed analyses of behavioral differences between populations.

Hirsch and his colleagues (Hirsch and Erlenmeyer-Kimling 1962) used artificial selection to change the phototactic and geotactic behavior of flies, and Hirsch's recognition of the importance of genetic studies in understanding the evolution of behavior can be seen by the invited papers in symposia that he organized (Hirsch 1967). His selected lines are now being used in evolutionary studies to evaluate ways in which gene interactions may have changed (Ricker and Hirsch 1988). Hirsch and Erlenmeyer-Kimling (1962) showed that the differences in behavior between a selected line and a control line were under the control of genes on several chromosomes, thus demonstrating the polygenic nature of behavioral traits. Hirsch and Erlenmeyer-Kimling's paper is the first demonstration for a behavioral trait of this fundamental assumption of quantitative genetics.

The application of quantitative genetics to behavioral ecology has grown substantially over the last ten years. This corresponds, in large part, to the appearance of Lande's models and Arnold's illustrations of the value of a quantitative genetic approach to behavioral biologists. Their treatment of sexual selection (Lande 1981b; Arnold 1983b) has been particularly influential, stimulating both research and debate (e.g., Bradbury and Andersson 1987). However, it is perplexing that evolutionary applications of behavioral genetics were so rare before about 1980.

Many behavioral traits vary quantitatively and are thus appropriate for study by the methods of quantitative genetics. Some examples discussed in this book are the amount of insulation a mouse puts into its nest (Lynch, chap. 13) and the tendency of a bird to display migratory behavior (Dingle, chap. 7). On the other hand, some species' differences, such as courtship signals in *Schizocosa* spiders, appear to show simple Mendelian inheritance (Stratton and Uetz 1986), as do certain inherited behavioral defects in mice, such as waltzing (Plomin, DeFries, and McClearn 1980). However, even apparently qualitatively varying traits can be under polygenic control, and can be successfully studied with the techniques of quantitative genetics. For example, Maynard Smith and Riechert (1984) proposed a two-locus model to explain differences in territory size between two populations of the spider *Agelenopsis aperta*, but their more recent data suggest that the two genetically based behavioral tendencies ("fear" and "aggression") of their original model are probably better explained as due to polygenic inheritance (Riechert and Maynard Smith 1989). In all animal species, the vast majority of behavioral acts are probably under polygenic control, and thus quantitative genetic approaches to their study are appropriate.

In contrast to other genetic analyses, quantitative genetic approaches focus on the phenotype rather than on allele frequencies. This focus has several consequences. First, one need not identify the genes influencing an action pattern, a behavioral strategy, or any other trait. Second, the study of the phenotype leads to an analysis of evolutionary processes as being the result of the combination of selection and inheritance (Fisher 1958). Selection acts on the phenotype within a generation, and inheritance allows the transmission of change to the next generation. This book emphasizes the study of inheritance, but the interaction between inheritance and selection will be seen in many of the chapters, and Partridge's discussion (chap. 6) of the issues involved in measuring sexual selection can be generalized to selection on many kinds of behavior.

The Scope of Quantitative Genetic Analyses of Behavioral Evolution

The formal goals of those using theoretical quantitative genetics in evolutionary biology are to predict phenotypic evolution and to infer rates and directions of past evolutionary processes (Lande 1976a; Heisler, chap. 5). This body of theory is being applied to studies of behavioral evolution by numerous scientists, using the techniques outlined in Arnold's review (chap. 2). Arnold describes how both heritability analyses and the measurement of the direction and magnitude of genetic relationships between traits can be used to predict the rate and direction of evolution. Roff (chap. 3) discusses recent theory that shows how complicated the interpretation of genetic correlations can get (Charlesworth 1990). Studies of the magnitude of genetic variation and covariation within a population, illustrated by work by Dingle (chap. 7), Hoffmann (chap. 9), Hedrick (chap. 11), and Garland (chap. 12), are the beginning of empirical searches for indications that a population has the potential to evolve in response to selection. Studies of variation between populations in the nature and magnitude of genetic parameters can lead to interpretations regarding adaptation, described in this book in the chapters by Dingle (chap. 7), Travis (chap. 8), Hoffmann (chap. 9), Stevens (chap. 10), and Lynch (chap. 13).

Since the publication of quantitative genetic models for behavioral evolution in the 1980s, the term "quantitative genetics" has often been used to refer to these models. The authors in this book discriminate between two classes of models. First are "quantitative genetic" models of evolution, which incorporate estimates of both selection on and inheritance of characters in order to predict evolutionary trajectories. Second are "phenotypic selection" models, which focus on selection within a generation and which do not make evolutionary predictions. Quantitative genetic models have been a source of novel and sometimes controversial ideas about the processes of behavioral evolution; some of their results are described by Arnold (chap. 2), Cheverud and Moore (chap. 4), and Heisler (chap. 5). Serious debates continue concerning the validity of certain assumptions of quantitative genetic models. In particular, the assumption that selection is sufficiently weak for ongoing mutation to maintain genetic variation in a population is controversial (Barton and Turelli 1989; Arnold, chap. 2; Roff, chap. 3). Yet if the models eventually prove to be uninformative for studies of long-term behavioral evolution, quantitative genetics will still be valuable to behavioral ecologists. Important applications of quantitative genetics that are independent of the models include the study of adaptation (Stevens, chap. 10; Lynch, chap. 13), the understanding of the genetic bases of variation within and between populations (Travis, chap. 8; Hoffmann, chap. 9; Stevens, chap. 10; Lynch, chap. 13), the testing of hypotheses derived from other kinds of models (Roff, chap. 3; Partridge, chap. 6), and the elucidation of mechanisms underlying behavioral variation (Travis, chap. 8; Hoffmann, chap. 9; Garland, chap. 12).

Skeptics recurrently raise the question, "Why study the inheritance of behavior?" because they believe that Fisher's fundamental theorem obviates the necessity. The argument goes as follows: Fisher stated as his fundamental theorem of natural selection that "the rate of increase in fitness of any organism at any time is equal to its genetic variance in fitness at that time" (Fisher 1958, 37). This statement means that if a population is at equilibrium (i.e., fitness is not increasing), there will be no genetic variance in fitness. It has been extrapolated to assert that furthermore there will be no genetic variance in components of fitness, and that because many behavioral traits such as foraging behavior and mating behavior obviously contribute to fitness, they should also show no genetic variance. It has also been argued that if genetic variance for a trait is low, that trait must be a major component of fitness. However, such a straightforward interpretation of Fisher's conclusion is usually unrealistic.

First, a variety of conditions, such as mutation or variation of selection in space or time, may allow total fitness to maintain some genetic variation (Charlesworth 1987). Although it may be convenient to assume that a population is in equilibrium, this assumption needs to be tested each time it is applied. Second, a great difficulty with applying Fisher's idea to behavioral characters is that even if total fitness does indeed show a vanishingly small amount of genetic variation, components of fitness may maintain substantial additive genetic variation (Rose 1982; Lande 1982a; Mousseau and Roff 1987). One mechanism for the maintenance of additive genetic variation of a component of fitness is antagonistic pleiotropy, in which a gene has a positive effect on fitness through one trait and a negative effect through another trait (Rose 1982). Furthermore, unless components of fitness are on average positively correlated with each other, any one component may have a zero correlation with total fitness, and the underlying genes will not be correlated with fitness (Wallace 1991).

The converse of Fisher's fundamental theorem, the idea that low genetic variance for a trait implies that the trait must be a major component of fitness, is also extremely weak. The reason is that low heritabilities may be caused by factors other than the continued action of selection (Price and Schluter 1991). In particular, every step between a DNA sequence and the external trait that it affects is also influenced by environmental variation. Behavioral and life history traits, which are influenced by morphological (and other) characters, will receive additional input from environmental variation compared with those characters, and thus will have lower heritabilities. Weak support for behavioral characters having lower heritabilities than morphological characters is given by Mousseau and Roff (1987) and Roff and Mousseau (1987). The message from these arguments is that measuring genetic variation in behavioral characters is indeed worthwhile, although interpreting the results may require sophistication. A number of the chapter authors address problems related to Fisher's fundamental theorem; these points are discussed in the final chapter.

The use of quantitative genetics for studying adaptation is exemplified by the research discussed in Lynch's and Dingle's chapters. If a trait has been selected in order to adapt to particular circumstances, then additive genetic variation for the trait may be very low in the population (Fisher 1958). However, behavioral differences between populations may have a genetic basis. Genetically based population differences can be detected if individuals from both populations that are reared in the same environment (a "common garden") maintain the differences that were observed in the wild (Dingle, chap. 7). The nature of the genetic control of these differences, including the number of genes involved and the degree of dominance, can be used to infer past selection on the traits (Broadhurst and Jinks 1974; Mather and Jinks 1982). This genetic architecture can be evaluated by using sophisticated versions of Mendelian crosses (Mather and Jinks 1982; Stevens, chap. 10; Lynch, chap. 13). These methods have been applied to studies of behavior for several decades (van Abeelen 1974; Plomin, DeFries, and McClearn 1980). At a fundamental level, such analyses describe the past evolution of a population: traits with low heritability but which show strong directional dominance are thought to have been under directional selection in the past (Mather and Jinks 1982; discussed by Lynch, chap. 13). A thorough analysis of genetic architecture sidesteps the difficulties of interpreting Fisher's fundamental theorem, because so much data is available in addition to the additive genetic variation.

The understanding of the genetic bases of variation within and between populations may involve studies of adaptation, as in Lynch's chapter. However, an analysis of the nature of the genetic control of phenotypic variation can be valuable for testing other kinds of hypotheses. Male mating strategies in sailfin mollies are size-dependent, and Travis (chap. 8) has been using quantitative genetic methods to examine variation between populations in absolute body size and its influence on conditional behavior. Stevens' (chap. 10) analyses of variation between populations in cannibalism provides insight into models of the evolution of sociality and of infanticide.

Many people (as argued by Cheverud and Moore, chap. 4) consider optimality models to be more limited than quantitative genetic models, since optimality models make assumptions about adaptation. Roff (chap. 3) shows that both kinds of models can be valuable for understanding evolution, and describes particular kinds of measures that make the models interdependent. Hedrick's research (chap. 11) was designed to evaluate an assumption common to both quantitative genetic and optimality models: that genetic variation must exist in order for a trait to evolve. Hoffmann (chap. 9) has been testing hypotheses about the costs and benefits of territoriality by examining the behavior of lines artificially selected for increased or decreased territorial behavior.

Finally, quantitative genetic techniques can be used to elucidate mechanisms underlying behavioral variation, as Lynch and her colleagues are doing to relate neuroanatomy to thermoregulatory behavior (chap. 13) and as Dingle (1992) is doing to examine life history variation in relation to body size. Garland's (chap. 12) studies of physiology could be applied to discussions of the evolution of prey capture and predator avoidance, and to energy expenditure during mating. Although studies of mechanisms are not directly evolutionary, they are important for understanding possible limitations on the direction and rate of evolution. Mechanisms can also evolve, but the study of the evolution of control of behavior so far has been usually based on comparative, rather than experimental, techniques (but see Huntingford 1993).

A decision to undertake a quantitative genetic study is far from trivial, and anyone who is contemplating such research must understand the limitations of the discipline as well as its potential benefits. Several sticky points will be raised in the following chapters. First, the value of the theory in explaining evolution has been questioned (Barton and Turelli 1989), as discussed by Arnold (chap. 2) and Roff (chap. 3). Second, the theory is not necessarily the best way to understand the evolution of behavior, although it provides a powerful additional tool (Roff, chap. 3; Partridge, chap. 6). Finally, the methods of quantitative genetics are suitable for only some studies: major limitations include the sample sizes needed and the question of how well laboratory studies model natural processes (Arnold, chap. 2; Partridge, chap. 6; Boake, chap. 14). These reservations are presented in the text because although the authors of this book all see value in using quantitative genetics, they also understand that the only valuable experiments will be those with adequate sample sizes and clearly defined goals.

Aspects of two major debates will be found in this book. The first, concerning the mutation-selection balance (Arnold, chap. 2), is being conducted by evolutionary geneticists. A detailed description of it is beyond the scope of this book, and interested readers are encouraged to consult the references cited by Arnold and by Roff, starting with Barton and Turelli's (1989) review. The second debate, concerning quantitative genetic versus optimality models, is presented in this book (Roff, chap. 3; Cheverud and Moore, chap. 4). The deep differences between the viewpoints indicate that this second debate deserves a major forum because of its central importance to behavioral ecologists.

Structure of the Book

The insights given by theory and the problems with applying it are the focus of chapters by Arnold (chap. 2), Roff (chap. 3), Cheverud and Moore (chap. 4), Heisler (chap. 5), and Partridge (chap. 6), but many of the other chapters include explicit discussions of data in relation to particular aspects of theory. Theoretical treatments can lead to predictions that are counterintuitive, such as those for the evolution of female mating preferences (Heisler, chap. 5), for the evolution of maternally influenced characters (Cheverud and Moore, chap. 4), and for evolutionary trade-offs (Roff, chap. 3). The concept of trade-offs in behavioral and ecological circumstances is incorporated in the evaluation of genetic correlations between different traits; these correlations will not necessarily be negative where trade-offs occur (Roff, chap. 3). The theory of genotype-environment interactions leads us to realize that individuals with different phenotypes may have the same genetic background, and that individuals with different genetic backgrounds may have similar phenotypes (Arnold, chap. 2; Travis, chap. 8).

Empirical studies make up the rest of the book, in chapters by Dingle (chap. 7), Travis (chap. 8), Hoffmann (chap. 9), Stevens (chap. 10), Hedrick (chap. 11), Garland (chap. 12), and Lynch (chap. 13). In each case, the authors address issues that arise directly from quantitative genetic theory, or from the distinctions between quantitative genetic and other models for the evolution of behavior. Quantitative genetic approaches have allowed behavioral biologists to formalize analyses of selection and to predict the response to selection (Dingle, chap. 7; Garland, chap. 12; Lynch, chap. 13), thus addressing the major issue of identifying those traits that are truly adaptations (Gould and Lewontin 1979). In some cases, quantitative genetics can be used to understand nonintuitive results and to develop new ways of looking at problems: for example, the maintenance of cannibalism in populations can be elucidated (Stevens, chap. 10). Besides giving estimates of heritability, the practice of artificial selection can provide a source of material for further research: selected lines can be examined with respect to ecology, behavior, or physiology (Hoffmann, chap. 9; Lynch, chap. 13). These studies can provide us with an understanding of evolutionary plasticity.

Table 1.1. Cross-listing of chapters in this book, showing topics and animals

Chapter	Author	Type	Breeding design	Behavioral problem	Organism
2	Arnold	Conceptual		Introduction to methods	
3	Roff	Conceptual		Optimality	
4	Cheverud and Moore	Conceptual		Maternal effects, sociality, parental care	
5	Heisler	Conceptual		Sexual selection	
6	Partridge	Conceptual	Measuring selection	Sexual selection	*Drosophila melanogaster*
7	Dingle	Conceptual and empirical	Selection, population crosses	Migration, life histories	Insects, birds
8	Travis	Empirical	Father-son regression	Size-dependent mating behavior	Sailfin mollies
9	Hoffmann	Empirical	Artificial selection	Territoriality	*Drosophila melanogaster*
10	Stevens	Empirical	Population crosses	Cannibalism, sociality	*Tribolium confusum*
11	Hedrick	Empirical	Father-son regression	Sexual selection	Crickets
12	Garland	Empirical	Sib analysis	Speed and stamina	Snakes
13	Lynch	Empirical	Population crosses	Behavioral thermoregulation	*Mus domesticus*

Each empirical chapter includes statistical references, so that a reader can go to the primary literature and learn about the methods in as much detail as necessary. We assume that readers know the fundamentals of regression analysis and analysis of variance; for most purposes in this book, reference to Sokal and Rohlf (1981) should be sufficient to explain statistical principles. For details of the interpretation of statistical results in genetic terms, we refer readers to Falconer's (1989) text, which includes many numerical examples. The computation of causal components of variance and relevant statistical tests are laid out extremely clearly and thoroughly by Becker (1984); unfortunately, in the spring of 1993, Becker's book was out of print. The concept of genetic architecture and the methods for evaluating it, which are used by Stevens (chap. 10) and by Lynch (chap. 13), are developed by Mather and Jinks (1982) and applied to behavioral evolution by Broadhurst and Jinks (1974) and Jinks and Broadhurst (1974).

Two chapters contain material that might be valuable for reference purposes. Arnold (chap. 2) has written an appendix that outlines the basics of matrix algebra. Models of maternal effects have been developed with two different notations; a key to these notations is provided by Cheverud and Moore (chap. 4, table 4.1).

The chapters included were chosen to demonstrate the application of a variety of breeding designs to a variety of behavioral problems in a variety of organisms. Table 1.1 is a guide to the topics in the book, which may help readers who have particular interests. We hope that this book illustrates both the limitations and the scope of quantitative genetics as applied to behavioral evolution.

Acknowledgments

All of the chapter authors discussed the topics in this chapter with me. I particularly thank Ary Hoffmann, Carol Lynch, and two anonymous reviewers for helpful advice.

1

Quantitative Genetic Theory in Relation to Behavioral Evolution

2

Multivariate Inheritance and Evolution: A Review of Concepts

Stevan J. Arnold

A female dusky salamander (*Desmognathus fuscus*), uncovered while she was brooding her eggs. The mass of yolk provided by the female affects the size of the hatchling larvae. Female behaviors that affect yolk size (e.g., feeding behavior) are said to exert maternal effects on hatchling size. Females also protect their developing young by attacking small predators (Forester 1978). Such antipredator behaviors are said to experience maternal selection (Kirkpatrick and Lande 1989).

The aim of this chapter is to review basic concepts in quantitative genetics. Most readers have heard of heritability and the axiom that genetic response to selection is a function of heritability and selection intensity. That axiom is now at least fifty years old, and the discipline of quantitative genetics has matured considerably in the intervening years. One important recent development has been the ability to predict genetic responses to selection acting simultaneously on multiple traits. The development of a useful multivariate theory has attracted the attention of evolutionary biologists, because natural selection inevitably acts on multiple traits. Curiously, heritability no longer holds center stage in the new multivariate theory of inheritance and response to selection. Instead, the most general theory of quantitative genetics is cast in terms of genetic variances and covariances, or in terms of the more basic concept of covariance between additive genetic value and phenotypic value. Heritability's fall from grace and the emergence of a new pantheon of concepts is easiest to understand from a historical perspective.

In the following sections I give a simplified account of conceptual developments in evolutionary quantitative genetics. (For more advanced overviews, see Lande 1988; Barton and Turelli 1989; and Arnold 1992b). Heritability and its antecedents are a good starting point, but my main goal is to outline the spirit and content of recent theoretical developments. Some equations are required to succinctly outline these results; I have tried to keep the number of equations at a minimum. Derivations are referenced and are not produced here. Matrix equations are given in boldface, and some useful rules for interpreting them are given in appendix 2.1.

Short-term Response to Selection

A major goal in quantitative genetics is to predict how selection acting in one generation will produce genetic change in the next generation. A series of response to selection equations has evolved in the discipline. These equations always have the same form: response to selection is a function of selection and inheritance. *Selection* is a phenotypic concept that describes statistical change in the means of traits in the parental generation. *Inheritance* is a genetic concept that describes how the effects of selection are transmitted from the parental generation to the offspring generation. In the sections below I outline a series of progressively more general equations. The whole series can be written in either of two ways: (1) with inheritance described as a covariance and selection described as a regression or (2) with inheritance described as a regression and selection as a covariance. Both formats have their merits, but I have adopted the first format because it makes the continuity between the various models somewhat clearer.

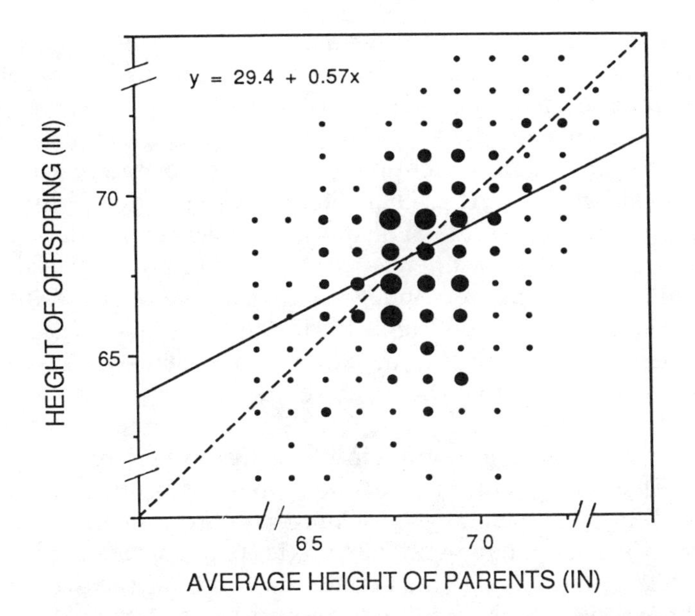

Fig. 2.1 Galton's (1889) data showing the height of adult human offspring (N = 928) as a function of the average height of their parents (N = 205 sets) in a British population. The dashed line describes perfect inheritance. The calculation of the regression slope (solid line) excluded 36 offspring for which the exact heights of offspring and/or parents were not reported. These 36 extreme points are for parents with average heights less than 63.5 inches or greater than 73.5 inches and for offspring less than 61.2 or greater than 74.2 inches in height.

Response to Selection on a Single Trait

Resemblance between offspring and their parents is the fundamental concept in quantitative genetics because it enables us to predict responses to selection. Galton (1889) introduced a graphical portrayal of offspring-parent resemblance that persists to the present day. An example of a Galton plot is shown in figure 2.1, which shows for a human population the height of adult offspring plotted against the average height of their parents. Such data plots convinced Galton's contemporaries and succeeding generations that there were regularities in the inheritance of traits with continuous distributions. Aside from an enduring graphical format, Galton's contribution was to show that a least squares line fitted to his data deviated from the line representing perfect inheritance (fig. 2.1). Offspring of exceptionally tall parents regressed downward toward the offspring mean, whereas offspring of exceptionally short parents regressed upward toward the mean. Consequently, Galton referred to his best fit line as a regression, a term which later took on a more general meaning in the field of statistics. Regression in its original incarnation as a descriptor of offspring-parent resemblance is still the fundamental concept in quantitative genetics.

The problem facing the generations that succeeded Galton was to reconcile the regression of offspring on parents with the emerging laws of Mendelian inheritance. Mendelian roots for Galton's plots and regressions were independently unearthed by Weinberg (1908), Fisher (1918), and Wright (1921). The basic idea is that many genes affect a trait such as human height, but each gene has only a small effect. If we allow the genes to display dominance in their effects on height and work through an algebraic model, we obtain a disconcerting result: it is not simply the genotypic properties of parents that produce resemblance with their offspring. Instead, we need to define a particular genetic property of parents that encodes the transmission of height characteristics to offspring. At the level of a single genetic locus, this property is known as the *average effect* of the gene, which, perhaps not surprisingly, is defined by a regression of phenotypic value on number of gene copies (zero, one, or two). Falconer (1989) gives a lucid account of the algebra. We can now define the elusive genetic property of parents that produces resemblance with offspring. That genetic property is simply the sum across gene loci of the average effects of alleles on height. Because of the summation, the key genetic property of parents has come to be known as *additive genetic value*. In other words, the recurrent adjective "additive" in the field of quantitative genetics heralds the early twentieth-century triumph in grafting Mendelian roots onto Galton's observations, while accounting for dominance in gene action.

Armed with the concept of additive genetic value, we return to Galton's plot. We can now show that Galton's best fit line is the regression of additive genetic values (observed as the average heights of offspring) on the phenotypic values of parents (observed as their measured heights). The regression slope is known as *heritability.* The slope turns out to be equivalent to the ratio of variance in additive genetic values (*additive genetic variance*) to variance in phenotypic values (*phenotypic variance*). We have now assembled the basic vocabulary used in textbook accounts of quantitative inheritance. Our goal, however, is to understand contemporary extensions of quantitative genetics to evolutionary biology.

To achieve those goals it will be helpful to have an explicit model of the concepts we have reviewed.

Our model for the inheritance of a behavior, or any other attribute affected by many genes, is that an individual's phenotypic value is composed of an additive genetic value (which is transmitted to offspring) and a part which is not transmitted to offspring. The nontransmitted part is conveniently referred to as the individual's environmental value, but it is composed of nonadditive genetic effects (due to dominance and epistasis) as well as purely environmental effects. If we denote the phenotypic value as **z**, the additive genetic value as **a**, and the environmental value as **e**, then our model is

$$\mathbf{z} = \mathbf{a} + \mathbf{e} \tag{2.1}$$

(see appendix 2.1). The mean phenotypic value in the population is similarly composed of two parts, and consequently the population variance in phenotypic values, **P**, is the sum of an additive genetic variance, **G**, and a environmental variance, **E**, which includes nonadditive genetic effects. In other words,

$$\mathbf{P} = \mathbf{G} + \mathbf{E}, \tag{2.2}$$

with the assumption that there is no correlation between the additive genetic and environmental values of individuals (see appendix 2.1). We shall return to this assumption later. If we also assume that there is no correlation between the environmental values of parents and their offspring, the slope of the regression line relating offspring values to parental values is $h^2 = G/P$, (h^2 is the standard symbol for heritability).

The primary importance of heritability is that it enables us to predict the genetic consequences of selection that acts on a single trait (e.g., a behavioral attribute). We have discussed the concept of heritability. How shall we conceptualize selection? A useful way to think about selection is to focus on its phenotypic manifestations. Consider the idealized Galton plot shown in figure 2.2a. The projection of the points onto the *x*-axis represents the distribution of phenotypic values for all the potential parents of the next generation. Suppose that only a subset of potential parents actually produce offspring (fig. 2.2b). The actual parents are a biased subset of all potential parents by virtue of selection acting on the phenotypic trait graphed on the *x*-axis. The difference between the mean value of actual parents and the mean value of all potential

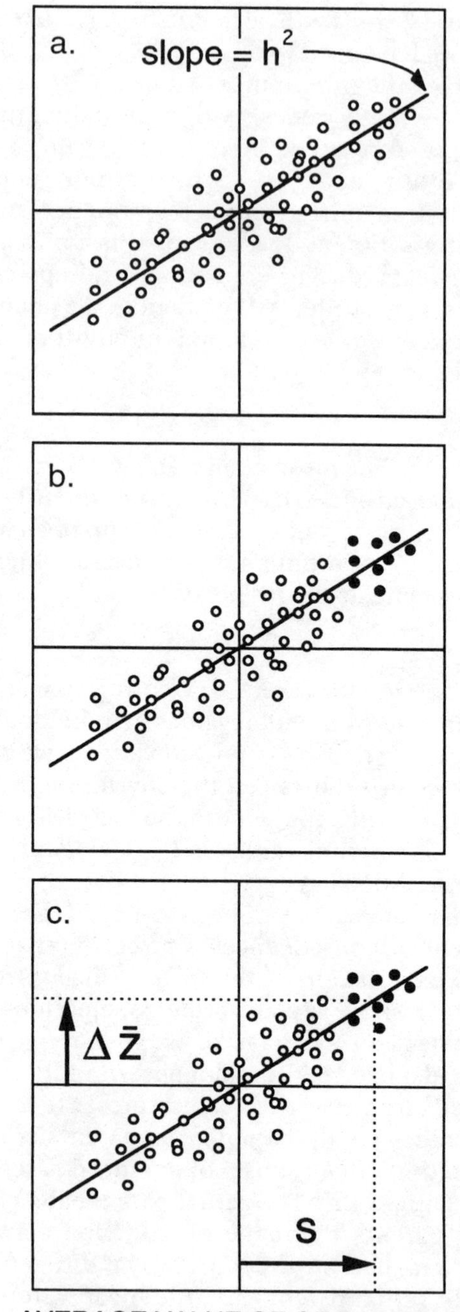
a.
slope = h^2
b.
c.
$\Delta \bar{z}$
S
OFFSPRING VALUE
AVERAGE VALUE OF OF PARENTS

parents is a useful measure of the strength of selection. This difference in means is known as the *selection differential.* By graphing the selection differential on Galton's plot, we see that the slope of the regression line enables us to predict the genetic consequences of selection (fig. 2.2c). In making this prediction we implicitly assumed that the regression for actual parents is the same as the regression for all potential parents. Pearson (1903) showed that this is the case. As Lush (1945) put it, heritability is the fraction of the selection differential that is transmitted to the next generation. Symbolically, we have the following result: the change in mean from one generation to the next when selection acts on a single trait is

$$\Delta\bar{z} = h^2 s, \tag{2.3}$$

where s is the selection differential. Equation 2.3 is a powerful tool for an experimentalist or a plant or animal breeder who imposes selection on a single trait and wants to know how much change (improvement) to expect. In the natural world, however, selection undoubtedly acts simultaneously on many traits, so we will need an even more powerful tool.

Response to Selection on Multiple Traits

To predict the evolutionary response to selection acting on multiple traits it is convenient to use a new concept of selection, the *selection gradient* (Lande 1979). The selection gradient measures the direct force of selection on a trait. It is equivalent to the partial regression of relative fitness on the trait, holding other traits constant (Lande and Arnold 1983). In contrast, the selection differential measures both the direct force of selection on a trait and the

Fig 2.2 A graphical portrayal of response to selection on a single trait. (*a*) Resemblance between offspring and their parents. The set of all potential parents and their offspring is shown. The slope of the regression of offspring values on parental values estimates heritability (h^2). The vertical line indicates the mean of parental values (before selection) and the horizontal line indicates the mean offspring value when there is no selection in the parental generation. (*b*) Same conventions as *a* but with the set of actual parents and their offspring values indicated with solid symbols. (*c*) The selection differential, s, is the difference between the mean values of actual parents and all potential parents. The response to selection, $\Delta\bar{z}$, is the difference between the mean offspring values of actual parents and all potential parents. The response to selection is proportional to the selection differential and heritability. (Adapted from Lush 1945 and Falconer 1989.)

indirect effects of selection acting on other traits. The selection differential is equivalent to a covariance between a trait and relative fitness (Robertson 1966).

The only other new concept we need in order to deal with selecting acting on multiple traits is the idea of *additive genetic covariance.* The idea here is that traits may be genetically coupled. Consequently, when selection acts directly on one trait, it may produce reverberations in traits that are genetically coupled to it. Genetic coupling will be revealed in a Galton plot in which we plot offspring values for one trait on the y-axis and parental values for some other trait on the x-axis. A nonzero regression in such a plot is due to covariance between additive genetic values for the two traits. This covariance is known as additive genetic covariance. Such covariance or coupling arises from pleiotropy (individual genes exerting effects on both traits) or linkage disequilibrium. By linkage disequilibrium we do not mean the physical linkage of genes on chromosomes, but rather statistical associations between alleles at different loci (which may or may not be on the same chromosome). Now, the additive genetic covariance between the two traits may be positive, zero, or negative. In the last case, there might be a preponderance of genes with positive pleiotropic effects on one trait, but negative effects on the other trait. Standardized additive genetic covariances are known as *genetic correlations* and can take values between +1 and −1. When dealing with multiple traits it is convenient to arrange the additive genetic covariances in a table known as the *additive genetic variance-covariance matrix,* **G**. In this table the first row and column refer to the first trait, the second row and column refer to the second trait, and so on. For p traits, the **G**-matrix looks like this

$$\begin{bmatrix} G_{11} & G_{12} & \cdots & G_{1p} \\ G_{21} & G_{22} & \cdots & G_{2p} \\ \cdot & \cdot & \cdots & \cdot \\ \cdot & \cdot & \cdots & \cdot \\ \cdot & \cdot & \cdots & \cdot \\ G_{p1} & G_{p2} & \cdots & G_{pp} \end{bmatrix}$$

where G_{11} is the additive genetic variance for the first trait, G_{12} ($= G_{21}$) is the additive genetic covariance between the first and the second trait, and so forth.

We can use the **G**-matrix and the set of selection gradients for the traits to predict responses to selection acting simultaneously on many traits. In tabular (matrix) form our equation looks like this,

$$\begin{bmatrix} \Delta\bar{z}_1 \\ \Delta\bar{z}_2 \\ \cdot \\ \cdot \\ \cdot \\ \Delta\bar{z}_p \end{bmatrix} = \begin{bmatrix} G_{11} & G_{12} & \cdots & G_{1p} \\ G_{21} & G_{22} & \cdots & G_{2p} \\ \cdot & \cdot & \cdots & \cdot \\ \cdot & \cdot & \cdots & \cdot \\ \cdot & \cdot & \cdots & \cdot \\ G_{p1} & G_{p2} & \cdots & G_{pp} \end{bmatrix} \begin{bmatrix} \beta_1 \\ \beta_2 \\ \cdot \\ \cdot \\ \cdot \\ \beta_p \end{bmatrix}$$

in which $\Delta\bar{z}_1$ is the change across generations in the mean of the first trait, β_1 is the selection gradient for the first trait, and so forth. The set of changes in means and the set of selection gradients are known as column vectors. The whole set of equations is conveniently represented as

$$\Delta\bar{\mathbf{z}} = \mathbf{G}\boldsymbol{\beta}, \tag{2.4}$$

in which $\Delta\bar{\mathbf{z}}$ and $\boldsymbol{\beta}$ are the column vectors given above. We can now let the equations do some work for us. Using the rules of matrix algebra (see appendix 2.1), we can write out the equation for the change in mean of the first trait, which is

$$\Delta\bar{z}_1 = G_{11}\,\beta_1 + G_{12}\,\beta_2 + \ldots + G_{1p}\,\beta_p.$$

We see before us several important consequences of multivariate selection. The evolutionary response of the first trait to selection, i.e., $\Delta\bar{z}_1$, is composed of several possible contributions. One of these is the direct response to selection, $G_{11}\,\beta_1$, representing the change in the mean of trait 1 due to selection acting directly on trait 1. The other contributions are correlated responses to selection. For example, $G_{12}\,\beta_2$ represents change in the mean of trait 1 due to selection acting directly on trait 2 and inducing change in trait 1 via the additive genetic covariance G_{12} (fig. 2.3). It is possible that the change in the mean of trait 1 is wholly or largely due to direct response to selection on trait 1. But, it is also possible that the many possible correlated responses to selection will dominate or swamp the direct response. In the extreme case, selection might favor high values of the first trait (positive β_1), yet the trait will evolve toward a lower mean value because of negative additive genetic covariances (fig. 2.3c) or because of selection for low values of other traits. In general, if we think of selection as a pool cue ($\boldsymbol{\beta}$) and

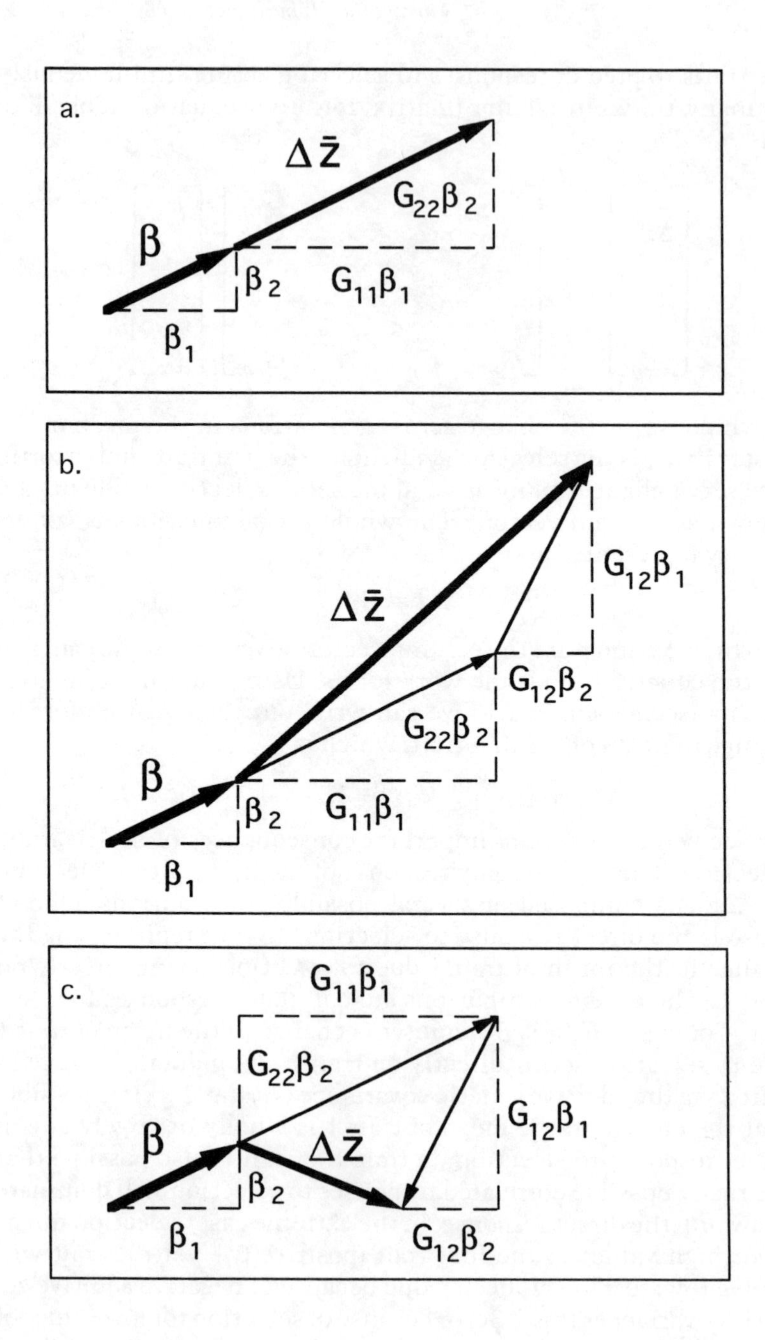
a.
Δz̄
G22β2
β
β2
G11β1
β1
b.
G12β1
Δz̄
G12β2
G22β2
β
β2
G11β1
β1
c.
G11β1
G22β2
G12β1
β
Δz̄
β2
β1
G12β2

response to selection as the path of the ball ($\Delta\bar{\mathbf{z}}$), the effect of genetic covariance is to throw the ball's path out of alignment with the cue (fig. 2.3).

Estimation of Genetic Variances and Covariances

Defining behavioral traits. How to define a behavior is an important decision that needs to be made before launching a program of genetic parameter estimation. Two important considerations to keep in mind are relevance to ecological circumstances or evolutionary history and simplicity of scoring. If a study of inheritance is conducted with traits that have straightforward ecological or evolutionary significance, the results can add a genetic dimension to ecological or evolutionary issues. Simplicity of scoring is crucial because the behaviors must be scored on many individuals (and sometimes many times on each individual) to estimate genetic parameters. The study of behavioral inheritance is especially challenging because the expression of a particular behavior by an individual may fluctuate over time, show maturational change, or be influenced by experience or condition.

1. Fluctuation without a temporal trend can be handled by making repeated scores on each individual. The issue of how many scores to make can be decided in a pilot experiment in which the behavior of each individual is scored two or more times. The among-individual component of variance, expressed as a fraction

---

Fig. 2.3 A graphical portrayal of response to selection on two traits. Our response to selection equation for the first trait is, $\Delta\bar{z}_1 = G_{11}\,\beta_1 + G_{12}\,\beta_2$, and for the second trait, $\Delta\bar{z}_2 = G_{22}\,\beta_2 + G_{12}\,\beta_1$. In each of the three cases shown, direct selection on the first trait, β_1, is positive and twice as strong as direct selection on the second trait, β_2, and the additive genetic variances for the two traits are equal, $G_{11} = G_{22} = 1$. The vector representing direct selection, $\boldsymbol{\beta}$, is shown as a heavy arrow. The vector representing response to selection, $\Delta\bar{\mathbf{z}}$, is shown as a medium arrow. The vectors representing direct and correlated responses are shown as light arrows. (*a*) No genetic covariance between the two traits, $G_{12} = 0$. The response to selection is composed only of direct responses to selection, $G_{11}\,\beta_1$ and $G_{22}\,\beta_2$. Because the two genetic variances are equal and genetic covariance is zero, the response is aligned with the direction of selection. (*b*) Positive genetic covariance between the two traits, $G_{12} = 0.75$. The two correlated responses to selection, $G_{12}\,\beta_1$ and $G_{12}\,\beta_2$, throw the total response out of alignment with the direction of selection. (*c*) Negative genetic covariance between the two traits, $G_{12} = -0.75$. The genetic covariance, acting through the correlated responses, now produces an extreme departure from alignment. Notice that trait z_2 evolves in the opposite direction to the selection acting directly on it.

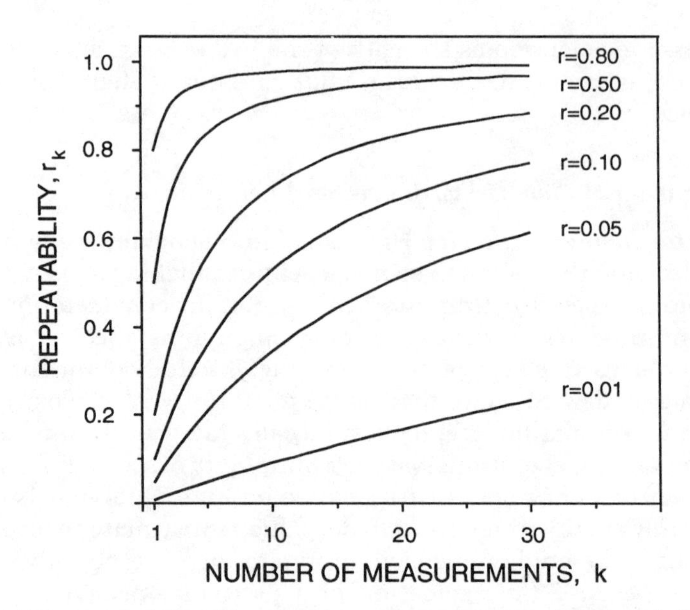

Fig. 2.4 The repeatability of k measurements or trials, r_k, as a function of single-trial repeatability, r, and the number of measurements or trials, k. Using formulas given in Winer (1971) and Falconer (1989), $r_k = kr/(kr - r + 1)$. In Falconer's (1989) notation, r_k is also equal to $(V_G + V_{Eg})/V_{P(k)}$, where $V_G + V_{Eg}$ is the among-individual component of variance and $V_{P(k)}$ is the total phenotypic variance for the average of k trials.

of total phenotypic variance, is known as the intraclass correlation, reliability, or *repeatability* (Winer 1971; Falconer 1989). Such a repeatability refers to individual consistency of single-trial scores, but we can also compute the repeatability of the average of two or more trials or measurements (fig. 2.4). We see from figure 2.4 that repeatability increases with the number of measurements, k. Thus, if a pilot experiment indicates that single-trial repeatability is 50%, by making four measurements on each individual and using the four-trial averages, we can increase repeatability to 80%. Because repeatability places an upper bound on heritability (Boake 1989b; Falconer 1989), we are more likely to be able to detect statistically significant heritability if our behavioral score is a four-trial average rather than a single-trial score. The use of multiple-trial averages can be justified on biological as well as statistical grounds. An animal has only one left eye, so measuring the repeatability of the width of the left eye is an exercise in assessing measurement error. But an animal may perform a courtship display hundreds or even thousands of times during its life. Consequently, the repeatability of a behavior, such as a courtship display, depends on the time scale over which behavioral averages are assessed. The proper time scale in turn depends on the question being asked. If the issue is response to lifetime selection, then life span is the proper time scale, and k is the average number of times the behavior is expressed in a lifetime. If we are interested in female discrimination of male behavior, then k might be the average number of male displays given to an attending female. Houck, Arnold, and Thisted (1985) give an example of time scale for repeatability as an ecological issue. An important technical point is to keep estimates of repeatability and heritability in register. If repeatability is computed using the sum or average of k trials, then the heritability for k trials should be computed, rather than the single-trial heritability. Arnold and Bennett (1984) give an example.

2. Behaviors that change with age can be treated as age-specific attributes. Examples of genetic analysis of behavioral ontogenies can be found in Hahn et al. (1990).

3. Behaviors that can be affected by experience call for special experimental procedures. One simple approach is to score behavior in neonates that are either completely naive or all have identical, limited experience (e.g., Arnold 1981a). If the inheritance of learning is the issue of interest, the difference in behavioral scores before and after experience can be treated as a trait. Repeated measures or profile analysis of variance are useful tools for data analysis (Winer 1971; Bock 1975; Harris 1975; Simms and Burdick 1988).

4. Techniques for handling behaviors that vary with nutritional condition or body size are discussed by Travis (chap. 8).

Reference population. The choice of a reference population is also a critical decision. Genetic variances and covariances may vary from one population to the next, even within a species. Therefore, if we draw sets of relatives from more than one population, the genetic parameters we estimate may not be characteristic of any natural population. Likewise, genetic parameters may change over several generations of captive propagation (e.g., due to inbreeding and/or drift), so that estimation based on a long-maintained laboratory line may not be characteristic of any natural population. If the aim of genetic parameter estimation is to make inferences about inheritance and responses to selection in nature, the best procedure generally is to sample directly from a single natural population. A second reason for having a reference population is that the genetic results will apply to a particular population with definable ecology and phylogenetic history. The ecology of the reference population can be studied along with its behavioral inheritance with synergistic results. Grant and Grant's (1989) study of Galápagos finches, for example, illustrates the power of focusing both genetic and ecological studies on the same reference population.

Overview of parameter estimation. Genetic variances and covariances (or their standardized analogues, heritability and genetic correlation) can be estimated in two ways: by conducting selection experiments or by assembling replicated sets of relatives. Falconer (1989) gives an elementary account of selection experiments; Hill (1980) discusses design considerations. I will focus on the use of sets of relatives. The classic approach in using sets of relatives is to assemble replicated sets of first-degree relatives (e.g., sets of parents and offspring, full sibs, or paternal half sibs) or to produce them in a breeding design. The basic tools of data analysis are regression and analysis of variance (Cockerham 1963; Becker 1984; Falconer 1989). A more recent approach is to use maximum likelihood to estimate the genetic parameters. This approach can be used with sets of first-degree relatives (Thompson 1977; Shaw 1987) or sets of individuals related by complex pedigrees (Lange, Westlake, and Spence 1976; Hopper and Matthews 1982; Lange, Weeks, and Boehnke 1988).

The basic idea in using sets of relatives to estimate genetic parameters is that observable phenotypic resemblance among different sets of relatives can be used to isolate and estimate the additive genetic and other interesting parts of the total phenotypic variance in behavior. Think of a stick representing the total variance, with marks indicating the additive genetic variance and other parts of the total. By observing the phenotypic resemblance among parents and offspring, among full sibs, among paternal half sibs, etc., we can break the stick at well-defined points and so isolate the segment that interests us most—the segment representing additive genetic variance. The same analogy applies to a stick representing the total phenotypic covariance between a pair of traits. Some complicated breeding designs have the goal of breaking the stock into as many interesting pieces as possible. I will focus on some simple sets of relatives and their use in estimating additive genetic variances and covariances.

Parents and offspring. We have already discussed the correspondence between the regression of parental on offspring scores and heritability. Two important assumptions were made in equating the regression slope with heritability. One assumption was that there is no correlation between the environmental values of parents and their offspring. This assumption can be violated in many ways. For example, if territories are passed from parents to their offspring, then the properties of territories, rather than genes, may promote resemblance between parents and their offspring. In a laboratory setting, care can be taken to standardize environments, but in sets of free-ranging parents and offspring the possibility of cross-generational correlation in environmental effects should be considered in the interpretation of resemblance between parents and offspring. If parents directly affect the behavior of their offspring (e.g., by tutoring or by nutritional effects), a more complicated model of phenotypic value should be considered (see below). A second critical assumption was that additive genetic values are not correlated with environmental values. Falconer (1989) gives the example of larger individuals, with higher additive genetic values for body size, also getting more to eat and so inducing a positive correlation between genetic and environmental values for size. In many instances such covariances between environmental values of parents and offspring or between genetic and environmental values within a generation may be implausible, but they should be considered.

In situations where the complications just discussed can be put aside, regression analysis can be used to estimate genetic variances and covariances from parent-offspring data. Estimates can be made using data on both parents or data on just one parent (Falconer 1989, chap. 9). When *family size* (i.e., the number of offspring in a brood) varies, the technical problem of how to weight families arises. Unfortunately the best weight is itself a function of the genetic parameters being estimated. Bulmer (1985) discusses an iterative solution to the problem. A shortcut is to compute two regressions, weighting families by their sizes and weighting them equally. The correct weighting is somewhere in between these extremes. If the regression slopes found using the two extreme weightings do not differ, it is probably not worth resorting to the more exact iterative approach.

Full sibs. Sets of offspring each with a different mother and father (full sibs) produce more ambiguous estimates of genetic parameters than any of the other sets of relatives. The procedure with sets of full sibs is to estimate the within- and among-sibship components of variance using a random effects analysis of variance (Sokal and Rohlf 1981). Unfortunately, in this case the stick representing total phenotypic variance is broken in a rather jagged way, so we do not get a clean estimate of additive genetic variance. The among-sibship variance component estimates additive genetic variance plus part of the genetic variance due to dominance and environmental variance arising from the common conditions experienced by sibs. One can minimize the common family environmental variance by standardizing as much as possible the environment during pregnancy and/or rearing. Even so, one needs to qualify the conclusions made from full sib data because the estimates of additive genetic parameters are inflated (or depressed) by some unknown amount.

Paternal half sibs. In contrast to using sets of full sibs, using sets of paternal half sibs enables us to break the stick of variation in a very clean way. Paternal half sibs are usually produced using a deliberate breeding design in which each of a series of males (sires) is mated to a different set of females (dams). Consider first the analysis when phenotypes are scored on only the offspring. The data structure is a nested one, with dams nested within sires. The data are analyzed using a two-level analysis of variance, with the goal of estimating the among-sire component of variance (Falconer 1989, chap. 10). Unless there are paternal effects (see below), the among-sire variance component estimates one-fourth of the additive genetic variance. Likewise, a two-level analysis of covariance is conducted to estimate the among-sire components of covariance and hence the additive genetic covariances between traits. If phenotypic scores are available for parents as well as for offspring (and if both generations were reared under equivalent conditions), all of the data can be used to estimate genetic parameters. The best approach is probably to use maximum likelihood estimation (Shaw 1987).

Maternal half sibs. Another way to cleanly break the stick of variation is to produce both maternal and paternal half sibs. The primary gain from the extra work is an estimate of the contribution of maternal effects. The production of maternal half sibs is usually possible only in animals with external fertilization (e.g., most anurans, some fishes). The most feasible design is to randomly assign breeding animals to blocks, with d dams and s sires in each block.

The ova from each dam in a block is divided into *s* groups and each group is fertilized with sperm from a different sire in that block. The offspring are reared individually and their phenotypes are scored. The data in a block have a factorial structure, sires crossed with dams. The data can be analyzed with a two-way analysis of variance (e.g., Simms and Rausher 1989). The among-sire component of variance (or covariance) can be used to estimate additive genetic variance (or covariance). The contrast between the among-sire and the among-dam components of variance can be used to isolate the contribution of maternal effects (Willham 1963, 1972).

Sample size. When heritability is high (>70%), a statistically significant effect may be detected with only a dozen or two parent-offspring sets or sibships. But when heritability is low (<30%), many dozens or scores of parent-offspring sets or sibships will be needed to show a statistically significant effect. Mousseau and Roff's (1987) survey suggests that behavioral heritabilities are usually in the low range. Under these circumstances, investigators often adopt one of two postures: "genetics without tears," or the more ambitious parameter estimation approach. The goal of the "genetics without tears" approach is to see whether a genetic effect can be detected with a modest sample of families or sibships (Antonovics 1982). This approach is one of exploration and can be justified if the phenotype or behavior in question has never been scrutinized from the heritability point of view. The gamble is that statistically significant genetic effects might be detected. If they are not, the investigator is in no position to argue that heritability is nonexistent, because the experimental design lacks power. An advantage of "genetics without tears" is that family structure can be incorporated into an experimental design and permit qualitative tests for genetic differences as part of a larger set of issues (e.g., Newman 1988; Trexler and Travis 1990; Arnold 1992a). In contrast, the estimation of parameters approach focuses on making standard errors around estimates of genetic parameters as small as possible. More effort is expended and more precision is the reward. A disadvantage of parameter estimation is that the thorough pursuit of one or a few genetic issues may cause the investigator to sacrifice other pursuits.

The potential gains from the additional effort can be gauged by a study of sampling properties. Klein, DeFries, and Finkbeiner (1973), for example, give tables of expected standard errors for various values of parametric heritability and numbers of families in the sample. Their tables, however, are constructed for students of primates and other organisms with small numbers of offspring (1–3). When family sizes are substantially larger, standard errors are considerably smaller than those given in the tables. For example, suppose one wishes to estimate heritability by regressing offspring phenotypic values on the phenotypic value of one parent. How many parent-offspring sets should be assembled to give a good heritability estimate? The answer depends on magnitude of parametric heritability and family size. A good solution is to proceed on a worst-case basis, imagining that parametric heritability might be as low as 20%. We can see from figure 2.5a that if there is only one offspring per parent we would need nearly 400 families to bound our estimate of h^2 away from zero. But, if we had ten offspring per family, we would need only about 70 families to achieve a comparably small standard error. A conscientious investigator can produce tailor-made tables of expected standard errors in less than an hour using a spreadsheet such as Lotus 1-2-3. If a full sib-half sib breeding design is being contemplated, Robertson (1959a,b) should be consulted.

The inefficiency of small family sizes can also be appreciated by considering the power of a heritability estimate. Roughly speaking, statistical power is the probability of correctly concluding that heritability is greater than zero when parametric heritability is in fact greater than zero. A number of authors and reviewers have erroneously cited Klein (1974) in reaching the conclusion that hundreds of families are needed to achieve good statistical power in estimating heritability. Klein, however, considered only the case of small family sizes ($n = 1–4$). We see from figure 2.5b that substantial statistical power can be achieved with less than a hundred families, if family size is large.

Other Models for Phenotypic Value and Variance

Genotype-environment correlation. Under some circumstances it is conceivable that environmental effects are correlated with additive genetic value. For example, in territorial species, larger individuals (which may tend to have large genetic values for body size) may obtain better territories and hence more food. The model for phenotypic values is the same as that given in equation 2.1, but the model for phenotypic variance becomes,

$$\mathbf{P} = \mathbf{G} + \mathbf{E} + 2\mathrm{COV}(\mathbf{a},\mathbf{e}), \qquad (2.5)$$

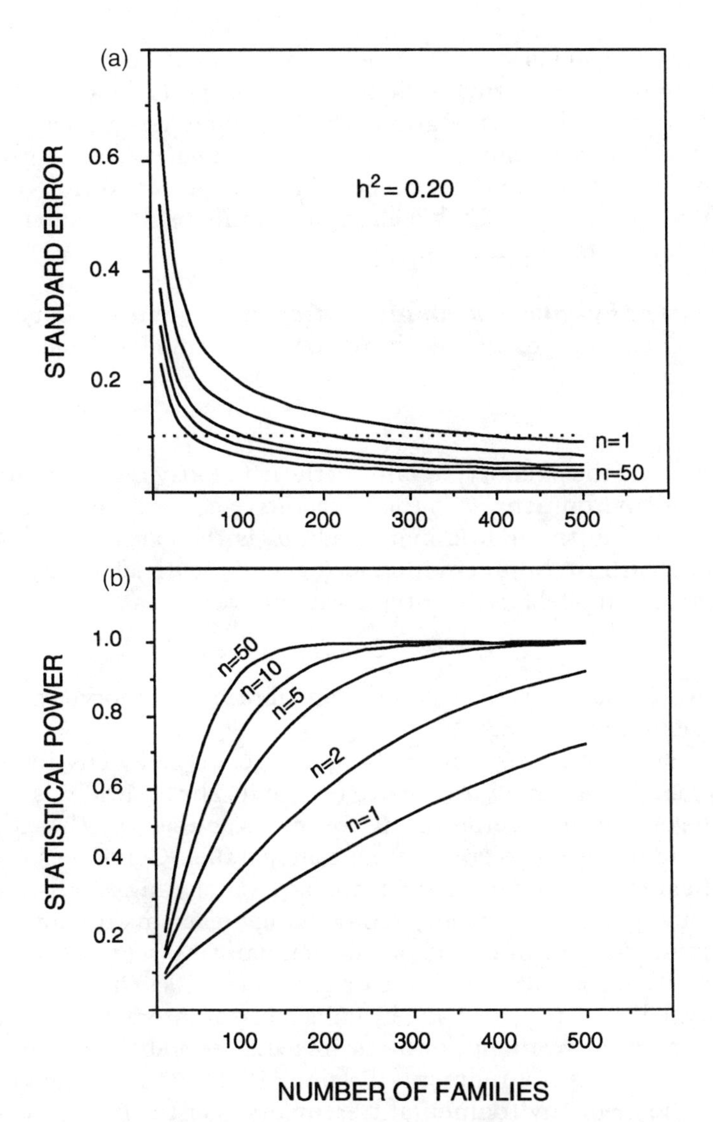

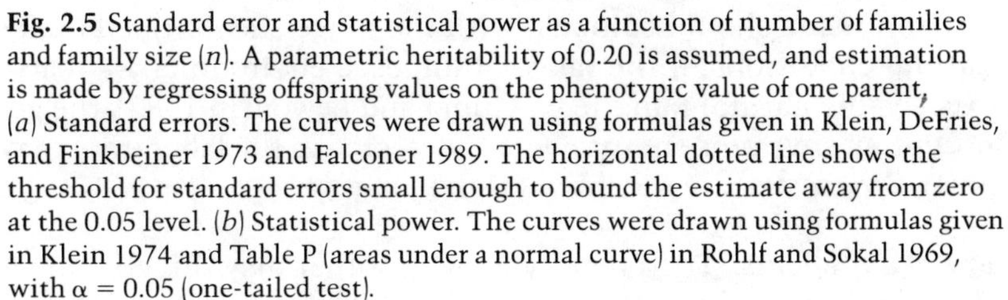

Fig. 2.5 Standard error and statistical power as a function of number of families and family size (n). A parametric heritability of 0.20 is assumed, and estimation is made by regressing offspring values on the phenotypic value of one parent. (a) Standard errors. The curves were drawn using formulas given in Klein, DeFries, and Finkbeiner 1973 and Falconer 1989. The horizontal dotted line shows the threshold for standard errors small enough to bound the estimate away from zero at the 0.05 level. (b) Statistical power. The curves were drawn using formulas given in Klein 1974 and Table P (areas under a normal curve) in Rohlf and Sokal 1969, with $\alpha = 0.05$ (one-tailed test).

where COV(**a**,**e**) is the covariance between additive genetic values and environmental effects. Such covariance pulls the rug out from under standard interpretations of phenotypic covariance among relatives, which assume zero covariance between additive genetic values and environmental effects. When rearing conditions can be controlled, however, COV(**a**,**e**) can be eliminated by standardizing environmental conditions.

Genotype by environment interaction. If the environment affects different genotypes differently, we have a new model of phenotypic value,

$$z_{ij} = a_i + e_j + ae_{ij}, \tag{2.6}$$

where z_{ij} is the phenotypic value of the ith additive genetic value in the jth environment, a_i is the ith additive genetic value, e_j is the effect of the jth environment, and ae_{ij} is the extra contribution from the jth environment interacting with the ith additive genetic value. Our model for phenotypic variance becomes

$$\mathbf{P} = \mathbf{G} + \mathbf{E} + \mathrm{VAR}(\mathbf{a} \times \mathbf{e}), \tag{2.7}$$

where VAR(**a** × **e**) is the variance contribution of genotype by environment interaction.

Most investigators routinely ignore genotype by environment interaction, but there are ways to test for it. The basic idea is to rear each genotype in a series of environments and see whether the genotypic reactions to environment are parallel. Genotypes can be replicated if inbred lines are available or if cloning is possible (as in some plants). In sexually reproducing organisms it is more appropriate to distribute sets of paternal half sibs (representing sets of additive genetic values) across environments. The data can be analyzed by testing for a sire by environment interaction in a two-way analysis of variance or by estimating the additive genetic covariance across environments (Falconer 1960; Via and Lande 1985). The choice of environmental treatments is critical. The range of environments should coincide with the range encountered under natural conditions. If the range of imposed environments greatly exceeds the natural range (e.g., Gupta and Lewontin 1982), the genotype by environment interaction detected in the laboratory may greatly overestimate the interaction in nature.

Maternal effects. The phenotype of a mother may directly affect the phenotypes of her offspring (Cowley 1990; Mousseau and Dingle

1990, 1991; Reznick 1990; Bernardo 1991; Cheverud and Moore, chap. 4). Such maternal effects complicate the interpretation of maternal-offspring resemblance. Consider the following linear model for the offspring's phenotypic value for the *i*th trait,

$$z_{o_i} = a_{o_i} + e_{o_i} + \sum_{j=1}^{n} M_{ij} \, z^*_{m_j} \, , \qquad (2.8)$$

where a_{o_i} is the offspring's additive genetic value for the *i*th trait, e_{o_i} is the environmental value for the offspring's *i*th trait, $z^*_{m_j}$ is the mother's phenotypic value for her *j*th trait (with the asterisk denoting that her phenotypic value is evaluated after selection), and M_{ij} is a maternal effect coefficient describing the effect of the mother's *j*th trait on the offspring's *i*th trait (Kirkpatrick and Lande 1989). Our earlier model for the composition of phenotypic variance (eq. 2.2) no longer holds. Furthermore, when selection acts on a single trait, response to selection is no longer a simple function of heritability and selection differential (eq. 2.3) (see Cheverud and Moore, chap. 4). And when selection acts on multiple traits, response to selection is no longer a simple function of the **G**-matrix and the selection gradients (eq. 2.4). Instead, we have a new response to selection equation which involves a new matrix, $\mathbf{C_{az}}$, representing the covariances between additive genetic values and phenotypic values. Our new equation for response to multivariate selection in generation *t* is

$$\boldsymbol{\Delta}\bar{\mathbf{z}}(t) = \mathbf{C_{az}}\, \boldsymbol{\beta}(t) + \mathbf{M}\boldsymbol{\Delta}\bar{\mathbf{z}}^*_{\mathbf{m}}(t-1), \qquad (2.9)$$

where $\mathbf{M}\boldsymbol{\Delta}\bar{\mathbf{z}}^*_{\mathbf{m}}(t-1)$ is a sum of terms representing change in the mean phenotypic values of mothers after selection in the preceding generation (i.e., from generation $t-2$ to generation $t-1$) (Kirkpatrick and Lande 1992). Thus, the response to selection in the current generation is partly due to selection acting in that generation and partly due to selection acting in the mothers' generation and then affecting offspring values via maternal effects. The important message of equation 2.9 is that in the presence of maternal effects the key genetic parameters are no longer heritabilities or even the **G**-matrix. Instead, the main targets for estimation are two matrices, $\mathbf{C_{az}}$ and **M**.

The estimation of genetic parameters when there are maternal effects, but no paternal effects (eq. 2.8), is discussed by Lande and Price (1989). The basic idea is that maternal effects contribute to the phenotypic resemblance between mothers and offspring but they do not contribute to the resemblance between fathers and offspring. Consequently, if we have data on the phenotypic values of mothers (including the attributes that exert maternal effects), fathers, and offspring, we can estimate the matrix of maternal effects, $\mathbf{M}$, as well as the matrix $\mathbf{C}_{az}$. Alternatively, one can use a breeding design that generates a large set of different kinds of relatives and/or cross-fostering (Eisen 1967; Hanrahan and Eisen 1973; Cheverud et al. 1983; Riska, Rutledge, and Atchley 1985). (See Cheverud and Moore, chap. 4, for more details.)

Maternal and paternal effects. In some fishes, birds, and mammals both sexes care for the offspring (Clutton-Brock 1991). In such species, paternal as well as maternal effects are possible. Our model for phenotypic value becomes

$$\mathbf{z}_o = \mathbf{a}_o + \mathbf{e}_o + \mathbf{M}\mathbf{z}^*_m + \mathbf{F}\mathbf{z}^*_f, \quad (2.10)$$

where $\mathbf{M}\mathbf{z}^*_m$ is algebraic shorthand for the sum of terms on the right side of equation 2.8, $\mathbf{z}^*_f$ represents the phenotypic values of fathers (after selection), and $\mathbf{F}$ represents a matrix of paternal effect coefficients that translate values of fathers' phenotypes into effects on offspring traits. The response to selection equation is analogous to equation 2.9 but involves an additional term representing the change in mean values of phenotypes of fathers in the preceding generation. The key target for genetic parameter estimation is again the matrix $\mathbf{C}_{az}$, but there are now two matrices of parental effect coefficients that need to be estimated, $\mathbf{M}$ and $\mathbf{F}$.

The appearance of a new inheritance matrix, $\mathbf{C}_{az}$, in our response to selection equations is at first a disconcerting result. The reason we need $\mathbf{C}_{az}$ instead of $\mathbf{G}$ becomes clear, however, when we consider our underlying model. To predict offspring values from parental values we need the covariance between additive genetic values (revealed as the average values of offspring) and phenotypic values of parents (Dickerson 1947). This covariance is $\mathbf{C}_{az}$. We see from our model (applying eq. 2.10 to the parental generation) that the phenotypic value of a parent is composed of its additive genetic value, a value transmitted directly from its mother, a value transmitted directly from its father, and an environmental deviation (The latter deviation or value will not concern us, because we have assumed that it is uncorrelated with additive genetic value). Because phenotypic value is composed of three relevant parts, when we consider its covariance with additive genetic value, we find that the covariance, $\mathbf{C}_{az}$, is composed of three parts: $\mathbf{G}$, representing the covariance between additive genetic values; $\frac{1}{2}\,\mathbf{C}_{az}\mathbf{M}^{T}$, representing the covariance between additive genetic value and maternally transmitted value; and $\frac{1}{2}\,\mathbf{C}_{az}\mathbf{F}^{T}$, representing the covariance between additive genetic value and paternally transmitted value. In other words,

$$\mathbf{C}_{az} = \mathbf{G} + \frac{1}{2}\,\mathbf{C}_{az}\,\mathbf{M}^{T} + \frac{1}{2}\,\mathbf{C}_{az}\,\mathbf{F}^{T} \qquad (2.11)$$

The superscript T denotes matrix transposition (see appendix 2.1). With maternal effects but no paternal effects, we drop the last term on the right in equation 2.11 (Kirkpatrick and Lande 1989). In the absence of both maternal and paternal effects, $\mathbf{C}_{az} = \mathbf{G}$, and we could have substituted $\mathbf{C}_{az}$ for $\mathbf{G}$ in our response to selection equation (eq. 2.4). Thus, $\mathbf{C}_{az}$ is the fundamental inheritance concept that runs through all of our response to selection equations (eqs. 2.3, 2.4, 2.9, 2.12, 2.13).

Estimation of genetic parameters when both sexes exert parental effects has been little explored, and there is virtually no empirical work. One possible experimental route is to rear broods allowing only mothers or fathers to exert care, but cross-fostering is probably the most efficient approach.

Selection Exerted by Relatives

Maternal selection. In species with maternal care, the attributes of the mother may directly affect the fitness of offspring. Such maternal selection is different from the maternal effects we considered in the preceding section. A maternal attribute exerts a maternal effect by affecting the expression of a trait in the offspring. A maternal attribute exerts maternal selection by directly affecting the fitness of the offspring (Kirkpatrick and Lande 1989). In many cases, the maternal traits exerting maternal selection will be behaviors. When a mother bird performs a display that distracts predators from her nest, the distraction display directly affects her offspring's fitness. In other words, the display exerts maternal selection. Such maternal selection constitutes an additional force promoting the evolution of distraction displays in the offspring generation. In addition, maternal selection on the distraction display can evoke correlated responses in traits that are genetically coupled to the distraction display.

The nature of response to selection when maternal selection prevails depends on the system of inheritance. In the absence of maternal effects (i.e., eq. 2.1), the evolutionary response to multivariate selection is relatively simple,

$$\Delta \bar{\mathbf{z}} = \mathbf{G}\boldsymbol{\beta}_{o} + \frac{1}{2}\ \mathbf{G}\boldsymbol{\beta}_{m}, \tag{2.12}$$

where $\boldsymbol{\beta}_o$ denotes the offspring selection gradient and $\boldsymbol{\beta}_m$ denotes the maternal selection gradient (Kirkpatrick and Lande 1992). In the presence of maternal effects (i.e., eq. 2.8), the evolutionary response to multivariate selection is somewhat more complicated,

$$\Delta \bar{\mathbf{z}}(t) = \mathbf{C}_{az}\boldsymbol{\beta}_{o}(t) + \frac{1}{2}\ \mathbf{C}_{az}\boldsymbol{\beta}_{m}(t) + \mathbf{M}\Delta \bar{\mathbf{z}}^{*}_{m}\ (t-1) \tag{2.13}$$

(using equations 2 and 3 of Kirkpatrick and Lande 1992). (See Cheverud and Moore, chap. 4, for more discussion of the first two terms on the right side of equation 2.13 for the case in which selection acts on a single trait.)

Maternal and paternal selection. When both sexes perform parental care, both maternal and paternal selection are likely to prevail. The multivariate response to selection equations are analogous to equations 2.12 and 2.13 but with additional terms representing contributions from paternal selection, $\boldsymbol{\beta}_f$, and in the case of equation 2.13, an additional term representing a contribution from selection acting on fathers in the preceding generation.

Nonlinear maternal and paternal selection. In the preceding sections we considered directional parental selection in which the attributes of parents have linear effects on the fitness of their offspring. The attributes of parents can also have nonlinear effects on offspring fitness. The simplest nonlinear effects are quadratic effects, which involve the squared phenotypic values of maternal or paternal attributes and the products of phenotypic values of mothers, fathers, and offspring. Such nonlinear parental selection does not directly contribute to the evolutionary response to selection (the change in trait means across generations), but it does affect the maintenance and evolution of genetic variances and covariances. As in the case of directional parental selection, nonlinear parental selection will often involve behavioral attributes that shelter or protect the offspring from hazards.

Overview of Short-term Selection Response

The genetic concept that runs through these various cases of response to selection is the covariance between additive genetic value and phenotypic value, $\mathbf{C}_{\mathbf{az}}$. In the breeder's response equation (eq. 2.3), this covariance cryptically resides in the numerator of heritability. In the equation for response to multivariate selection (eq. 2.4), $\mathbf{C}_{\mathbf{az}}$ masquerades as the **G**-matrix. When we consider parental effects, we are forced to make a distinction between $\mathbf{C}_{\mathbf{az}}$ and **G** (eq. 2.11). In this revealing case, we find that it is $\mathbf{C}_{\mathbf{az}}$ (rather than heritability or additive genetic variance/covariance) that plays a central role in predicting response to selection.

Evolutionary Models (Long-term Response to Selection)

For many years response to selection equations were used to predict only one or a few generations of genetic change. Lande (1976a, 1980a, 1984, 1988) broadened the scope of applications by arguing that genetic variances and covariances might be maintained in equilibrium by a balance between input from mutation/recombination and loss to selection. Or, as Kurtén (1959) put it, "the replenishing of genetic material keeps pace with the appetite of selection." In this view, relative constancy of the **G**-matrix (eq. 2.4) might be maintained while the phenotypic mean undergoes a substantial evolution over many generations. Lande's argument for constancy of the **G**-matrix has been challenged by Turelli (1984, 1985, 1986, 1988a), who argued that the amount of genetic variation maintained at equilibrium depends on untested assumptions about the statistical distribution of allelic effects on phenotypic traits. Empirical tests of the constancy issue have been made by comparing **G**-matrices after varying intervals of phylogenetic separation. (Methodologies for comparing matrices are discussed by Flury 1988; Shaw 1991, 1992; and Cowley and Atchley 1992.) These tests have yielded results ranging from rough structural similarity of **G**-matrices (Arnold 1981a; Atchley, Rutledge, and Cowley 1981; Lofsvold 1986; Kohn and Atchley 1988) to striking constancy (Wilkinson, Fowler, and Partridge 1990; Arnold 1992b). Thus, the jury is still out on the general issue of **G**-matrix constancy as well as on the related issue of how much **G**-matrix similarity is required to model evolution with quantitative genetics. Meanwhile, a series of evolutionary models has been pursued, predicated on the constancy of the **G**-matrix (reviewed by Lande 1988).

The range of evolutionary problems that can be treated using equation 2.4, under the assumption of **G**-matrix constancy, is very broad and includes the evolution of allometric relationships (Lande 1979), behavioral ontogenies (Arnold 1990), plasticity (Via and Lande 1985), sexual dimorphism (Lande 1980b), life history (Lande 1982a), and evolution by sexual selection (reviewed by Arnold 1987b; Pomiankowski 1988). In all these models, selection is described by a surface, as will be discussed below. One solves for the phenotypic composition of the population at equilibrium and for the trajectories of populations evolving toward that equilibrium. A selling point for these models is that they do not assume that populations are in equilibrium. Instead they give a picture of how populations might evolve toward equilibria, and they make predictions about how long the transit to equilibrium might take. A further attractive feature of the models is that they are cast in terms of observable properties of continuously varying traits (genetic variances and covariances, selection gradients). The models provide an agenda of issues to be tackled in natural populations, as well as insights about evolutionary dynamics. Heisler (chap. 5) provides more discussion of quantitative genetic models of evolution as they are applied to the evolution of mating behavior. Partridge (chap. 6) and Roff (chap. 3) discuss how other approaches can be used to complement a quantitative genetic study or model.

The concept of selection as a surface is a fundamental feature of quantitative genetic models of long-term evolution. The basic ideas can be grasped in a two-dimensional phenotypic space. Imagine phenotypic values for one trait plotted on the x-axis and phenotypic values for a second trait plotted on the y-axis. The bivariate values for the distribution in a population might take the form of a bivariate normal distribution. At each point in this distribution we plot the expected lifetime fitness of individuals in a third dimension. Or we might represent expected individual fitness as contours on the y- by x-space. The selection gradient in equation 2.4 is the direction of steepest uphill slope on this *surface of expected individual fitness*, averaged over the bivariate phenotypic distribution. In other words, at each point on the surface of individual fitness we can evaluate the slope or gradient of the surface. If we weight each of these slopes by the proportion of individuals in the population with the appropriate combination of phenotypic values, the weighted average of the slopes turns out to be the selection gradient. The selection gradient in an actual population can be estimated as the slope of a partial regression of individual fit-

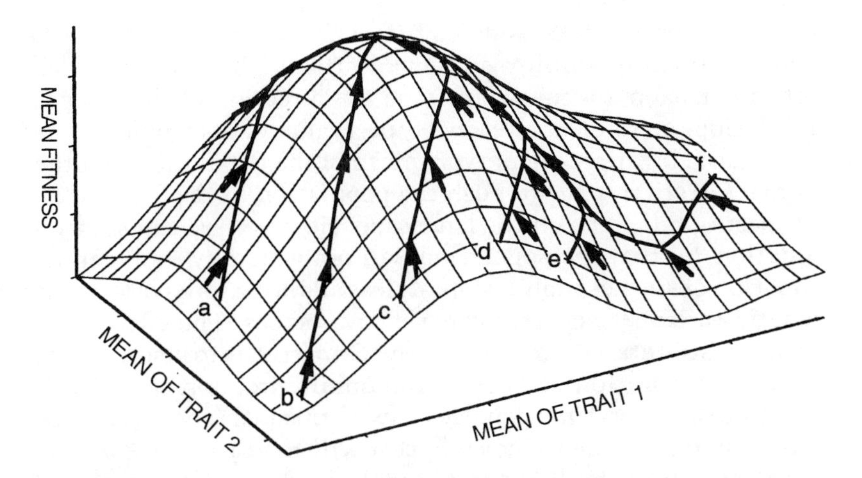

Fig. 2.6 Evolutionary trajectories of populations evolving on a hill-shaped adaptive landscape. The two horizontal axes represent the mean value of two quantitative traits, and the vertical axis represents mean fitness in the population. The genetic correlation between the two traits is strongly positive and is assumed to be constant. The additive genetic variances of the two traits are equal and constant. Heavy lines show evolutionary trajectories from six different starting points (*a–f*). Arrows show selection gradients (directions of steepest uphill slope) at representative points along the trajectories. Because of the genetic correlation between the traits, generally the evolutionary trajectories are not aligned with the selection gradients. However, selection gradients are aligned with evolutionary paths for the trajectory beginning at *b* and for the common trajectory of paths beginning at *e* and *f*. Those exceptional trajectories are known as the eigenvectors of the dynamical system. (Adapted from Lande 1980b.)

ness on a trait, holding the other traits constant (Lande and Arnold 1983).

The evolutionary movement of the population on an *adaptive landscape* is a useful way of summarizing the results of a quantitative genetic model of long-term phenotypic evolution. The adaptive landscape is closely related to the selection surface we just considered. Instead of plotting the phenotypic values of individuals, we plot the average phenotypic values for the two traits. The bivariate mean of the population is now represented as a point in the *x*- and *y*-space. The evolution of the population can be represented as the movement of the point in the *x*- and *y*-space. Instead of plotting expected individual fitness in the third dimension, we plot average fitness in the population. The resulting plot is called an adaptive landscape (fig. 2.6). It will generally be similar to the selection surface we considered first, but the landscape will have gentler curvature (Phillips and Arnold 1989). The selection gradient in equation 2.4 turns out to be the direction of the steepest uphill slope on the adaptive landscape, evaluated at the point representing the bivariate phenotypic mean of the population. In other words, selection tends to push the population in the direction of steepest uphill slope on the landscape. However, we can see from equation 2.4 that the population will not generally evolve in the direction of steepest uphill slope unless all the traits have the same genetic variances and zero genetic covariances. The effect of genetic constraints, summarized in the **G**-matrix, is to distort the direction of evolution away from the uphill direction specified by selection. In general, the population (with unequal genetic variances and/or nonzero genetic covariances) will tend to evolve along a curved trajectory (fig. 2.6), upward on the landscape but not straight toward the highest point of the landscape (Lande 1979). If the adaptive landscape does not change from generation to generation (e.g., if selection is not frequency dependent), a large population will eventually evolve to the highest local point on the landscape, but its trajectory will be more or less circuitous, depending on the constraints embodied in the **G**-matrix (Lande 1979, 1980b). Another message from equation 2.4 is that the rate of evolution in each generation will be proportional to the strength of selection. In other words, the population will evolve fastest when it is located on the steep flanks of a landscape. The evolutionary rate will gradually decelerate as the population climbs toward a gentle summit, until finally the population stops at the summit. The population is now said to be in evolutionary equilibrium. If the population drifts away from the summit, selection will tend to push it back toward the summit. (See Heisler, chap. 5, especially fig. 5.3, for more discussion of bivariate evolutionary trajectories.)

Thus, models for long-term evolution on an adaptive landscape allow us to visualize the joint operation of many themes in evolutionary biology: selection, inheritance, genetic constraints, maximization of fitness, evolutionary equilibrium, rates and paths of evolutionary change. The effects of small population size (e.g., drift in phenotypic mean) and alternative evolutionary outcomes can be incorporated as well. Because such dynamic models integrate so many themes, they provide a useful framework for quantitative genetic studies of behavior. Dynamic models can be used as a vehicle to connect an inheritance study with a wide variety of evolutionary issues. The possibilities of these multiple connections can also provide the motivation for mastering the elements of matrix algebra that are required to use the most powerful version of evolutionary genetic theory.

Summary

1. Heritability and the selection differential are the key concepts for predicting genetic response to selection on a single trait. Thus, heritability and the selection differential can be used to predict responses when deliberate selection is imposed on a single trait (e.g., in selection experiments or in plant and animal breeding). In nature, however, selection acts simultaneously on multiple traits, so other parameters of inheritance and selection are needed to predict response to selection.

2. Additive genetic variances and covariances and the selection gradient are the crucial concepts for predicting genetic responses to multivariate selection. This formulation of selection response assumes no covariance between additive genetic and environmental values and no maternal (or paternal) effects.

3. Additive genetic variances and covariances can be estimated by conducting selection experiments or by analyzing variation and covariation in replicated sets of relatives. Offspring-parent sets or sets of paternal half sibs give less ambiguous estimates of genetic variances and covariances than sets of full sibs.

4. Covariance between additive genetic and phenotypic values and the selection gradient are the critical concepts for predicting response to multivariate selection in the presence of parental (maternal and/or paternal) effects. By parental effects we mean the direct effects of parental phenotypes on offspring phenotypes (e.g., parental care that affects offspring size). The basic tools for estimating genetic parameters when parental effects prevail are measurement of traits in parents and offspring, the production of paternal and maternal half sibs, and cross-fostering.

5. When the attributes of parents directly affect the fitness of their offspring, we need an additional concept, parental selection, to predict genetic responses to selection. In addition to the partial regression of offspring fitness on offspring traits (the ordinary or offspring selection gradient), we need to consider the partial regression of offspring fitness on parental traits (the parental selection gradient).

6. Quantitative genetic models of long-term evolutionary response to selection are predicated on the somewhat controversial

assumption that genetic parameters equilibrate in a mutation-selection balance and remain constant. A wide variety of evolutionary processes have been modeled on this equilibration assumption (e.g., sexual selection, allometric evolution, plasticity, and life history evolution). An evolutionary trajectory on an adaptative landscape is the fundamental concept in these models.

7. These quantitative genetic models can also be used as bridges between behavioral inheritance and a wide variety of evolutionary issues.

Acknowledgments

I am grateful to P. Gowaty and L. Konigsberg for comments on the manuscript, to J. Gladstone for help with graphics, and to M. Kirkpatrick and R. Lande for consultations regarding parental effects and selection. The preparation of this manuscript was supported by NSF Grants BSR 89-06703, BSR 89-18581, and BSR 91-19588.

Appendix 2.1 Some Rules of Matrix Algebra

Many texts cover the basic elements of matrix (or linear) algebra (e.g., Campbell 1965; Finkbeiner 1966; Bradley 1975; Searle 1982). Only a few simple rules for matrix operations are reviewed here.

Addition of two column vectors. The addition of two column vectors is similar to the addition of ordinary numbers. Consider the two-trait case of equation 2.1, $\mathbf{z} = \mathbf{a} + \mathbf{e}$, which represents

$$\begin{bmatrix} z_1 \\ z_2 \end{bmatrix} = \begin{bmatrix} a_1 \\ a_2 \end{bmatrix} + \begin{bmatrix} e_1 \\ e_2 \end{bmatrix}.$$

The vector sum is

$$\begin{bmatrix} z_1 \\ z_2 \end{bmatrix} = \begin{bmatrix} a_1 + e_1 \\ a_2 + e_2 \end{bmatrix}.$$

or $z_1 = a_1 + e_1$ and $z_2 = a_2 + e_2$.

Addition of two matrices. Matrix addition is a simple extension of the rule for adding two vectors. Thus, the two-trait case of equation 2.2, $\mathbf{P} = \mathbf{G} + \mathbf{E}$, is

$$\begin{bmatrix} P_{11} & P_{12} \\ P_{12} & P_{22} \end{bmatrix} = \begin{bmatrix} G_{11} & G_{12} \\ G_{12} & G_{22} \end{bmatrix} + \begin{bmatrix} E_{11} & E_{12} \\ E_{12} & E_{22} \end{bmatrix}$$

or

$$\begin{bmatrix} P_{11} & P_{12} \\ P_{12} & P_{22} \end{bmatrix} = \begin{bmatrix} G_{11} + E_{11} & G_{12} + E_{12} \\ G_{12} + E_{12} & G_{22} + E_{22} \end{bmatrix}$$

Thus, the phenotypic variance for trait 1 (P_{11}) is the sum of the trait's additive genetic variance (G_{11}) and its environmental variance (E_{11}).

In general, the lower left-hand element in a 2 × 2 matrix would not be equal to the upper right-hand element and would be denoted P_{21}, G_{21}, or E_{21}. However, our **P**-, **G**- and **E**-matrices are *symmetric,* which means that $P_{12} = P_{21}$, $P_{13} = P_{31}$, etc.

Multiplying a matrix and a column vector. The basic rule to remember in multiplying a matrix and a vector or two matrices is "rows times columns." For example, consider the two-trait case of equation 2.4, $\Delta\bar{\mathbf{z}} = \mathbf{G}\boldsymbol{\beta}$, which represents

$$\begin{bmatrix} \Delta\bar{z}_1 \\ \Delta\bar{z}_2 \end{bmatrix} = \begin{bmatrix} G_{11} & G_{12} \\ G_{12} & G_{22} \end{bmatrix} \begin{bmatrix} \beta_1 \\ \beta_2 \end{bmatrix}$$

To evaluate the product $\mathbf{G}\boldsymbol{\beta}$, we first multiply the first row of **G** by the column $\boldsymbol{\beta}$. The product is the first element in the first row of **G**, G_{11}, times the first element of $\boldsymbol{\beta}$, β_1, plus the second element of the first row of **G**, G_{12}, times the second element of $\boldsymbol{\beta}$, β_2. In other words,

$$\Delta\bar{z}_1 = G_{11}\beta_1 + G_{12}\beta_2.$$

Likewise, the second element of the product $\mathbf{G}\boldsymbol{\beta}$ is

$$\Delta\bar{z}_2 = G_{12}\beta_1 + G_{22}\beta_2.$$

The rule is easily extended to the many-trait case.

Multiplying two matrices. Again, the rule is "rows times columns," but now there is more than one column. Consider the second term in equation 2.11, $\frac{1}{2}\mathbf{C}_{\mathbf{az}}\mathbf{M}^{\mathrm{T}}$, in the two-trait case. Before we take the product $\mathbf{C}_{\mathbf{az}}\mathbf{M}^{\mathrm{T}}$, we note that $\mathbf{M}^{\mathrm{T}}$ is the so-called *transpose* of the matrix **M**; the rows of **M** are the columns of $\mathbf{M}^{\mathrm{T}}$. In other words, if

$$\mathbf{M} = \begin{bmatrix} M_{11} & M_{12} \\ M_{21} & M_{22} \end{bmatrix},$$

then

$$\mathbf{M}^{\mathrm{T}} = \begin{bmatrix} M_{11} & M_{21} \\ M_{12} & M_{22} \end{bmatrix},$$

The product $\mathbf{C_{az}M}^{\mathrm{T}}$ is then

$$\begin{bmatrix} C_{az_{11}} & C_{az_{12}} \\ C_{az_{21}} & C_{az_{22}} \end{bmatrix} \begin{bmatrix} M_{11} & M_{21} \\ M_{12} & M_{22} \end{bmatrix}$$

The upper left-hand element (with subscript 11) of the product is the *first* row of $\mathbf{C_{az}}$ times the *first* column of $\mathbf{M}^{\mathrm{T}}$; the upper right-hand element of the product (subscript 12) is the *first* row of $\mathbf{C_{az}}$ times the *second* column of $\mathbf{M}^{\mathrm{T}}$; the lower left-hand element (subscript 21) of the product is the *second* row of $\mathbf{C_{az}}$ times the *first* row of $\mathbf{M}^{\mathrm{T}}$, and finally, the lower right-hand element (22) of the product is the *second* row of $\mathbf{C_{az}}$ times the *second* row of $\mathbf{M}^{\mathrm{T}}$. Our matrix product is

$$\begin{bmatrix} C_{az_{11}}M_{11} + C_{az_{12}}M_{12} & C_{az_{11}}M_{21} + C_{az_{12}}M_{22} \\ C_{az_{21}}M_{11} + C_{az_{22}}M_{12} & C_{az_{21}}M_{21} + C_{az_{22}}M_{22} \end{bmatrix}$$

The symbol $\frac{1}{2}$ in front of $\mathbf{C_{az}M}^{\mathrm{T}}$ in equation 2.11 denotes an ordinary number or scalar. Multiplying the scalar $\frac{1}{2}$ times the product $\mathbf{C_{az}M}^{\mathrm{T}}$ means that each element in $\mathbf{C_{az}M}^{\mathrm{T}}$ is to be multiplied by $\frac{1}{2}$. Thus,

$$\frac{1}{2}\mathbf{C_{az}M}^{\mathrm{T}} = \begin{bmatrix} \frac{1}{2}C_{az_{11}}M_{11} + \frac{1}{2}C_{az_{12}}M_{12} & \frac{1}{2}C_{az_{11}}M_{21} + \frac{1}{2}C_{az_{12}}M_{22} \\ \\ \frac{1}{2}C_{az_{21}}M_{11} + \frac{1}{2}C_{az_{22}}M_{12} & \frac{1}{2}C_{az_{21}}M_{21} + \frac{1}{2}C_{az_{22}}M_{22} \end{bmatrix}$$

Matrix multiplication is unlike ordinary or scalar multiplication in that the order of multiplication matters. Notice that $\mathbf{C_{az}M}^{\mathrm{T}} \neq \mathbf{M}^{\mathrm{T}}\mathbf{C_{az}}$!

3

Optimality Modeling and Quantitative Genetics: A Comparison of the Two Approaches

Derek A. Roff

In the last two decades, two approaches to the analysis of variation in life history traits, including morphological, physiological, and behavioral components, have been actively pursued: optimality modeling and quantitative genetic analysis. Optimality modeling is based on two premises: first, that there is some measure of fitness that is maximized by selection, and second, that there are trade-offs between traits. Suppose we wish to predict the value of some trait, such as the optimal age at first reproduction; the various trade-offs are combined to produce a predicted relationship between the trait, or traits, under study and the fitness measure. The predicted value of the trait(s) is that at which the fitness measure is maximized. Traits to which this technique has been successfully applied are the adult size of *Drosophila melanogaster* (Roff 1981), age and size at maturity in fishes (Roff 1984; Stearns and Koella 1986) and amphibians (Stearns and Crandall 1981; Kusano 1982), the relative call rate between neighboring toads (Arak 1988), male mating behavior of sunfish (Gross and Charnov 1980; Gross 1982), the seasonal breeding strategy of the deermouse, *Peromyscus maniculatus* (Fairbairn 1977), and territorial behavior in *D. melanogaster* (Hoffmann, chap. 9).

While the optimality approach is centered on the premise that some fitness measure is maximized, the genetic approach is based on a model that specifies the rules for the transmission of characters from parent to offspring. The number of offspring left by each individual is determined by the relative fitness of each genotype, which is a function of the survival rate of each genotype, its fecundity, and so forth. There is no overall measure of fitness to which the population is presumed to move, though for particular models such maxima may exist (Lande 1979; Charlesworth 1980, 1990). Though in principle a quantitative genetic model could be used to predict the value of a particular trait, the requirements of this approach, in terms of specifying the rules of transmission (i.e., the genetic variance-covariance matrix) are enormous (see below), and I know of only one successful application to a field situation (Morris 1971; discussed in detail in Roff 1990b).

Optimality modeling has come under criticism for several reasons. First, it has been accused of creating "just so stories" (Gould and Lewontin 1979). The gist of this argument is that the organism is broken into single traits and an adaptive story created for each. If this indeed were the approach taken then the method would be invalid. But, as pointed out by Mayr (1983), this interpretation of the method reduces it to a caricature of itself. The optimality approach does not assume that there necessarily exists an adaptive explanation for every trait. It begins with a trait or suite of traits and, recognizing the possibility that present structures or patterns may be influenced to a greater or lesser degree by history and genetic constraints (discussed later), proceeds by combining the observed or presumed trade-offs into a model from which the optimal trait value or combination of values can be calculated. It should be regarded primarily as a method for guiding research, for it serves to direct our attention to the critical factors driving evolution in a particular direction. Thus from a larger set of possible interactions it directs attention to that smaller critical subset on which further research can most profitably be focused.

A second objection to the optimality approach is that it sweeps all the genetic architecture under the rug, and thereby all the potentially confounding influences explicitly dealt with by population genetic theory, such as migration, mutation, genetic drift (Rose, Service, and Hutchinson 1987). But single-locus genetic theory cannot be used to predict most life history traits, for, with few exceptions, traits of ecological interest, particularly those relating to behavior, are not determined by single-locus mechanisms. To move from simple Mendelian models to quantitative genetic models requires simplifying assumptions that cast doubt on their general applicability, doubts that have been acknowledged by geneticists for over twenty years (Lee and Parsons 1968; Robertson 1977; Kempthorne 1977, 1983; Barton and Turelli 1987).

Given the uncertainties over an adequate genetic model for evolution, the optimality approach remains a viable method of analysis. The true test is, of course, the validity of its predictions, and as noted above this approach has been successful both in making predictions and in guiding experimental investigations. Optimality and genetic theory are not competitors but components of a holistic approach: optimality modeling addresses the question, "Where should the population be?" while quantitative genetic analysis deals primarily with the questions, "How does, and can, the population get there?" Ideally we would like a method that is capable of predicting the time course of evolution over both short and long time scales. Such a method may not be practical simply because of the amount of information that would have to be incorporated. The present statistical approach to quantitative genetics would almost certainly not be adequate; a more mechanistic approach utilizing molecular biology, physiology, and developmental biology would be required. At present we do not have sufficient information to tackle this problem. In the interim we can address and answer many relevant questions utilizing a combination of optimality and genetic modeling (see also Cheverud and Moore, chap. 4).

In this chapter I shall consider the elements that constitute an optimality model and ask to what extent these elements are supported by, or are themselves components of, a quantitative genetic model. The chapter is divided into three sections: (1) an overview of quantitative genetic and optimality models, (2) measures of fitness, and (3) the role of trade-offs in the evolution of traits.

An Overview of Quantitative Genetic and Optimality Models

In comparing the two approaches it is necessary to remember that there is no single quantitative genetic model. The one most frequently employed is that developed from the biometric approach and developed most fully within the framework of artificial selection (Falconer 1989; Bulmer 1985): for an alternative model applied to genetic variation in a heterogeneous environment see Gillespie and Turelli (1989). The biometric model has been popularized as a tool for understanding evolution in natural populations largely by the work of Lande (see particularly Lande 1979, 1982a), and, unless otherwise stated, discussion of quantitative genetic analysis in this chapter will assume the biometric model. (For a review of basic concepts in quantitative genetics see Arnold, chap. 2.)

To apply quantitative genetic theory one must assume a constant genetic variance-covariance matrix, which at the very least requires weak selection and a large population size so that mutation can replace genetic variation lost by selection (Kempthorne

1983; Lande 1982a). In general, traits that are analyzed with optimality models are likely to be under strong selection, and hence the assumption of a constant variance-covariance matrix must be doubted. Whether in reality the genetic variance-covariance matrix is constant over more than a few generations is unknown (Turelli 1988a), though there are good theoretical grounds for believing that it will not be stable (Bohren, Hill, and Robertson 1966). Under artificial selection genetic correlations have been found to be stable (Bell and Burris 1973; Cheung and Parker 1974) or to vary substantially and unpredictably (Mather and Harrison 1949; Sheridan and Barker 1974). The genetic variance-covariance matrix may vary at the species level (Lofsvold 1986), making the use of quantitative genetic theory to explore long-term changes questionable.

Another major problem with applying quantitative genetic theory to life history evolution is that estimation of the relevent parameters is extraordinarily difficult, requiring sample sizes that are possible for only a relatively few species (Robertson 1959a; Van Vleck and Henderson 1961; Bogyo 1964; Klein, DeFries, and Finkbeiner 1973). Even estimating whether the number of loci is large or small is an extremely difficult task (Thoday and Thompson 1976; Luckinbill et al. 1987).

Despite the theoretical and experimental shortcomings of quantitative genetic analysis, it is a valuable tool in the analysis of variation in life history, behavioral, and morphological traits. First, artificial selection experiments have demonstrated that, at least in the short term, it can be a useful predictive tool. Second, quantitative genetics provides the metric, heritability, by which the amount of relevant genetic variation can be assessed. This is important, for optimality modeling assumes that sufficient genetic variation exists to permit the organism to evolve the optimal combination of traits. Third, the trade-offs that form the core of any optimality model restrict the set of possible combinations; how rigid such restrictions are can be judged, in part, by the genetic correlations between traits. While such measurements cannot preclude the evolution to a particular predicted optimum, they can give us insights into how readily that optimum can be attained.

A primary purpose of optimality modeling is to establish how much information is required to predict a particular suite of traits. We begin with the minimum set of trade-offs and with these predict where the trait values should lie in parameter space. Finding that the prediction is incorrect does not demonstrate that optimality arguments are false; rather, it is taken as an indication that the model is deficient in some respects. The next task is to modify the model, and again check predictions against observations. In this way we build up a model comprising a minimal set of constraints and trade-offs that adequately predict the observed phenomenon. But this is not sufficient grounds for assuming that the model is correct. There are likely to be a variety of plausible biological models that give the same prediction. The actual test of the model resides in testing the component assumptions, i.e., trade-offs, of the model; the predictions are simply a logical outcome of these assumptions. This principle can be illustrated with the following example. Among a variety of taxa propagule size has been observed to increase with adult size (e.g., plants: Hendrix 1984; insects: O'Neill and Skinner 1990; fishes: Coates 1988). Parker and Begon (1986) and McGinley (1989) independently constructed models that make this prediction. A critical assumption of the model of Parker and Begon is that mortality *increases* with clutch size. In contrast, a critical assumption of McGinley's model is that mortality *decreases* with clutch size. The difference in the models lies not in their predictions but in their component assumptions. Verification of either model is made not solely by virtue of its prediction but by seeing whether the biological assumptions of the model are appropriate in a particular case.

One assumption that is not generally under test is the assumption that natural selection maximizes some measure of fitness, though it is possible that an incorrect fitness measure may be chosen. In the next section I examine different measures of fitness and the circumstances in which each is applicable.

Measures of Fitness

The analysis of a suite of characters is predicated on the assumption that some measure of fitness is maximized, the particular measure being dependent upon whether the population is constrained primarily by density-dependent or density-independent factors.

Density-Independent Measures of Fitness

In the density-independent case, the most obvious fitness measure is the intrinsic rate of increase, r. Suppose we have two clones with intrinsic rates of increase r_1 and r_2, where $r_1 > r_2$: it is obvious that, in an environment where growth is density independent, the clone growing at the instantaneous rate r_1 will eventually come to be numerically dominant in the population, regardless of its starting frequency. When there is age structure the rate of increase satisfies the relationship

$$\Sigma e^{-rt} l_t m_t = 1$$

where t is age, l_t is survival to age t and m_t is the number of female offspring produced at age t. If the population is stationary (i.e., $r = 0$), and both clones are equal with respect to competitive ability, the clone that has the highest expected lifetime fecundity ($R_0 = \Sigma l_{t,i} m_{t,i}$, where i denotes clone), will eventually replace all other clones, since it leaves proportionally more offspring per generation.

The above theory applies only to an environment that is not fluctuating in time. In a temporally fluctuating environment the long-term growth of clone i, r_i, is given by

$$r_i = (\Sigma r_{t,i})/n$$

where $r_{t,i}$ is the instantaneous rate of increase of clone i during the interval t, and n is the number of sequential environments. If the fitness measure is taken to be the finite rate of increase $\lambda_{t,i}$, ($= \exp(r_{t,i})$), fitness is the geometric mean of λ (Cohen 1966). The geometric mean can be increased by either increasing the arithmetic mean or decreasing the variance (Gillespie 1977; Lacey et al. 1983). Because of the latter fact, the most fit strategy may appear to be a conservative strategy, frequently referred to as "bet-hedging" (Slatkin 1974; Seger and Brockmann 1987). Since all environments fluctuate to some degree, bet-hedging is probably a common phenomenon.

A number of studies have attempted to incorporate the concept of intrinsic rate of increase into population genetic models (reviewed in Charlesworth 1980). In general, there is no single parameter that can be equated with the fitness of a genotype under arbitrary selection intensities (Charlesworth 1980, 153–54). However, the maximization of r can be justified by the fact that, in an age-structured, diploid, random-mating population, any mutant that decreases r will be eliminated, while any mutant that increases r will increase in frequency (Charlesworth 1973; Charlesworth and Williamson 1975; Charlesworth and León 1976). Furthermore, Lande (1982a) showed that under polygenic inheritance and weak selection, r will also be maximized.

The reproductive value, $V_x (= (\exp[r_x]/l_x) \Sigma l_t m_t \exp(-rt))$, is sometimes a more convenient parameter with which to work (Schaffer 1979, 1981; Taylor et al. 1974; Goodman 1982). Maximization of V_x is mathematically the same as maximizing r (or R_0) and hence the choice of which metric to use is simply a matter of preference.

In many instances, particularly in analyses involving behavior, a fitness component, such as energy intake per unit time (e.g., MacArthur and Pianka 1966; Fritz and Morse 1985), is maximized under the assumption that this also maximizes r. Such a "restricted analysis" is appropriate provided the suite of traits involved is independent of other traits contributing to the rate of increase.

Density-Dependent Measures of Fitness

Defining a fitness criterion in a population that is regulated by density-dependent interactions is more difficult than when density effects can be ignored. An intuitively appealing criterion is the equilibrium population size. Suppose we have two clones that in isolation equilibrate at population sizes K_1 and K_2, where $K_1 < K_2$, and both clones are individually identical in competitive ability. Suppose now we mix the two clones together in equal proportion to make a population initially of size K_1. Clone 2 will continue to increase, but clone 1 will now be above its equilibrium size, and hence must decline to the extent that the combined numbers of the two clones equal K_1. Since clone 2 can always increase at this level (i.e., population size = K_1), clone 1 will continue to decline, eventually being eliminated. Numerous studies using both simple clonal models and population genetic models have verified that in an age-structured population the number of individuals in the age group subject to density dependence will tend to be maximized (Charlesworth 1980, 168; Iwasa and Teramoto 1980; Desharnais and Costantino 1983). The appropriate measure of fitness is therefore the expected lifetime fecundity, R_0, with density dependence incorporated in the appropriate age group.

Further Complications

In some instances, particularly those involving social organisms, it is necessary to estimate inclusive fitness, which takes into account the increment in fitness accruing to an individual by virtue of interactions with related individuals (Hamilton 1964a). There has been considerable confusion over the correct definition of inclusive fitness (Grafen 1982); the best operational approach is to use Hamilton's rule (Grafen 1984; Creel 1990). Complications arise when the interaction between clones or genotypes is frequency dependent: in this case the optimal set of strategies can be found using the game theoretical approach (Maynard Smith 1982). Although it can be shown that, with a genetic mechanism consisting of a single locus with two alleles, an evolutionarily stable strategy (ESS) may not be possible (Maynard Smith 1981), more complex genetic models do permit populations to converge to the ESS predicted by game theory (Eshel 1982; Bomze, Schuster, and Sigmund 1983).

CONSTRAINTS AND TRADE-OFFS

Theoretical Considerations

If a trade-off is to have an evolutionary impact, it must be genetically based (Reznick, Perry, and Travis 1986). For example, if there is a phenotypic trade-off between number of offspring and their size, (see, for example, Olsson 1960; Hamilton 1962; Strong 1972; Mann and Mills 1979), then selection for increased offspring number must be accompanied by a decrease in offspring size if the trade-off is to have any evolutionary significance.

Optimality models assume that all organisms lie on a line defining the trade-off between any two traits (in principle they can also lie below the line, but such combinations are clearly suboptimal). Further, it is assumed that variance about the line is not genetically based; in other words, that an individual that deviates from the line in the direction of, say, increased offspring number and size will not pass on this tendency (individuals that deviate in the opposite direction will have decreased fitness and hence will not be favored, by arguments from either optimality or genetic theory).

In contrast, the quantitative genetic model examined by Via and Lande (1985) assumes that this variance is genetically based, and hence that there are some individuals capable of attaining any joint optima. The importance of the genetic correlation is in mediating the rate at which the joint optima are achieved. The assumption that joint optima are always possible is clearly not universally true: for example, if the amount of resources available for the production of eggs is strictly limited, there must be a functional trade-off between size and number of offspring. Charlesworth (1990) incorporated such functional constraints in the quantitative genetic framework and showed that these lead to convergence in the predictions of optimality and genetic analyses.

Table 3.1 Signs of the genetic correlations between the life history traits discussed in this chapter

	$s(2)$	$m(1)$	$m(2)$	$m(3)$
$s(1)$	<0	-1	>0	>0
$s(2)$		>0	-1	0
$m(1)$			<0	<0
$m(2)$				0

Source: Adapted from Charlesworth 1990.

With only a single trade-off that is functionally based, the genetic correlation at equilibrium will be -1 (Charnov 1989), but with more traits, "positive genetic correlations are mathematically possible between some pairs of components of fitness that are under constraints that imply negative trade-offs between them" (Charlesworth 1990, 525). This proposition can be illustrated with the following simple model. A hypothetical organism lives for three years, breeding in years 1, 2, and 3. Total egg production is fixed, so that $m(1) + m(2) + m(3) =$ constant, where $m(i)$ is the number of female offspring produced at age i. Survival between ages i and $i + 1$, $s(i)$, declines with reproductive effort at age i, according to the relationship, $s(i) = 1 - m(i)^2$. There are thus three trade-offs. Population size is assumed to be determined by density-dependent effects in the immature stage; hence, fitness is proportional to lifetime fecundity estimated from age 1, $m(1) + s(1)m(2) + s(2)m(3)$. Given the above constraints, the signs of the genetic correlations between pairs of traits can be determined (table 3.1). The genetic correlations between the two functionally defined trade-offs between survival and reproduction are both -1. But other possible genetic correlations that one might hypothesize to be negative are in fact positive. The important point is that in these cases no biological trade-offs are explicitly built into the model, and hence the a priori hypothesis of a negative trade-off is, in fact, incorrect.

Types of Constraints and Trade-offs

Constraints, or trade-offs, can be conveniently divided into five categories: genetic, phylogenetic, physiological, mechanical, and ecological. These categories are not mutually exclusive, and behavioral traits could fall into more than one category: for example, some physiological factor such as nutritional deficiency that causes an animal to forage more may increase its susceptibility to predation, an ecological factor.

The first four categories have been termed internal constraints; the last one, an external constraint (Gans 1989). As with the division into five categories, this division into two groups is by no means exclusive: external factors may play a role in defining internal constraints, as, for example, when wave action sets mechanical limits on the morphology of intertidal organisms (Denny, Daniel, and Koehl 1985).

Genetic constraints. Optimality models assume that there is always sufficient genetic variation to permit evolution to the optimal combination. In most laboratory examinations, an abundant amount of additive genetic variation has been found (Roff and Mousseau 1987; Mousseau and Roff 1987), suggesting that additive genetic variation is unlikely to be a factor preventing an organism from evolving under selection on one trait. However, genetic correlations among traits will at the very least slow the rate at which a particular optimum is approached, and hence traits may be far removed from their equilibrium values (Arnold, chap. 2). The question of equilibrium versus nonequilibrium conditions is very important and deserves greater study, both theoretically and empirically (Cheverud and Moore, chap. 4).

Phylogenetic constraints. Evolution of particular traits may constrain future evolution by eroding genetic variation. For example, the evolution of flightlessness in penguins and rheas has clearly led to the loss of genetic variation in wing form to the point that it is highly unlikely that future evolutionary branches of these avian families will become capable of flight. Phylogenetic constraints can be incorporated into optimality models by restricting attention to a particular taxon: for example, we might consider the optimal life history of penguins subject to the restriction that flight is not a permissible option.

Physiological constraints. By physiological constraints I mean those processes that act internally in the organism, but are above the level of the gene or genes. Development may be constrained within certain bounds, such that the likelihood of following one pathway may be greater than others (Maynard Smith et al. 1985). Demonstrating that a physiological constraint exists is very difficult, particularly as it might be exhibited only under specific conditions. It is important to recognize such interactions when designing experiments to demonstrate the existence of trade-offs. For example, most laboratory examinations of the cost of reproduction, such as studies of differences in longevity between virgin and mated females, or the correlation between longevity and fecundity (Reznick 1985; Bell and Koufopanou 1985), have been based on physiological constraints and have been carried out under relatively benign conditions; therefore, the results must be interpreted with caution. In this regard it is significant that reproductive costs have most frequently been demonstrated under conditions of stress such as starvation (Bell and Koufopanou 1985).

Mechanical constraints. Mechanical constraints set very clear limits within which organisms must operate. Growth in hermit crabs, for example, can be strongly constrained by available shell size, and when given access to larger shells individuals may switch from reproduction to growth (Bertness 1981).

Ecological constraints. I define an ecological constraint as one that is external to the organism and is dependent upon the particular environment the organism occupies. For example, female *Daphnia* that contain eggs suffer higher rates of predation than males or females not carrying eggs (Hairston, Walton, and Li 1983; Koufopanou and Bell 1984). Females are more visible when they contain eggs (Hairston, Walton, and Li 1983), but the increased susceptibility is eliminated when *Daphnia* are placed against a dark background (Koufopanou and Bell 1984). Predation risk may also create a trade-off between growth and survival by limiting foraging time or habitat (Wilson, Leighton, and Leighton 1978; Murdoch and Sih 1978; Sih 1980, 1982; Streams and Shubeck 1982; Werner et al. 1983; Weissburg 1986).

Physiological and mechanical constraints are themselves functions of the ecological setting, and in this respect ecological considerations may be of paramount importance in defining trade-offs: for example, the constraint on growth in hermit crabs due to small available shell size is a mechanical constraint contingent on the distribution of shell sizes, an ecological factor.

Measuring Trade-offs

Methods of measuring trade-offs can be divided into four categories (Reznick 1985):

1. Phenotypic correlations: measurement of the correlation between two traits at the level of the phenotype, and involving no manipulation of the organisms
2. Experimental manipulations: direct manipulation of a single factor while keeping all other factors constant, or at least randomly assigned
3. Sib analysis: quantitative genetic estimation of the correlation between the two traits using covariation between individuals within and between families
4. Correlated responses to selection: measurement of correlated changes in one factor in response to selection on another.

The first two categories measure only the phenotypic association between two traits, while the second two address the issue of whether the trade-off can produce evolutionary change.

Phenotypic correlations. There are many examples of correlations based on data from unmanipulated situations, but because of the problem of inferring causation from correlation, the interpretation of such data is fraught with difficulty (Partridge and Harvey 1985; van Noordwijk and de Jong 1986; Pease and Bull 1988). This problem can be illustrated with the following simple example in which there is a trade-off between egg number and egg size. Suppose that this correlation arises because of geometric constraints; for example, a limited space inside the mother. The relationship will then take the form

$$\text{Number of eggs} \propto \text{Size of mother/Size of egg}$$

Taking logarithms, we have

$$Y \propto B - X$$

where Y is log(egg number), B is log(body size), and X is log(egg size). Provided the body size of the mother is constant, there will be an observed trade-off between the number of eggs and their size. But suppose body size varies and this is not taken into account; what will be the observed correlation? If variation in body size is small there will still be an overall negative correlation between Y and X (upper panels, fig. 3.1), but this correlation will decline as variance in body size increases, and if the variance is moderate no trade-off may be discernible (middle panels, fig. 3.1). Furthermore, if there is a correlation between egg size and body size (such as has been ob-

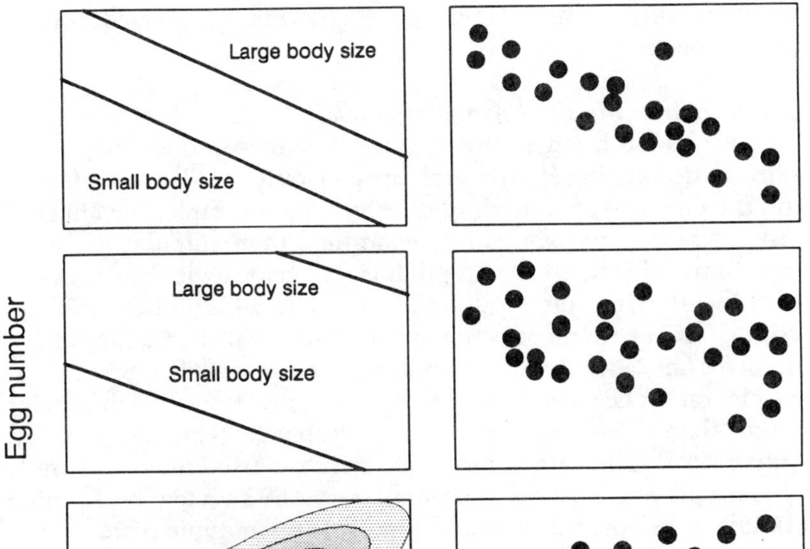

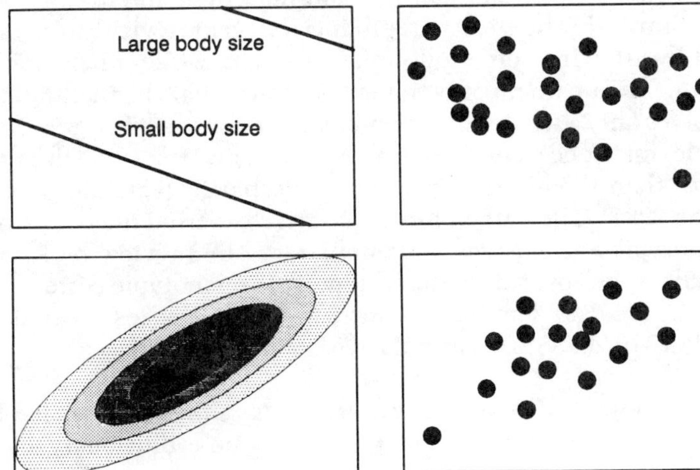

Fig. 3.1 Possible correlations between egg number and egg size when effects of body size are not taken into account. The panels on the left indicate the actual relationship; the panels on the right, what might be obtained by sampling. In all cases egg number decreases with egg size for a fixed body size (solid lines). For a fixed egg size, egg number increases with body size. *Upper panels:* The range in body size is small and the observed phenotypic relationship indicates a trade-off between egg size and egg number. *Middle panels:* The range in body size is large, and swamps the negative relationship between egg size and egg number. *Lower panels:* There is a positive correlation between egg size and body size (the darker the shading, the higher the probability of a particular body size/egg size combination): small females have small eggs and low fecundities, while large females have larger eggs and larger fecundities. As a consequence, a positive relationship between egg size and egg number is obtained. (Adapted from Roff 1992.)

served in Arctic char, *Salvelinus alpinus* [Jonsson and Hindar 1982], the wasp *Trichogramma embryophagum* [Klomp and Teerink 1967], the copepod *Pseudocalanus* sp. [Hart and McLaren 1978], and garter snakes, *Thamnophis sirtalis* [Garland, chap. 12]), the correlation between number of eggs (Y) and egg size (X) may be reversed (bottom panels, fig. 3.1).

Experimental manipulation. It is clearly better to examine a proposed trade-off by manipulating the variables in question, at the same time keeping all other factors constant or at least randomizing the effect of confounding variables. For example, the effect of brood size on survival can be examined by manipulating brood size. Similarly, the effect of reproduction on longevity can be examined by allowing some individuals to mate while others are kept virgin. Great care must be taken to ensure that the manipulation involves only a single change: increased longevity of virgins, for example, might be a consequence of a lack of interaction with males rather than the physiological stress of mating (Partridge and Farquhar 1981). The value of manipulative experiments over simple phenotypic correlations is demonstrated in a review by Reznick (1985) on the costs of reproduction: of 33 phenotypic correlations, only 22 (66.7%) demonstrated the presence of a cost or trade-off, while 17 of 20 (85%) manipulative experiments did so.

Sib analysis and selection experiments. Unless the trade-off in question has a genetic basis it will have no evolutionary significance. Thus it is essential to demonstrate the genetic basis of the traits, either by a comparison of variation among families or by selection. However, the finding of Charlesworth (1990) that at equilibrium positive genetic correlations may exist despite functional constraints that imply negative trade-offs suggests caution in the interpretation of genetic correlations considered in isolation from other interacting traits. A negative genetic correlation indicates a trade-off between two traits; a positive correlation indicates that the functional basis of any trade-off cannot directly involve the two traits alone. A functional analysis of trade-offs is an essential companion to a genetic analysis (Pease and Bull 1988). However, there is the possibility of getting stuck in the problem of infinite regress, with the search for the functional basis of the trait leading one into very detailed physiological investigations.

The most serious problem with the estimation of genetic correlations by a comparison of variation among families is a logistic one: enormous sample sizes are required to adequately estimate the genetic correlation (Van Vleck and Henderson 1961; Klein, DeFries, and Finkbeiner 1973). Selection experiments require fewer individuals per generation, but, depending upon the selection intensity that can be applied and the heritability, a large number of generations of selection may be required to observe a consistent response, making these experiments very labor-intensive. A more serious problem with selection experiments is that artificial selection is practiced, in general, on only a single character: for example, nesting behavior in mice (Lynch, chap. 13), migration propensity (Dingle, chap. 7), mating success in male *D. melanogaster* (Partridge, chap. 6), or territorial success in *D. melanogaster* (Hoffmann, chap. 9). However, natural selection operates on the suite of characters contributing to fitness. For example, there have been numerous attempts to select for changes in the phototactic behavior of *D. melanogaster.* These experiments have generally been successful, but detailed investigations of the mechanism producing changes in phototaxis have shown that the actual focus of selection is the eye pigment; that is, artificial selection has operated by producing flies that are visually impaired (Kohler 1977; Markow and Clark 1984). In the wild, visual impairment is likely to have serious consequences on other components of fitness. Selection for changes in behavior must be integrated into the life history, and artificial selection may give little insight into the malleability of the trait under natural conditions (see also Partridge, chap. 6).

Discussion

According to strict optimality arguments the trade-off between traits represents a barrier beyond which the organism cannot pass. Analysis of variation requires both a functional analysis of the mechanisms generating the trade-offs and an analysis of genetic correlations, the latter indicating the extent to which the presumed trade-offs are of evolutionary significance. The number of studies attempting to measure genetic correlations are relatively few, mostly concern morphological traits, and have standard errors on the estimates generally so large that all one can really say is that the traits are genetically correlated. Cheverud (1988a) has suggested that differences between phenotypic and genetic correlations may largely be a consequence of relatively small sample sizes, which if true would greatly reduce sampling problems since the phenotypic correlation is easily estimated. This proposition has been challenged by Willis, Coyne, and Kirkpatrick (1991), who criticized the method of analysis and the types of traits examined. Phenotypic correlations may reflect the sign of the genetic correlation when the two component traits are morphological, but may be inadequate guides for other types of comparisons (Roff and Mousseau 1987; Willis, Coyne, and Kirkpatrick 1991; Roff 1992).

In constructing an optimality model we begin with the premise that all trade-offs are contained within the model. This is also true of a quantitative genetic model, since in such a model one assumes that all correlated traits are included in the model. One of the purposes of constructing an optimality model is to test the hypothesis that a particular set of trade-offs is adequate to predict the character under study. Consider the relationship between growth and survival in insects. Fast growth may be favored because it increases *r* (Lewontin 1965), or because it enables the insect in an ephemeral environment to complete development before conditions in the habitat become inhospitable (Fairbairn 1984; Roff 1991). However, to achieve a high growth rate the immature insect must forage more frequently, and, as shown with several species, increased foraging can increase mortality rate (Weissburg 1986). Thus, there will be a trade-off between growth rate, foraging frequency, and survival. Selection for increased foraging frequency should cause the population to move along the observed trade-off function to a new position, *unless foraging frequency is also genetically correlated with some other behavior that changes the susceptibility of the insect to predation.* In all optimality models the range of responses is prescribed; the question is, "To what extent are such assumptions reasonable, and what addition must be made to the model if there is reason to believe that the assumption may not be valid?" Suppose the organism can vary its foraging behavior in such a manner as to reduce its susceptibility to predation. If such is the case, why do not all members of the population adopt the behavior? A reasonable conjecture is that a behavior that reduces susceptibility to predation also reduces foraging efficiency: for example, in the presence of potential predators, immature backswimmers (*Notonecta*) and small bluegills (*Lepomis*) select habitats that are relatively free of predation but are suboptimal for growth (Murdoch and Sih 1978; Streams and Shubeck 1982; Werner et al. 1983). Thus the trade-off between growth and survival resulting from foraging frequency can be extended by incorporating the proportion of time spent in each habitat. Because, in this instance, the costs of foraging are determined primarily by ecological factors, the assumption of fixed boundaries is probably reasonable. Therefore, while selection for an organism that spent more foraging time in the habitat favorable for growth but containing predators would lead to an increased growth rate, it would also necessarily lead to a decrease in survival, entirely predictable from the phenotypic correlations. This result is a logical outcome of the mechanism underlying the assumed trade-off. By basing analyses on the underlying mechanisms, we can generate models in which the trade-offs will be functionally assigned and thereby be reasonably sure that the combination predicted by optimality modelling will be congruent with that expected on the basis of genetic variation, as suggested by the analysis of Charlesworth (1990). Unfortunately, in many instances, particularly those involving physiological costs, our ignorance of mechanisms may not permit such models. Therefore, it is likely that at least some elements in a model will always be based on empirical relationships that cannot be demonstrated to be fixed constraints.

Optimality modeling serves as a means of generating hypotheses and to this end is useful even if it is later shown that one of the assumptions is incorrect: the model is then revised and further tests made. Optimality models and quantitative genetic models are not mutually exclusive methods of analysis. Both approaches have deficiencies and should be viewed as complementary rather than antagonistic (see also Garland, chap. 12). It is much easier to obtain phenotypic correlations than genetic correlations, and the former may be reasonable estimates of the latter, at least for morphological traits (Cheverud 1988a). Manipulative experiments can examine the degree to which trade-offs are functionally based and hence are likely to be genetically correlated. Optimality models are quickly and easily derived from phenotypic relationships; sensitivity analysis of such models indicates the appropriate experiments to be undertaken, which may well be selection experiments. Obviously, if a model does not fit the data, one can add components until a fit is obtained, but this is not a deficiency of the approach, for it should be viewed as a means of generating testable hypotheses, not as an end in itself. The fit of a model to data gives us reassurance that a sufficient number of factors have been taken into account, but the fit is not itself proof that the component relationships are correct (Feller 1940). A model is only the logical outcome of a set of interactions; the onus is on the modeler to demonstrate that these interactions are correct. This caution applies to those developing quantitative genetic models as well as those working from the optimality perspective.

Summary

Analysis of the evolution of traits may be done via optimality modeling or quantitative genetic analysis. Both methods have strengths and weaknesses, and should be viewed as being complementary to each other, rather than as exclusive alternatives. A study should, if feasible, adopt both methods, a marriage that is likely to strengthen conclusions. The basic assumptions of optimality modeling are that there is some measure of fitness that is maximized, and that trade-offs and constraints limit the possible set of parameter combinations. Although quantitative genetic theory does not make the assumption of fitness maximization, it does make a number of assumptions, such as weak selection, multivariate normality, and a constant variance-covariance matrix, which are highly suspect, particularly with traits that are likely to be under strong selection. Provided the trade-offs and constraints postulated for the optimality and quantitative genetic models have the same functional basis, the predicted optima at equilibrium will be the same by the two methods of analysis (Charlesworth 1990). Thus, analyses should center upon finding those constraints for which a functional basis can be assigned. Achieving this will require a combination of manipulative experiments and experiments designed to estimate the genetic correlations among traits.

Acknowledgments

I am most grateful for the critical comments of Drs. D. J. Fairbairn, M. Bradford, C. Boake, F. Breden, and B. Buchanan. This work was supported by an operating grant from the Natural Sciences and Engineering Research Council of Canada.

4

Quantitative Genetics and the Role of the Environment Provided by Relatives in Behavioral Evolution

James M. Cheverud and Allen J. Moore

Parental care in diverse taxa. (*a*) A family of rhesus macacques (*Macaca mulatta*) on Cayo Santiago. (Photograph by J. Cheverud.) (*b*) A mother scorpion (*Vejovis carolinus*) carrying her young. (Photograph by A. Moore.)

Behavioral Ecology and Interactions among Individuals

Behavioral ecologists have repeatedly shown that the actions of relatives in one generation can have a significant effect on traits expressed in subsequent generations. For example, the importance of both male and female parental care and its effect on offspring survivorship, fecundity, fertility, and behavior is well documented in a wide variety of organisms from arthropods to mammals. Biparental care of offspring is relatively common, especially in birds and mammals (Rubenstein and Wrangham 1986). Alternatively, just the male or just the female may invest in the offspring after fertilization (Ridley 1978; Gross and Shine 1981; Gross and Sargent 1985).

Relatives other than parents may also influence offspring characteristics. Helping behavior is an example in which the behavior of older siblings can have a significant effect on offspring survivorship (Emlen 1984). Eusociality is an extreme example of a social system in which the behavior of relatives (siblings, aunts) influences offspring fitness (Wilson 1971). Not all interactions within families are positive: siblings may compete for mates (Hamilton 1967) or other resources (Clark 1978). In all these examples both the form of the investment and the character that is affected are often behavior, including acts such as aggression, brooding, diapause, feeding, and defense.

These patterns in intrafamilial interactions and their effects are commonly interpreted in light of ecological influences (Gross and Shine 1981; Gross and Sargent 1985; Wrangham and Rubenstein 1986). Game theory, optimization models, the theory of parent-offspring conflict, behavioral reciprocity, and other ecologically or socially based explanations are typically cited to provide the underlying theoretical contexts (Krebs and Davies 1991; Roff, chap. 3). Evolutionary mechanisms such as kin selection, natural selection, or sexual selection may be invoked to provide an evolutionary cause of the observed pattern of behavior (although these are different levels of explanation and not alternatives to the ecological influences) but are often not investigated directly. Rarely are genetic mechanisms or consequences underlying these forms of social behavior considered, except to note that additive genetic variation is required for an evolutionary response to selection. A genetic basis underlying behavior is assumed, but not investigated. The role of genetic variation in constraining and guiding the evolutionary process is therefore not considered in many behavioral studies.

One consequence of the ecological, or "optimality," perspective is a focus by behavioral ecologists on the evolutionary outcome rather than the evolutionary process (e.g., chapters in Krebs and Davis 1991; Roff, chap. 3). Under the optimality approach populations are considered to be at or very near a stable intermediate optimum for the characters being studied. Dynamic programming, game theoretical, or other optimality models and approaches are used to identify an expected character state. The behavior of the animal under study is then judged in relation to this theoretical optimum. Thus, it is the outcome, or expected outcome, that is of interest using this approach. The outcome is of course predicted based on the current behavior of animals. While this approach has considerable appeal, and has been very powerful in understanding initially perplexing behavior (e.g., Maynard Smith 1982; Krebs and Davies 1991), it does not always explain the observed patterns of behavior (Rose, Service, and Hutchinson 1987; Riechert 1993). Deviation of the population from optimal behavior is usually seen as a failure of the specific optimality model employed, necessitating revisions of the model or tests of the assumptions (Krebs and Kacelnik 1991). Interestingly, the possibility that the population is not in equilibrium is rarely considered (but see Riechert 1993 for an exception).

Optimality approaches need not be evolutionary (Rose, Service, and Hutchinson 1987) since optimality theories do not specifically require that a process be considered. An optimality approach, by itself, is equally consistent with argument from design models and can be derived from nonevolutionary theories such as William Whewell's belief in an all-wise designer (Ruse 1979). The optimality approach is also consistent with non-Darwinian evolutionary theories, such as use and disuse. All that is required by optimality theory is that some mechanism be supposed; the actual mechanism is not specified by the theory itself. That an optimality approach can become nonevolutionary is perhaps not surprising since the basic mathematical underpinnings of optimality arguments derive not from biology, but from economics.

Optimality arguments become evolutionary when the process of evolution by natural selection is invoked to provide a mechanism by which a population reaches an adaptive peak. For this fusion of evolution and optimality to succeed, the trade-offs predicted by the optimality approach must have some underlying genetic basis (Rose, Service, and Hutchinson 1987; Smith 1991; Roff, chap. 3). Thus, without eventually incorporating genetic experiments, such optimality approaches ultimately have limited utility and may provide misleading results (Rose, Service, and Hutchinson 1987; Eshel and Feldman 1991).

Optimality theory also fails when populations are not near stable equilibria. The expectation of population equilibrium is influenced in part by the underlying model of fitness surfaces that is employed. In contrast to the assumption of stable single optima often used in interpretation of behavioral ecological models and data, Wright (1982, 1988) conceived of fitness surfaces that were in constant flux over evolutionary time, like waves on the ocean, with many local minor optima of lower fitness appearing and disappearing at a relatively high rate due to environmental fluctuations. Evolutionary models using these fitness surfaces, and other non-Gaussian surfaces (Wagner 1988; Bürger 1986), do not necessarily predict stable optimal character combinations. One might expect behavior, being so directly involved with the organism-environment interface at which selection takes place, to be especially subject to fluctuating selection pressures over evolutionary time. In this view, we do not commonly expect to find populations in evolutionary equilibrium.

To resolve the problem of whether populations typically reside at selection optima empirical studies are required. While we do not yet have data that bear directly on this question, we can infer the likelihood of population equilibrium from studies of selection. If populations reside at optima, empirical measurements of natural selection should discover stabilizing selection on individuals and a population selection gradient of zero. There are a growing number of empirical studies that have quantified the direct effects of selection, i.e., measured selection gradients. The most common finding is for directional selection to be present (Endler 1986; Conner 1988; Moore 1990a; Anholt 1991), although sometimes in combination with a stabilizing component (Endler 1986; Moore 1990a). Populations seem to deviate from local optima and to be subject primarily to directional selection. Unfortunately, studies of selection do not indicate how "near" or "far" from a local optimum a population may reside.

Evolutionary biologists therefore consider the evolutionary process itself, as well as evolutionary outcomes, a subject for empirical and theoretical consideration. Models of the adaptive evolutionary process all basically indicate that evolution proceeds by natural selection on heritable phenotypic variation, as first defined by Darwin (1859). Both the rate and direction of evolution are influenced jointly by the strength and direction of natural selection and the magnitude and pattern of heritable variation. If the pattern of natural selection is stable over long periods of evolutionary time, populations will eventually reach optima, although eventually may be a very long time if the direct path to the optimum is inconsistent with patterns of heritable variation (Lande 1981b).

For these reasons, it seems worthwhile to consider models of evolutionary process in behavioral ecology. Optimality models may define the relative performance of character sets in a given environment and thus define patterns of natural selection, but they must be combined with information on patterns of heritable variation to obtain a view of the process of evolution itself. If populations are not at an evolutionary equilibrium, a quantitative genetic approach is required. A quantitative genetic approach can therefore be considered complementary to the optimality approach by permitting a consideration of the process as well as the outcome of evolution under certain conditions (Charlesworth 1990, 1993; Smith 1991; Roff, chap. 3). However, when populations are far from an equilibrium, a genetic approach is the only alternative. The process of evolution is a rich and interesting phenomenon in itself and a better understanding of it will lead to a fuller understanding of the outcomes produced.

In this chapter we outline how interactions among relatives and the resulting evolutionary process can be considered by a quantitative genetic approach. A large body of theory and empirical research has shown that such interactions are prevalent and have important behavioral consequences. The models that we present suggest that interpreting the direction and, perhaps, the ultimate outcome of evolution without considering the evolutionary effects of the environment provided by relatives can be misleading. Negative responses to positive selection are possible because of the effects of relatives. At population equilibrium this may be of little consequence. However, for those populations still evolving, these nonadaptive responses are an interesting and potentially fruitful area of study for behavioral ecologists (Stamps 1991).

The Influence of Family on Fitness: Maternal Effects

An alternative to the optimality approach to the study of the effects of one individual on a related individual's fitness comes from quantitative genetic models of evolutionary processes. Consider, for example, the effect a mammalian mother can have on her offspring through suckling. Clearly, the quality and quantity of the resources provided to the offspring depends, in part, on the mother. Further, the nutritional quality of her milk is likely to have a large effect on many aspects of the offspring's biology. Such cross-generational effects on characters, or the effect of the environment provided by a parent on the characteristics of its offspring, are treated by quantitative geneticists in models that include "maternal effects" or "kin effects" (Cheverud 1984a). Maternal effects contribute an environmental source of variance to the offspring. This effect contributes a common familial environmental variance, V_{Ec}, to the total environmental variance contributing to the offspring phenotype (Falconer 1989, 158–60). However, variation among mothers related to the quality of the environment that they produce for their offspring (in the present example, maternal care through nursing) may be influenced by genetic differences among mothers. Because they are partly influenced by genetic factors, maternal effects themselves can be subject to selection and evolve.

Models of maternal or kin effects can include any genetic relationship. Thus, a quantitative genetic perspective is useful in understanding the broad topic of the role of environmental influences in evolution, especially the role of resources or care provided by relatives (Stamps 1991). The term "maternal effect" itself developed from a consideration of the evolution of characters in mammals, in which contributions from the mother are often substantial (Legates 1972; Willham 1972). Maternal effects were originally defined as the effects of the environment provided by the mother on the growth and development of her offspring (Dickerson 1947; Willham 1963, 1972; Legates 1972; Cheverud 1984a). However, this concept can be generalized to incorporate the effects of more than just mothers: the effects of interactions between fathers and their offspring, grandparents and their grandchildren, and siblings can just as easily be considered. "Kin effect" may be a more descriptive term than "maternal effect" (Willham 1963, 1972; Cheverud 1984a; Lynch 1987). Nonetheless, the term *maternal effect* is well established in the literature, so we will continue to employ it in our discussion, keeping in mind that other relationships can and will be considered as well.

In quantitative genetics, two types of maternal effects can be distinguished (Falconer 1989). In the first, resemblance among offspring is influenced by the common environment provided, typically by the mother. Whenever this type of maternal effect is important, there is an increased resemblance between the offspring, but the resemblance between the offspring and the relative providing the common environment is not affected. An example of this might be the effect of parental care on body weight. This type of maternal effect is often conceptualized as acting through a single character, usually referred to as maternal performance (e.g., Young and Legates 1965). Maternal performance is a composite phenotype composed of all of the (maternal) features that have an effect on a particular offspring phenotype. For example, maternal characters contributing to maternal performance for offspring body weight at weaning in mammals include nest site selection, preparation and form of the nest, milk production, and care of the young. Such characters combine as maternal performance to affect offspring body weight.

A second type of maternal effect can be identified when the value of the phenotype in the relative providing the environment influences the expression of the same phenotype in the offspring. Thus, both resemblance among the offspring and the resemblance between the offspring and the relative is influenced by this maternal effect. An example of this type of maternal effect occurs when the quality of parental care provided by an adult is influenced by the quality of parental care it received as a dependent. Another example is offspring body weight, which reflects the body weight of the mother, probably through the relationship between body weight and milk supply (Falconer 1989).

These two types of maternal effects are not mutually exclusive. The distinction arises, in part, from the way in which the maternal effects are detected. In the first case, the maternal effect is measured by relating differences in the offspring character(s) of interest to differences among mothers (hence maternal *performance*). There is no requirement that the *cause* of the maternal effect be known. In the second instance, the effect of a specific maternal phenotype is investigated. Thus, in this second approach, one is interested in both the maternal character itself and its effect on the *same* offspring character.

Maternal effects have long been of interest in agricultural genetics, and most of our theoretical and empirical knowledge of these effects comes from the animal breeding literature (Dickerson 1947; Willham 1972; Falconer 1965; Hanrahan 1976). In this tradition, maternal effects "are a frequent, and often troublesome, source of environmental resemblance, particularly among mammals" (Falconer 1989, 159). Maternal effects confound traditional estimates of heritability because the maternal environment causes phenotypic similarities among siblings raised by the same mother.

Yet when estimating heritability, we usually assume that the environment is randomly distributed with respect to the genetic relationship used in our estimation procedure. Thus a variety of experimental designs have been generated to eliminate this confounder of heritability estimates. The most familiar is the paternal half-sib design, in which several dams are mated to each sire, and the relationships through the sires alone are used to estimate heritability (Arnold, chap. 2; see below).

This view of maternal effects as contributing to resemblance among relatives through a common environment has relegated them to the theoretical sidelines in most evolutionary genetic studies. Why focus our attention on a confounding factor that must be controlled to obtain accurate estimates of important evolutionary parameters, such as heritability? Yet agricultural geneticists were well aware, both theoretically and empirically, that maternal performance and its component maternal characters are themselves inherited and, unlike other environmental factors, evolve under natural selection acting directly on maternal performance itself and as a correlated response to selection on offspring characters (Dickerson 1947; Willham 1963, 1972; Falconer 1965). Also, as noted above, behavioral ecologists have long been interested in the role of the parental or sibling environment on offspring fitness. Thus, a consideration of maternal effects, and an incorporation of maternal effects into evolutionary models, would seem to be timely despite past neglect.

A few recent evolutionary models have begun to include maternal effects. Cheverud (1984a) considered maternal effects as a special form of kin effects, using Dickerson's (1947) and Willham's (1972) models. Riska, Rutledge, and Atchley (1985) combined this model with persistent environmental effects, where both genes and environments are transferred across generations. Lynch (1987) also considered the extension of Willham's models to multiple kin. Finally, in a series of papers, Kirkpatrick and Lande (1989; Lande and Kirkpatrick 1990) generalized and combined the previous models into a single theoretical framework for the evolution of offspring characters affected by maternal environment and of the maternal characters composing maternal performance itself. These papers have all highlighted the potential importance of maternal effects under a wide variety of evolutionary conditions.

The models developed in agricultural genetics for the evolution of characters affected by maternal environment predict several counterintuitive results that help explain some practical difficulties encountered in artificial selection programs (Dickerson 1947; Willham 1972; Hanrahan 1976). They predict (1) that selection for an increased value of an offspring character could result in an expected evolutionary *decrease* in that character (i.e., opposite to the expected result) and (2) that selection for increased offspring phenotypic value may result in an expected *decrease* in maternal performance. For example, selection for increased early growth rate could lead to a decrease in milk production (see below). These results are counterintuitive given commonplace evolutionary thinking in which natural selection is thought to result in optimal phenotypes and be invariably adaptive; the optimality expectation is that growth should be related to quality and quantity of milk received. However, in the recent quantitative genetic models of maternal effects developed within an evolutionary framework, it has been shown that evolution may proceed in opposition to adaptation. This occurs because selection, by itself, does not produce evolution. It is filtered through the genetic system of inheritance, which we ignore at some risk.

Examples of Maternal Effects

The best example of the potential importance of maternal effects is provided by studies of body weight in mammals, the phenotype we know the most about. Maternal effects play a predominant role in preweaning offspring phenotypes and are significant even in adulthood (Young and Legates 1965; Young, Legates, and Farthing 1965; El Oksh, Sutherland, and Williams 1967; Ahlschwede and Robison 1971; Rutledge et al. 1972; Kuhlers, Chapman, and Furst 1977; Herbert, Kidwell, and Chase 1979; Atchley and Rutledge 1980; Cheverud et al. 1983; Riska, Rutledge, and Atchley 1985). In mammals generally, the proportion of phenotypic variance in offspring body weight due to maternal environment ranges from a high of about 50% at weaning to a low of about 10% in the adult, while heritability tends to remain stable across ages (Cheverud et al. 1983). The decline in percentage of variance due to maternal effects during development is countered by an increase in percentage of variance due to nonmaternal environment. The strong effect of maternal environment early in life is especially significant given the large opportunity for selection that exists at this life stage and the relatively high mortality rates typically experienced.

Although maternal effects are a well-known influence on body weight (above) and other characters in mammals (below), fewer studies have considered maternal effects outside of mammals or maternal effects on the expression of offspring behavior. In insects, maternal effects have been shown to influence several aspects of life history, including pupation, dispersal, foraging, sex ratio, fecundity, development, and coloration, as well as body size (DeFries and Touchberry 1961; Bondari, Willham, and Freeman 1978; Bauer and Sokolowski 1988; Janssen et al. 1988; Mousseau and Dingle 1991; Mousseau 1991). Perhaps the best-studied system with respect to maternal effects on behavior is diapause in insects (Mousseau and Dingle 1991; Mousseau 1991). One well-studied maternal effect is the influence of maternal age. Maternal age influences the frequency of diapause in offspring: offspring of older mothers typically enter diapause more often than offspring of younger mothers. However, there is variation among females in the effect on offspring diapause, and this variation is, in part, due to genetic differences among females. Mousseau (1991) has shown that the geographic pattern of maternal age effects in the striped ground cricket supports the hypothesis that maternal age effects can evolve as a mechanism for life history regulation in bivoltine populations. Unfortunately, Mousseau's work is one of the few evolutionary studies of behavior that explicitly considers maternal effects.

These evolutionary oddities would be just curiosities of models if maternal effects were rare and unimportant for most other offspring characters. However, empirical studies have shown the opposite to be true. Maternal effects on characters other than behavior are widespread throughout the plant and animal kingdoms. Schaal (1984) and Roach and Wulff (1987) have reviewed maternal effects in plants, and we have little more to add in a chapter concerned primarily with animal behavior. In animals, examples of maternal effects have been found in insects (reviewed by Mousseau and Dingle 1991), fishes (Reznick 1981, 1982), reptiles (Bull 1980), birds (Price and Grant 1985), and mammals (see references in Bradford 1972; Cundiff 1972; Legates 1972; Robison 1972; Hanrahan and Eisen 1973; Cheverud 1984a; Riska, Rutledge, and Atchley 1985; Southwood and Kennedy 1990). This is only a partial list, and many more examples could be produced. We have no reason to suspect that maternal effects on behavior are less common; in fact, the opposite may well be true (Stamps 1991). Thus, maternal effects are ubiquitous. While the assumption is rarely tested, it seems likely that most maternal effects, like most behavior, are also quantitative characters affected by both genes and environment.

Maternal Effects and the Consideration of Fitness

Perhaps one reason that behavioral ecologists have not considered maternal effects more closely is the view of fitness that is prevalent in this field. Typically, the survival of offspring to maturity and the values of offspring phenotypes that contribute to this fitness are considered entirely as a maternal character, and offspring survivorship is seen as an aspect of maternal fitness. Thus, measures of fitness that include "*surviving* and *reproducing* offspring" confound fitness across two (or more) generations. Such a consideration is reasonable only when the genetic (inheritance) component of the evolutionary process is ignored and we assume that there is no direct effect of genes in the offspring on their own characters. Empirical evidence has shown that this assumption is not true. The direct effects of genes carried by offspring on their own phenotypes (hereafter referred to as direct effects: Fisher 1918) typically account for about 30% of the behavioral differences among offspring (Mousseau and Roff 1987). While this is not as large as some maternal effects (e.g., 50% of the differences among offspring at weaning in mammalian body weight), it is still quite significant and should not be ignored.

In contrast, quantitative and population geneticists have typically assigned fitness and characters to the individual displaying them, and secondarily allowed for the effects of the mother on offspring characters and hence on offspring fitness. Fitness is defined with respect to a particular phenotype and is measured as the expected number of offspring for a given phenotypic value. In this approach, fitness in its entirety belongs to an individual and is assigned at birth; others can affect it but it is never reassigned to them. This approach contrasts with the common usage of inclusive fitness in which individuals are assigned portions of their relative's fitness (Grafen 1982). A personal view of fitness has also been advocated by Grafen (1988). He discusses definitions of fitness with respect to which individuals control fitness; thus, whatever measure is used for fitness, it is assumed that there is independence of control. To be independent, the measure itself must not be influenced by the phenotypes being counted.

The issue of how to score fitness is best illustrated by an example. Consider (as does Grafen) offspring number as a measure of parental fitness. This is perhaps one of the better estimates of fitness, but this measure still requires that we make assumptions. By counting the number of offspring produced, we implicitly assume (among other things) that offspring phenotypes do not influence the clutch size; that is, that control of offspring number depends on the fertility and fecundity of the parents and not on characteristics of the offspring. Thus, the number of offspring is a good measure of the parent's fitness, but not of the offspring's fitness. Including offspring survival to reproduction in definitions of parental fitness violates the assumption of independent control since aspects of the offspring's phenotypes also influence their own survival. Thus, survival is a measure of offspring fitness, not parental fitness. In the quantitative genetic models that include maternal or kin effects, independence of control is not assumed because the interdependence of fitness among kin is specifically included in the evolutionary model.

An individual approach to fitness does not require that we ignore the effects of relatives on fitness; as we show below, that is where maternal effect models come into play. A complementary method for keeping fitness assigned to individuals while studying kin selection is provided by the partitioned covariance models of Wade (1985), Cheverud (1985), and Queller (1992). Breden (1990) provides a review of this approach and its applications.

Evolutionary Models of Maternal Effects

As noted above, maternal effects can be considered a special case of kin effects (Cheverud 1984a; Lynch 1987) and can therefore be tied into kin and group selection theory. Quantitative genetic models of kin selection as a result of maternal effects have the advantage of focusing on the individual as the unit of selection, and thus overcome ambiguities associated with defining and measuring inclusive fitness (Grafen 1982; Queller 1989; Nee 1989). Furthermore, maternal effect models are rather general. As we imply above, the same models and approaches used for maternal effects can be fruitfully applied to paternal effects, sibling effects, grandparent effects (Cheverud 1984a) and their combinations (Lynch 1987). To illustrate this, we present one quantitative genetic model for evolution under maternal effects (from Cheverud 1984a), generalize it to any kind of kin or combination of kin, and then discuss various strategies for estimating the parameters contained in the models. In this way we hope to demonstrate the value of understanding something about the genetic factors that influence interactions among individuals.

Partitioning the Phenotype and Maternal Effects

Ordinarily in quantitative genetics, the phenotype (P) is modeled as the sum of the additive, or heritable, effects of genes carried by an individual (A_o) and the environment (E) (fig. 4.1a),

$$P = A_o + E. \tag{4.1}$$

In other words, the phenotype of an individual reflects the sum of the additive effects of its genes and the nonadditive genetic and environmental effects. In maternal effect models, the environmental effect is further divided into maternal performance (M), which exerts the maternal effect, and a nonmaternal or direct environmental effect (E_o) (fig. 4.1b)

$$P = A_o + M + E_o. \tag{4.2}$$

Maternal performance, being a phenotype of the mother, is also composed of her additive genetic effect (A'_m) and an environmental component (E'_m)

$$M = A'_m + E'_m, \tag{4.3}$$

where the $'$ indicates that the effects are in the maternal, or previous, generation ($t - 1$). Note that we assume in equation 4.3 that there is no maternal effect from a previous generation on maternal performance. We discuss relaxing this assumption below. In this simple maternal effect model, the phenotype of the offspring is composed of a direct additive genetic effect (A_o), a direct environmental effect (E_o), an indirect maternal genetic effect, referred to as maternal performance, exerted in the previous generation (A'_m), and a maternal environmental effect (E'_m) (fig. 1c),

$$P = A_o + A'_m + E_o + E'_m. \tag{4.4}$$

Selection, Evolution, and Maternal Effects

In quantitative genetics, the response of a trait to selection (R_P) is ordinarily considered as

$$R_P = h^2 S_P = V_A \beta_P, \tag{4.5}$$

where h^2 is the heritability (proportion of phenotypic variance that is inherited), and S_P is the selection differential (covariance between relative fitness and the phenotype). In the alternative formulation, V_A is the additive genetic, or heritable, variance, and β_P is

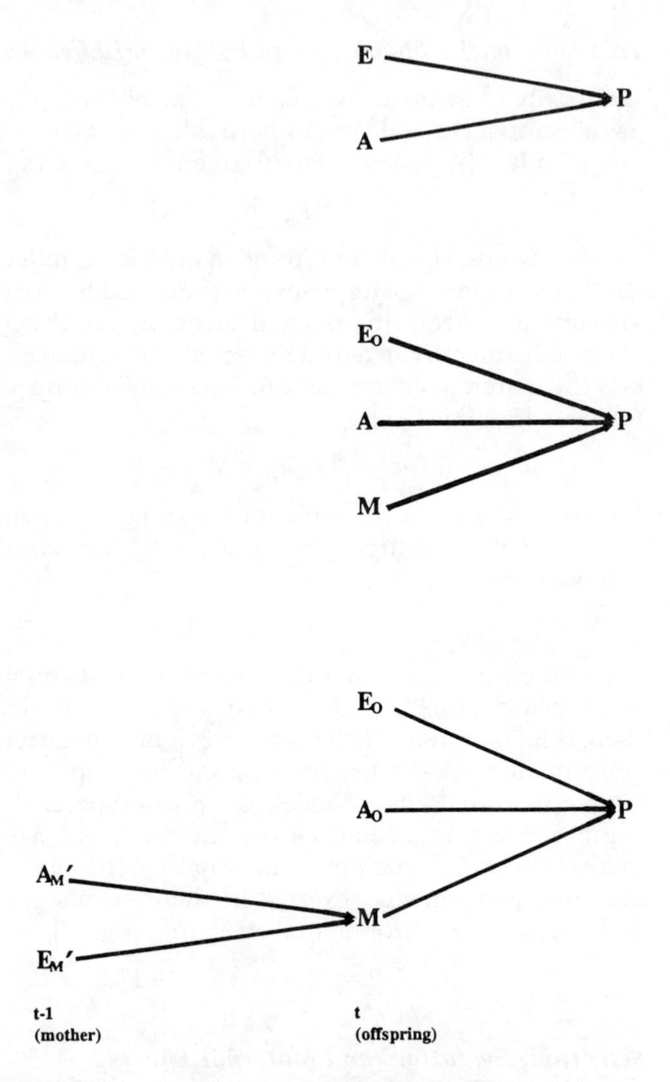

Fig. 4.1 Path diagram demonstrating how the phenotype can be decomposed into additive components. In this example, a parent (e.g., mother) and her offspring are considered. See Lynch 1987 for examples of other relationships. Wright (1968) or Li (1975) give a complete description of path analysis; Lynch (1988) provides a brief and less technical introduction to the application of path analysis in evolutionary biology and quantitative genetics. (*a*) Linear components of the phenotype assuming no maternal effects. *A*, additive effects of genes expressed in the individual; *E*, environmental and nonadditive effects on the phenotype. (*b*) As (*a*), except with latent variable *M* representing the effect of the environment provided by relatives on the phenotype of individual *o*. E_o now represents all nonmaternal

the selection gradient (regression of relative fitness on the phenotype: Lande and Arnold 1983). However, equation 4.5 is really a special case of the more general relationship between selection and its response (Falconer 1989). Actually the response to selection is given by

$$R_P = \mathrm{cov}(A, P)\, \beta_P, \tag{4.6}$$

where cov (A, P) is the covariance between the breeding value and the phenotypic value. The breeding value of an individual is the average phenotype of its offspring when it mates at random in the population (Falconer 1989, 117–18). The breeding value is equal to the sum of the average effects of all alleles carried by an individual within a given population. The breeding value, A, is therefore composed of the direct additive genetic effect, A_o, and the direct maternal genetic effect, A_m (Willham 1963). From equation 4.4 the phenotypic value can likewise be decomposed into direct additive genetic (A_o) and indirect maternal genetic (A'_m) effects (fig. 4.2). Thus, when maternal effects exist, cov(A, P) will include both direct and maternal components of each, so

$$\begin{aligned}\mathrm{cov}(A,P) &= \mathrm{cov}(A_o + A_m, A_o + A'_m)\\ &= \mathrm{cov}(A_o,A_o) + \mathrm{cov}(A_o,A'_m) + \mathrm{cov}(A_m,A_o) + \mathrm{cov}(A_m,A'_m).\end{aligned}$$

assuming no correlation between environmental effects in the parents and offspring (i.e., cov$(E_o, E_m) = 0$). This last equation can be further simplified by noting that the phenotype in the offspring reflects, in part, the influence of the parental additive genetic values (fig. 4.2). Thus, based on the theory of Mendelian inheritance with no assortative mating,

$$\mathrm{cov}(A_m,A'_m) = (1/2)\mathrm{cov}(A_m,A_m) = (1/2)V_{A_m}, \text{ and}$$

$$\mathrm{cov}(A_o,A'_m) = (1/2)\mathrm{cov}(A_o,A_m).$$

environmental influences. (*c*) A more complete path diagram showing how genetic effects in the mother (A'_m, maternal additive genetic effects) and E'_m, the environmental component of maternal performance, influence the expression of the maternal effect (M). This illustrates the point that the maternal effect is an environmental effect for the offspring but may be, in part, genetically determined in the mother. A_o symbolizes the direct additive genetic effects in offspring. Symbols marked with a prime indicate that these effects occur in the parental $(t - 1)$ generation.

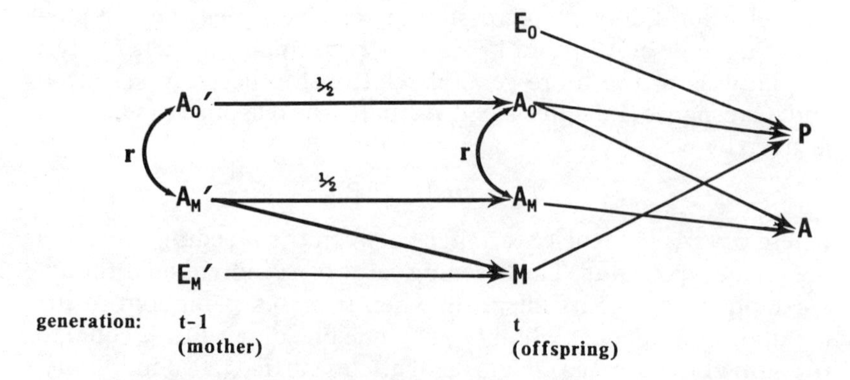

Fig. 4.2 Path diagram reflecting causal relationships between genetic and maternal effect influences on the phenotype of an offspring. A_o, direct additive genetic effects influencing the offspring phenotype (P); E_o, environmental and nonadditive genetic effects within the offspring affecting the phenotype; A_m, maternal additive genetic effects influencing maternal performance (M). Symbols marked with a prime indicate that these effects occur in the parental ($t - 1$) generation; thus, E'_m, environmental and nonadditive effects within the mother affecting the maternal performance; A'_m, maternal additive genetic effects contributing to the maternal performance experienced by the offspring. Transmission of additive genetic effects from mother to offspring assumes Mendelian inheritance; hence, ½ of the direct additive effects (A_o) in the offspring are inherited from the mother. The contributions of the additive genetic effects on the breeding value (A; the sum of all the additive genetic effects influencing all traits in the offspring) is included so that the paths contributing to the covariance between A and P can be illustrated. The correlation between the direct and maternal additive genetic effects, reflecting pleiotropy, is given by r. Note that the maternal additive genetic effects in the mother (A'_m) affect the offspring phenotype through maternal performance (M). The maternal additive genetic effects in the offspring (A_m) do not necessarily affect the offspring's phenotype (except to the extent that $|r| > 0$), but they do contribute to the offspring's breeding value. The magnitude and sign of the correlation between direct and maternal additive genetic effects (r) thus influences the response to selection on the offspring trait—and as a result, adaptive evolution.

These equalities are based on the theoretical result (again assuming Mendelian inheritance) that the covariance between parent and offspring breeding values is ($^1/_2$) and that the covariance of a trait with itself is its variance. Collecting terms, we get the equation

$$\text{cov}(A, P) = V_{A_o} + (^3/_2)\text{cov}(A_o, A_m) + (^1/_2)V_{A_m}. \qquad (4.7a)$$

This equation holds for any relative with a coefficient of relationship of $^1/_2$ (mother-offspring, father-offspring, full sibs). Other relationships such as half sibs, cousins, or grandparents can be considered by changing the coefficients of the paths to reflect the correct coefficient of relationship, since the covariance between relatives' breeding values is a function of their coefficient of relationship (Willham 1972), i.e.:

$$\text{cov}(A, P) = V_{A_o} + (1 + r)\,\text{cov}(A_o, A_q) + (r)\,V_{A_q}, \qquad (4.7b)$$

where the subscript q indicates the kind of relative and r is the appropriate coefficient of relationship.

From equations 4.6 and 4.7a, the response to selection with maternal effects is then

$$R_P = [V_{A_o} + (^3/_2)\text{cov}(A_o, A_m) + (^1/_2)V_{A_m}]\,\beta_P \qquad (4.8a)$$

or the more general (from equations 4.6 and 4.7b)

$$R_P = [V_{A_o} + (1 + r)\text{cov}(A_o, A_q) + (r)V_{A_q}]\,\beta_P \qquad (4.8b)$$

Note that the inheritance term (in square brackets) in equations 4.8a and 4.8b is not the heritability and actually includes a direct-maternal covariance term. This covariance accounts for the pleiotropic effects of single genes directly on the offspring character and directly on maternal performance. When the direct-maternal genetic covariance is highly negative, the inheritance term can also be negative, leading to evolution in the opposite direction of selection.

The direct-maternal genetic covariance is frequently negative (Dickerson 1947; DeFries and Touchberry 1961; Ahlschwede and Robison 1971; Hohenboken and Brinks 1971; Vesely and Robison 1971; Hanrahan and Eisen 1973, 1974; Kuhlers, Chapman, and Furst 1977; Bondari, Willham, and Freeman 1978; Nagai et al. 1978; Burfening, Kress, and Friedrich 1981; Cheverud 1984a; Janssen et al. 1988; Southwood and Kennedy 1990). A negative direct-maternal genetic covariance indicates that genes which directly increase the value of the offspring character will also decrease the maternal performance for that character, and thus indirectly decrease the offspring phenotypic value. Returning to our now familiar example of mammalian weaning weight, maternal performance for offspring weight at weaning typically has a negative genetic correlation with offspring weight itself (see references above). Thus, genes that result in a faster growth rate in the offspring (perhaps through their action on the physiology of early growth) typically have a negative effect on maternal performance characters that influence offspring weight (e.g., volume of milk production). Studies on ontogenetic change in direct-maternal genetic covariance are rare (Ahlschwede and Robison 1971; Kuhlers, Chapman, and Furst 1977; Cheverud 1984a; Riska, Rutledge, and Atchley 1986) but usually indicate a decrease in covariance from a positive value at birth to a negative value at weaning, after which the covariance increases again. The covariances occasionally reach levels where the predicted response to selection is opposite the direction of selection (Cheverud 1984a). Even when this negative covariance fails to reverse evolution, it can seriously limit response to selection.

The overall response to selection given in equation 4.8 can be divided into correlated responses in the direct (CR_o) and maternal (CR_m) components of the phenotype,

$$CR_o = [V_{A_o} + (1/2)\mathrm{cov}(A_o,A_m)]\,\beta_P \tag{4.9}$$

and

$$CR_m = [(1/2)V_{A_m} + \mathrm{cov}(A_o,A_m)]\,\beta_P. \tag{4.10}$$

Note that if $\mathrm{cov}(A_o,A_m)$ is negative and its absolute value exceeds that of one-half the additive genetic variance for maternal performance, maternal performance will decrease as a correlated response to selection for increased offspring phenotype. For example, selection for increased offspring growth in weight to weaning may lead to a decrease in milk production and other aspects of maternal performance in mice (Cheverud 1984a). Furthermore, if $(1/2)\mathrm{cov}(A_o,A_m)$ is greater than $(V_{A_o} + 1/2 V_{A_m})$, offspring weight is also expected to decline.

Thus far we have only considered selection on the offspring phenotype, but it is also possible to extend these models to include direct selection on maternal performance or its components (Cheverud 1984a). With selection on maternal performance, the direct response of maternal performance (R_m) and the correlated response of the offspring phenotype (CR_P) is

$$R_m = V_{A_m}\,\beta_M\,(1/2) \tag{4.11}$$

and

$$CR_P = [V_{A_m} + \mathrm{cov}(A_o, A_m)]\, \beta_M\, (^1/_2). \tag{4.12}$$

The maternal performance selection gradient (β_M) is multiplied by ($^1/_2$) in equations 4.11 and 4.12 to reflect the fact that selection on maternal performance can only occur in females, one-half of the parental population. Thus the total response of the offspring phenotype and maternal performance to selection on both characters is

$$R_m = [(^1/_2)V_{A_m} + \mathrm{cov}(A_o, A_m)]\, \beta_P + V_{A_m}\, \beta_M\, (^1/_2) \tag{4.13}$$

and

$$R_P = [V_{A_o} + (^3/_2)\mathrm{cov}(A_o, A_m) + (^1/_2)V_{A_m}]\, \beta_P + [V_{A_m} + \mathrm{cov}(A_o, A_m)]\, \beta_M\, (^1/_2). \tag{4.14}$$

This maternal effect model can be generalized to any form of kin by simply substituting the appropriate kin effect and coefficient of relationship (r) into the equations (Cheverud 1984a; eqs. 4.7b and 4.8b above). For example, the equations for paternal effects ($r = ^1/_2$) would be the same as for maternal effects, just substituting the subscript f for m. Full sib effects ($r = ^1/_2$) would substitute the subscript s for m and drop the ($^1/_2$) from the selection gradient, since selection would be acting on both sexes. If selection involves half siblings ($r = ^1/_4$), the response of the sibling "performance" character is

$$R_s = [(^1/_4)V_{A_s} + \mathrm{cov}(A_o, A_s)]\, \beta_P + V_{A_s}\, \beta_s$$

and the response of the target character is

$$R_P = [V_{A_o} + (^5/_4)\mathrm{cov}(A_o, A_S) + (^1/_4)V_{A_S}]\, \beta_P + [V_{A_S} + \mathrm{cov}(A_o, A_S)]\, \beta_s,$$

where the subscript s refers to the sibship effect. Note that selection occurs in both sexes in this model. This process can be further generalized to include several kinds of kin at the same time (Lynch 1987).

In addition to the usual assumptions of quantitative genetic models (that there are many genes, each with a small effect on the phenotype, and no dominance or epistasis) the simplifying assumptions used in our model are that: (1) only genes are inherited, not environments, so that there is no maternal effect on maternal performance; (2) only two characters are considered, an offspring phenotype and maternal performance for that phenotype; (3) selection remains constant from one generation to the next. Riska, Rutledge, and Atchley (1985) developed a model that relaxes the first assumption. Falconer (1965) also addresses the first assumption in his discussion of a model that considers maternal effects on the evolution of maternal performance itself.

A Generalized Maternal Effect Model

Equations 4.13 and 4.14 above represent a special case of a more general maternal effect evolutionary model, relaxing all three assumptions, recently developed by Kirkpatrick and Lande (1989; Lande and Kirkpatrick 1990). In this general model, the phenotype is defined as (see equation 4 in Kirkpatrick and Lande 1989; equation 1 in Lande and Kirkpatrick 1990; see also Arnold, chap. 2)

$$P = A_o + E_o + M_{ij}P'_j$$

We have changed their symbols somewhat to be consistent with the symbols we have used in this chapter; table 4.1 presents a translation. The term M representing maternal effects in our equation 4.2 has now been generalized by Kirkpatrick and Lande, being replaced by $M_{ij}P'_j$. In this more general model P'_j is any maternal character while M_{ij} is the partial regression of the phenotypic value of character i in the offspring on character j in the mother. In most earlier models of maternal effects, the symbol M stands in for Kirkpatrick and Lande's whole last term, with maternal performance being the combination or sum of all the effects of maternal characters ($M_{ij}P'_j$). Alternatively, if maternal performance is itself the maternal character, Kirkpatrick and Lande's M_{ij} equals one and P'_j is then the maternal performance. Thus, the sum of their product, $M_{ij}P'_j$, is the maternal effect (M from equation 4.2).

There are several advantages to the Kirkpatrick and Lande approach. First, it combines the two types of maternal effects, in which the maternal character(s) affect either the same character when expressed in the offspring (Falconer 1965) or other offspring characters (Dickerson 1947; Willham 1972). The coefficient M_{ij} includes the possibility of maternal effects on maternal performance. Thus, the environmental (maternal) effects can also be passed across generations through inheritance. Second, it is multivariate, allowing consideration of many offspring and maternal characters jointly. This is important when separate aspects of maternal performance are of interest and when maternal performance

Table 4.1 Correspondence between symbols used in maternal effect models

"Classic" Symbol[a]	K&L[b]	Definition
P_o	z_i	The phenotypic value (for character i) as measured in the individual *o* (generation $t + 1$; K&L).
A_o	$a_i(t + 1)$	The additive genetic component of the phenotypic trait in individual *o* $(t + 1)$.
E_o	e_i	The nonmaternal, direct environmental component of the phenotypic trait in individual *o* $(t + 1)$. This term includes nonadditive genetic effects.
P'_j	$z^*_j(t)$	Any maternal character, *j*, expressed in the mother (generation m $[= t]$).
	M_{ij}	The partial regression of the phenotypic value of character *i* in the offspring $(t + 1)$ on character *j* in the mother (t) holding genetic effects (a_i) constant. This measures the strength of the maternal effect of character *j* on character *i*.
A_m	a_m	The additive genetic component contributing to the maternal performance (M).
E_m	e_i	The direct environmental component of the maternal performance.
M	$M_{ij}z^*_j(t)$	Maternal, indirect, environmental effects on the phenotypic trait in individual *o* $(t + 1)$.
$\mathrm{cov}(A,P)$	$\mathbf{C_{az}}$	The covariance between the breeding value and the phenotypic value.
V_A		Additive genetic variance.
R_p	$\Delta\bar{\mathbf{z}}(t)$	Phenotypic response to selection (i.e., evolution).
R_m		Response to selection on the maternal component of the phenotype.
CR_p		Correlated response in the offspring phenotype due to selection on the maternal performance.
CR_o		Correlated response in the direct components of the phenotype due to selection on the overall phenotype.
CR_m		Correlated response in the maternal (indirect) components of the phenotype due to selection on the overall phenotype.

[a]Cheverud, 1984a; this chapter.
[b]Kirkpatrick and Lande 1989; Lande and Kirkpatrick 1990; Arnold chap. 2.

is a different character for different offspring features. Although correlations between maternal performance for a variety of body size characters are essentially unity, indicating that maternal performance for body growth affects all offspring somatic characters alike (Cheverud and Leamy 1985), the same may not be true for nonmorphological characters or mixtures of morphology, behavior, and life history. These beneficial features of the model combine the various aspects of previous models (Dickerson 1947; Willham 1963; Falconer 1965; Cheverud 1984a; Riska, Rutledge, and Atchley 1985) into a unified framework. However, because of these advantages one is also required to assume that all of the individual maternal characters affecting the offspring phenotypes are included in the analysis in order to use this model. In contrast, models using maternal performance, like the one we have described in some detail above, have the advantage of treating the maternal characters jointly and measuring them only through their effects on the offspring. This facilitates the study of the evolution of offspring characters but is less useful for studying the evolution of individual maternal characters.

A major finding of Lande and Kirkpatrick's (1990) model, not replicated in our model due to the assumption of no maternal effect on maternal performance, is that there can be a time lag in evolutionary response to selection and that evolution will continue even after selection has stopped, a form of evolutionary momentum being induced by the presence of maternal effects. This is considered further by Arnold (chap. 2).

Applications of Maternal Effect Models

Maternal Effects and Kin Selection

Models of evolution by kin selection are a special case of the evolutionary maternal effect model given in equation 4.13. The evolution of maternal performance can be considered altruistic when the direct selection on the offspring phenotype (β_P) and on maternal performance (β_M) are in opposite directions. For example, selection for increased offspring weight combined with selection against high maternal contributions to individual offspring weight, as may occur when high maternal performance lowers future survivorship, results in the evolution of a form of altruism (sacrifice of future reproduction by the mother). Recall that in the quantitative genetic approach everyone's fitness belongs to themselves, so that once offspring are born their future survival and reproduction no longer contribute to maternal fitness. Any reduction in the mother's future survivorship or fertility due to increased maternal care or performance can be considered altruism, especially when it increases offspring survival.

Maternal performance will evolve in an altruistic direction when the direct response of maternal performance to selection against increased performance is not sufficient to counterbalance the correlated response of maternal performance to selection for increased offspring phenotypes (Cheverud 1984a). Given that β_P is positive and that β_M is negative, altruism will evolve when

$$R_M + CR_M > 0$$

or when

$$(1/2) + [\mathrm{cov}(A_o, A_m)/V_{A_m}] > (|\beta_M|/\beta_P)\,(1/2), \qquad (4.15)$$

where the (1/2) on the left-hand side of the inequality is the coefficient of relationship between mother and offspring and the (1/2) on the right-hand side of the inequality is due to selection on maternal performance occurring only in females. This result can be easily modified to consider any kind of kin effect, such as that of siblings, fathers, or grandparents. In general, when only one relative is considered at a time, evolution proceeds toward more altruistic states when

$$(r + [\mathrm{cov}(A_o, A_q)/V_{A_q}]) > |\beta_Q|/\beta_P, \qquad (4.16)$$

where r is the coefficient of relationship between the altruist and the recipient, A_q is the additive genetic value for kin performance, V_{A_q} is the additive genetic variance for the kin performance, and B_Q is the selection gradient against increased altruism.

A concrete example may be of some benefit. Consider alarm calls, such as those provided by birds and small mammals. One could measure the selection against alarm calling in parents (β_Q, the regression of calling versus not calling on survival using the methods of Lande and Arnold [1983]) and the positive selection for a response to an alarm call in the offspring (β_P). Standard parent-offspring techniques (Falconer 1989) could also yield estimates of genetic covariances for these traits if relatedness is known. It is then a fairly simple matter to plug these values into equation 4.16, incorporating the appropriate coefficient of relationship (1/2 for parents and offspring). A similar approach could be adopted to measure the potential for evolution of alarm calls (or any other potentially altruistic behavior) among siblings. Although there will undoubtedly be difficulties in conducting such studies, given the problems of estimating relatedness in nature, the conceptual difficulties are no greater than those in any study of altruism.

These equations for the evolution of altruism are consistent with other polygenic models of kin selection (Yokoyama and Felsenstein 1978; Boyd and Richerson 1980; Aoki 1982; Engels 1983; Cheverud 1985). However, the unique feature of this model is the inclusion of pleiotropic genetic effects on traits expressed in the donors and recipients of altruism.

Equation 4.16 is a quantitative genetic formulation of Hamilton's rule, specifying the conditions under which evolution will proceed toward more altruistic states (Hamilton 1964a). The ratio of selection gradients on the right-hand side of the equation represents the ratio of the costs of altruism to its benefits. The left-hand side of the inequality differs from the usual Hamilton's rule by the addition of the genetic regression of offspring phenotype on maternal performance (the term in square brackets). This term is included here because we have *not* made the common, but often unstated, assumption that maternal performance exerts the sole effect on offspring characters. Instead, we have allowed genes carried by the offspring to have a direct effect as well. Our avoidance of this peculiar assumption of behavioral ecology is a direct result of assigning fitness to the individual surviving and reproducing rather than to a relative, such as its mother. The concept and approach of inclusive fitness is mathematically consistent with the one presented here because if we assume no direct effects of genes on offspring characters (or no covariance between direct and maternal effects), the usual form of Hamilton's rule is obtained. However, other versions of the concept of inclusive fitness, with a complex accounting of fitness in which fitness is shuffled between individuals, tend to cloak the assumption of no direct effects. Although the inclusive fitness concept tends to lead to the assumption of no direct effects, this assumption is not a requirement of the concept. Empirically we know that direct effects do exist and are important and that they genetically covary with maternal effects. Thus, the usual assumptions made in inclusive fitness models and derivations of Hamilton's rule are often not biologically tenable.

Inspection of equation 4.16 indicates that the maternal genetic regression, $[\mathrm{cov}(A_o, A_m / V_{A_m}]$, can have an overwhelming effect on the potential for and realization of altruistic evolution. This genetic regression ranges from negative to positive infinity and can thus swamp the effects of relatedness in considerations of altruistic evolution. If the regression is relatively large and positive (greater than 0.50), the left-hand side of the inequality can exceed one, allowing altruistic evolution to proceed even when costs exceed benefits. Thus a positive direct-maternal covariance enhances altruistic evolution. If the regression is a highly negative value (less than −0.50), the left-hand side of the equation can be negative, in which case altruistic evolution is prevented, despite minimal costs and tremendous benefits. The evolution of altruism can either be vastly facilitated or absolutely prevented by the structure of the system of inheritance. Ignoring this could lead to gross errors in inference from measurements of costs, benefits, and relatedness alone. Cheverud (1984a) provides several empirical examples in which maternal performance may evolve altruistically even when costs exceed benefits and in which maternal performance would be prevented from evolving altruistically regardless of the cost-benefit ratio. Only rarely is the genetic regression unimportant for altruistic evolution. These counterintuitive results apply to all forms of kin effects on altruism (Cheverud 1984a).

Maternal Effects and Parent-Offspring Conflict

The maternal effect models described above are particularly relevant to parent-offspring conflict theory (Trivers 1974; Wilson 1975; Cheverud 1984a). Trivers (1974) viewed parent-offspring relations at various ages as a result of selection operating in opposite directions on mothers and offspring. We return, once again, to the example of maternal care provided by nursing mammals. Improved maternal performance (increased quality or quantity of the milk supply) is selected for because it leads to improved offspring characters (such as higher body weight) which in turn lead to higher offspring survival. However, improved maternal performance is also selected against due to the consequent decrease in maternal fertility and survivorship (due to the cost of lactation). When the two selection regimes are equally balanced, weaning occurs. However, as noted above in considering the maternal effect models, selection for increased offspring phenotype, especially at the age of weaning, may often lead to an evolutionary *decrease* in maternal performance for that phenotype. Apparently opposed selection pressures may often result in mutually reinforcing, rather than opposing, evolutionary responses. While Trivers's analysis of parent-offspring conflict may be correct with regard to the opposing selective forces and their ecological causes, it may not be sufficient for predicting the evolutionary outcome of these contrary selection pressures. Prediction of evolutionary outcome requires a knowledge of the evolutionary processes that lead to it or requires that assumptions be made about these processes. Thus information, or assumptions, about patterns of inheritance, genetic variances, and covariances are required. Trivers's assumption that opposing selection pressures result in opposing responses to selection may not be appropriate in this instance, although other models have supported Trivers's ideas (Bull 1985). Information currently available in the literature strongly suggests that the direct-maternal genetic covariance is a significant factor in evolutionary responses when maternal effects are considered.

Maternal Effects and Mate Choice

Another area in which maternal, or paternal, effects are thought to be important is in the area of mate choice. Trivers (1972; see also Darwin 1871; Fisher 1958; Williams 1966; Alexander and Borgia 1979; Halliday 1983; Maynard Smith 1985; Thornhill 1986) analyzed the action of sexual selection with respect to parental investment and suggested that females might choose males that provide the greatest investment in offspring. Parental investment, which is an environmental influence on offspring characteristics, can be considered a form of "parental performance." Thus we can restate Trivers's hypothesis as suggesting that females should choose males with high paternal performance (see Trivers 1972).

Females may discriminate among males based on resources that males provide and which influence offspring characters (Thornhill and Alcock 1983; Thornhill and Gwynne 1986). One particularly complete example is the mating behavior of scorpionflies. Thornhill has shown that scorpionfly females discriminate among males based on the prey resources they are able to offer and has further suggested (based on an analysis of repeatability) that variation among males in ability to capture the prey used as nuptial gifts (an aspect of paternal performance) has a genetic basis in at least one species (Thornhill 1976, 1983, 1986; see also Boake 1989b). Another example of paternal performance that may influence mate choice is provision of territories providing protection and/or food to both mates and offspring (Thornhill and Gwynne 1986). Thus, both theoretical considerations and empirical data suggest that the degree of parental investment, or parental performance, can influence mating decisions by the opposite sex. Evolutionary models of this process need to include genetic variation in and covariation among paternal performance, female choice based on the male performance character, and offspring phenotypes.

Recently, Hoelzer (1989) presented a haploid population genetic model of the "good parent process" of sexual selection under which females choose among males based on a character that reflects *nonheritable* paternal quality. The assumption that differences in parental quality do not have a genetic basis is not supported by the few empirical data available (Sakaluk and Smith 1988; references above). Trivers's analysis and the data provided by Thornhill and by Sakaluk and Smith suggest that a more reasonable model of the process would consider the effects of mate choice on heritable parental differences influencing offspring characteristics (heritable parental performance).

It is also useful to adopt a quantitative genetic perspective in considering whether an "indicator" trait can evolve to signal parental quality or performance to potential mates. The suggestion that females should select mates based on their ability to provide for their offspring is well supported, but it is less clear that some trait will therefore be selected for that will "signal" this ability to provide quality resources. Much as in the example of parent-offspring conflict given above, the effects of selection cannot be predicted without some knowledge of the genetic basis of parental performance, the indicator trait, and female choice. If the epigamic indicator trait is genetically related to the ability to provide parental investment, selection for an increased value in the offspring phenotype can lead to a correlated *decrease* in parental performance for that phenotype, as discussed above for mammalian weight at weaning. This is the opposite of the prediction of the good-parent process of sexual selection based purely on optimization arguments. This genetic reasoning also applies to the "truth in advertising" hypothesis of sexual selection presented by Kodric-Brown and Brown (1984). An important, and apparently unrecognized, parameter in this model is the genetic covariance between the parental behavior (parental performance) and the offspring trait. Unless this value is zero, selection alone will not predict the evolutionary response. The value of the quantitative genetic reasoning in this example is that it provides information on the sorts of measures that must be obtained to evaluate the model.

It should be noted that Fisher (1915) first considered mate discrimination to be based, initially, on a character that provides "a fairly good index of natural superiority" (187). However, he supposed that this indicator would soon be elaborated beyond its value as an index due to a "runaway" process (Fisher 1915, 1958). This runaway process has been the subject of a great deal of attention, and has been modeled using a quantitative genetic framework by

Lande (1981b). Heisler (1984b, 1985; Heisler, chap. 5) develops quantitative genetic models of the origin of mating preferences based on initially adaptive characters. Incorporating maternal effects into these situations may be difficult, but illuminating.

Future Studies

The models of Cheverud (1984a), Lynch (1987), Kirkpatrick and Lande (1989), and Lande and Kirkpatrick (1990) highlight the utility of using quantitative genetic principles when considering the effects of interactions among related individuals. However, as with most issues in evolutionary biology, empirical data are needed to evaluate these models critically. The quantitative genetic models incorporating maternal (parental, kin) effects do not disprove the predictions derived from parent-offspring conflict theory, or invalidate the good-parent or truth-in-advertising models of sexual selection, but they do emphasize that verbal reasoning is often misleading because evolution cannot be predicted solely from selection. Further, quantitative genetic models identify additional parameters that could be measured in order to obtain valuable data. Information, or at least realistic assumptions, about the patterns of genetic inheritance, genetic variances, and covariances is required for an understanding of the potential evolutionary responses to selection and thus for an understanding of evolutionary outcome.

It is obvious that a large number of additional examples incorporating maternal effects could be developed, since a great number of studies of behavior include interactions among relatives. We suggest that researchers consider measuring maternal (kin) effects in any area within the study of behavior in which a common environment provided by a relative may influence offspring success. Studies of helping, social dominance, and other interactions that could potentially involve relatives might benefit from an examination of maternal and other kin effects.

Methods of Estimating Parameters of the Maternal Effect Models

The basic problem in estimating maternal effects parameters lies in the observation that mothers affect their offspring in two different ways, by passing on their genes and by providing a common environment for their offspring. In order to separate the environmental from the genetic effects of mothers on offspring it is necessary to experimentally or statistically separate inheritance and environment. This can be accomplished in two basic ways. One can construct an experimental design that separates inheritance from environment, or, if the two are not completely confounded, one can separate them statistically (just as partial regression coefficients separate the effects of correlated independent variables).

Experimental designs that utilize specific breeding methods to separate maternal environment from inheritance have been those most commonly employed (Eisen 1967; Bauer and Sokolowski 1988). The most popular one of these has been the paternal half-sib design. In this design, each of a sample of independent sires is mated to an independent set of two or more dams (Falconer 1989). The design is then analyzed by using a nested analysis of variance with dams nested within sires. The variance between sires (V_s among half-sib families) is then equal to $(1/4)V_{A_o}$, allowing estimation of V_{A_o} and heritability without confounding maternal effects. The variance between dams within sires ($V_{d(s)}$ among full-sib families) is $(1/2)V_{A_o} + (1/4)V_{D_o} + V_m$, where V_{D_o} is the direct dominance variance and V_m is the phenotypic variance in maternal performance. If we assume that V_{D_o} is zero, we can obtain an estimate of V_m by subtraction:

$$V_m = V_{d(s)} - V_s. \tag{4.17}$$

However, from this design we cannot estimate the important parameters V_{A_m} and $\text{cov}(A_o,A_m)$. Therefore, while this design is good at removing maternal effects from heritability estimates, it is not valuable for estimating maternal effect parameters themselves.

A second, less popular but more powerful, design is cross-fostering (Rutledge et al. 1972; Atchley and Rutledge 1980; Cheverud et al. 1983; Cheverud 1984a; Riska, Rutledge, and Atchley 1985). In a cross-fostering design, independent full-sib families are paired and a random half of the litter (or clutch, brood, etc.) is exchanged between dams. Half the offspring are thus reared by their own, genetic, dam while half are reared by a foster dam. A large number of cross-fostered pairs are desirable, and it is also desirable for the litters to be standardized in number at birth to avoid including confounding litter size effects in the analysis (Riska, Rutledge, and Atchley 1985).

With the cross-fostering design, it is possible to estimate the phenotypic variance of maternal performance (V_m) and the direct-maternal genetic covariance ($\text{cov}(A_o,A_m)$). In this design, the entire analysis is nested within cross-fostering pairs. The variance between dams (V_d, among full-sib families born of a single dam) is $(^1/_2)V_{A_o} + (^1/_4)V_{D_o}$. If we assume that V_{D_o} is zero, then V_{A_o} is twice V_d. The variance between foster mothers (V_n, among young reared by a single dam) is V_m and thus provides a direct estimate of the phenotypic variance in maternal performance. The genetic covariance between direct and maternal effects is contained within the phenotypic variance of unfostered individuals, but not in the phenotypic variance of fostered offspring (Riska, Rutledge, and Atchley 1985). Thus it can be obtained by subtracting the variance among fostered offspring (within pairs) reared by other mothers from the variance among offspring (within pairs) reared by their own mother (Riska, Rutledge, and Atchley 1984; Cheverud 1984a). If the design is modified so that the original dams are genetically related (preferably through their fathers) and related dams are not included in single cross-fostering pairs, it is also possible to estimate the additive genetic variance for maternal performance (V_{A_m}).

Cross-fostering experimental designs have not been used to study certain maternal environments, such as egg constituents and the mammalian uterus, because transfer of offspring has been difficult to carry out. However, technological advances such as embryo transfer have allowed this design to measure previously obscure prenatal mammalian maternal effects (Cowley et al. 1989; Cowley 1991; Atchley et al. 1991).

An alternative approach to estimating maternal effect parameters is based on the general theoretical approach of Kirkpatrick and Lande (1989; Lande and Kirkpatrick 1990). Lande and Price (1989) demonstrate how maternal and direct effect coefficients, as described in the Kirkpatrick and Lande (1989) model, can be obtained from the regression of offspring values on mothers' and fathers' values. Using this approach, one must directly measure the maternal phenotypes that contribute to maternal performance for the offspring phenotypes considered and assume that all of the maternal characters affecting the offspring characters of interest are measured and included in the analysis (Lande and Price 1989). Under this restrictive assumption, the differences in mother-offspring and father-offspring regressions and asymmetries in the pairs of parent-offspring cross-covariances (covariance between trait X in the parent and trait Y in the offspring versus the covariance between trait Y in the parent and trait X in the offspring) can be used to obtain the parameters for evolution with maternal effects. This approach has the advantage of directly addressing the maternal characters affecting offspring phenotypes and thus avoiding the complex designs described above, but carries the restrictive assumption that all maternal features contributing to maternal performance are included in the analysis. This approach is likely to be most useful with natural populations where the experimental designs discussed above cannot be implemented.

A final possibility is the use of maximum likelihood methods (Thompson 1976) such as the animal model (Southwood et al. 1989; Southwood and Kennedy 1990) or REML (Shaw 1987). These methods, like most quantitative genetic methods, have grown out of an applied animal breeding background. Nonetheless, maximum likelihood approaches are being increasingly applied in evolutionary quantitative genetics (Shaw 1987). This approach has several advantages and attributes that facilitate the analysis of field data (Lande and Price 1989; Southwood and Kennedy 1990). First, relationships other than parent-offspring or sibship can easily be incorporated into the analysis. Second, because several generations can be included, the data need not be obtained from experimental breeding studies (e.g., half-sib or cross-fostering), but can reflect a long-term field study of a population with a known genealogy. Finally, restricted maximum likelihood methods often have statistical advantages over least squares methods (Shaw 1987; Lande and Price 1989). Therefore, when data are available from several generations (e.g., from pedigrees), the use of a maximum likelihood approach (Thompson 1976) such as the animal model or REML (Shaw 1987; Southwood et al. 1989; Southwood and Kennedy 1990) should be considered. Shaw (1987) gives a particularly lucid introduction to these methods (see also references in Arnold, chap. 2).

Measurements That Can (or Should) Be Taken

Future studies should begin to estimate direct genetic variance (e.g., heritability of the offspring trait), the proportion of phenotypic variance due to maternal effects, maternal genetic variance (e.g., the heritability of the trait(s) in the parent that influence the offspring phenotype), and direct-maternal genetic covariance wherever possible. As reviewed above, the first two are sometimes measured, but we have almost no estimates of the last two. Further, it would be highly desirable to produce measures that were sufficiently precise to warrant discussing relative values. However, due mostly to sample size limitations (Klein, DeFries, and Finkbeiner 1973; Klein 1974), this ideal is rarely achieved in estimates of relatively simple parameters such as heritability of the offspring character, let alone estimates of maternal genetic variance. Nonetheless, large sample sizes and the techniques outlined above should be used wherever possible.

Providing values for these parameters is a worthy goal, but should not preclude the more tractable studies that simply demonstrate that variation among individuals is due in part to genetic differences among individuals. We need studies that simply demonstrate that the variation in traits such as maternal and/or paternal quality, helping, parental investment, or kin effects, is influenced at least in part by genetic differences among individuals. The study by Sakaluk and Smith (1988) on the inheritance of paternal investment, or performance, is a start. Many of the studies outlined in this book show how a genetic basis for a behavior can be demonstrated without providing specific values for genetic variances. The research of Mousseau (1991) is a good example of a demonstration of a genetic influence on maternal performance. Such studies directed at demonstrating a genetic basis of variation in maternal performance are still valuable, as they demonstrate the ubiquity (or scarcity) of the influence of genetics on the behavior of interest.

Finally, it should be noted that field studies do not preclude the use of a quantitative genetic perspective. One of the advantages of quantitative genetics is that it is phenotype oriented; given accurate measures of a phenotype and some knowledge of genetic relationships among individuals, all the information needed for quantitative genetic analyses is available. Behavioral ecologists are often involved in long-term field studies, often utilizing a banded or marked population. It would be useful if their data were analyzed, or made available for analysis, with respect to some of the characters of interest. The maximum likelihood methods developed by animal breeders could be especially valuable in analyzing such data (see Arnold, chap. 2).

Implications for Behavioral Ecology and Genetics

The value of models such as these is in stimulating further research. We, and others (Lynch 1987; Stamps 1991; Arnold, chap. 2), argue that most characters in parents constituting maternal effects will be behavioral because they involve social interactions. Many of the characters being influenced will also be behavioral traits. Investigation of social interactions constitutes one of the largest areas of study in behavioral ecology, focusing on such diverse organisms as slime molds, birds, arthropods, mammals, reptiles, and amphibians. Not all species can yield data that are suitable for consideration of maternal effects. However, in a number of groups, it should be possible to collect the relevant data. Breeding studies or genealogies will be necessary for estimating most of the parameters. Nonetheless, given the excitement and research generated by such complicated concepts as inclusive fitness, we expect that ideas derived from models such as these should prove valuable in testing a variety of concepts related to the evolution of behavior.

Summary

1. Research in behavioral ecology has a rich tradition of considering the effect of relatives on the evolution of behavior. Typically, behavioral ecologists use optimality models to generate testable predictions concerning the effects of animals' behavior on the behavior of others. However, optimality models do not consider the evolutionary process in which behavior is so often intimately involved. In this chapter, we argue that a consideration of inheritance will yield important insights into the evolutionary effects of behavioral interactions among relatives.

2. The class of quantitative genetic models that have typically considered the effects of relatives are called "maternal effect" models. However, these models can be and have been generalized to include the effects of any class of relative. Thus, they may be more accurately referred to as "kin effect" models. The term *maternal effect* is too well established in the literature to be replaced and we therefore use it to refer to all models incorporating the genetic effects of interacting relatives.

3. A consideration of these models in an evolutionary framework indicates that maternal, or kin, effects have important and counterintuitive consequences for short-term evolution. In maternal effect models, the response to selection depends in part on the sign of the covariance between direct additive genetic effects and the maternal additive genetic effects. Thus, genes in offspring that increase the value of a character are counteracted by genes that decrease the maternal performance influencing the offspring's character. If this covariance term is highly negative, short-term evolution can proceed in a direction opposite that of selection. Empirical studies have shown that this covariance is often negative.

4. In addition to highlighting the potential for nonadaptive evo-

lution, maternal effect models force an explicit consideration of fitness. Fitness is assigned to individuals in these models and is never partitioned, as it is in kin selection models. Quantitative genetic models have the advantage of separating selection and transmission in describing evolutionary changes.

5. We review an evolutionary maternal effect model originally developed by Cheverud (1984a), and then generalize this model to include any kind of kin or combination(s) of kin. Throughout this section we emphasize the need to study genetic factors that influence interactions among individuals. After developing the models, we show how this approach can be applied to a variety of topics, including studies of altruism, parent-offspring conflict, or even mate choice based on parental care.

6. Finally, we review potential methods that can be used to measure the parameters relevant for testing models that consider the effects of relatives on the evolution of behavior. Two approaches are possible: experimental separations of inheritance from environmental effects through paternal half-sib breeding designs or cross-fostering, or statistical separations involving regression or maximum likelihood approaches. We also outline the types of measurements that should be taken. In conclusion, we suggest some of the potential implications that this approach may have for research in behavioral ecology.

Acknowledgments

We thank John Alcock, William Atchley, Chris Boake, Deborah Clark, Sandra DeBano, Jennifer Fewell, Ken Haynes, Ann Hedrick, Diana Hews, Bruce Riska, Jack Rutledge, Donald Sade, and Ken Yeargan for their (sometimes vigorous, always enlightening) discussions of this chapter and/or the ideas on which it is based. Our views do not always agree with those expressed by our colleagues, but we have always benefited greatly from the care they have taken to patiently explain our differences. We have also benefited from the support of NSF grants BSR-8906041 to JMC and BSR-9022012 and BSR-9107078, plus state and federal Hatch support, to AJM during the writing of this chapter. Our interest in paternal performance has been invigorated by Bryan, Caitlin, Eirik, Kevin, and Mary Grace.

5

Quantitative Genetic Models of the Evolution of Mating Behavior

I. Lorraine Heisler

Fig. 52. Side view of male Argus pheasant, whilst displaying before the female. Observed and sketched from nature by Mr. T. W. Wood.

The courtship display of the male Argus pheasant, *Argusianus argus,* as drawn by T. W. Wood. From Charles Darwin, *The Descent of Man and Selection in Relation to Sex* (London: John Murray, 1871.)

Nowhere has the impact of quantitative genetics on behavioral ecology been as evident as in studies of the evolution of mating behavior. Before 1980 the subjects of mate choice, sexual selection, and the evolution of courtship displays were virtually untouched by population genetic theory. Since then, however, sexual selection has become one of the most active areas of genetic modeling, and theoretical results now play an influential role in guiding empirical research. This change in the status of sexual selection theory coincided with the introduction of quantitative genetic approaches to modeling mate choice (Lande 1981b). Earlier work by O'Donald (1967) had clearly established that genetic models were essential to understanding the evolution of mating behavior. However, traditional population genetic approaches yield models that are not very realistic biologically and are nearly impossible to analyze mathematically (see, however, Gomulkiewicz and Hastings 1990). Quantitative genetic models overcome the analytic difficulties of modeling mate choice while at the same time providing biological realism and an easier avenue for empirical application.

This chapter provides a brief introduction to quantitative genetic models of phenotypic evolution, using examples from recent models of the evolution of female "choice" of attractive male courtship displays. It is not intended as a comprehensive review of current theory; several excellent reviews already exist (Arnold 1983b, 1985; Kirkpatrick 1987b). My aims are to outline the biological assumptions and mathematical structure of quantitative genetic models of mating behavior, to describe the models' advantages and limitations, and finally, to discuss some of the issues that arise in testing the results of quantitative genetic theory.

The Attractions of Quantitative Genetic Models

If there were a perfect mathematical model for the evolution of behavior, it would be simultaneously realistic, applicable, and interpretable. The model would capture all of the qualities and consequences of behavioral traits that affect their evolution. It would represent genetic transmission accurately, so that evolutionary change could be extrapolated over many generations. Parameters and variables would be measurable in real organisms, making it easy to apply the model to empirical data. Taken together, these qualities would provide both the confidence that the model was a good imitation of nature and the ability to demonstrate that this was indeed the case.

However, if a model were completely realistic, and thus perfectly predictive, its behavior would be no easier to fathom than that of nature itself. Models that help us understand the causes of evolution are unrealistic by choice: they omit some factors for the sake of identifying the effects of others (Levins 1966). Our "perfect" model would therefore be exactly as realistic as required: more realistic if it were intended to provide accurate predictions; less realistic if its purpose were to gain insights into specific causes of evolution.

Quantitative genetic models have achieved great popularity because they provide what many see as the best balance yet reached between the conflicting needs of realism, applicability, and ease of interpretation. These models assume that characters vary continuously owing to the action of many small genetic and environmental effects. This is clearly more realistic for most behavioral, morphological, and life history traits than the alternative assumption of discrete variation with single-gene determination employed in classical population genetic theory as well as in most optimality modeling approaches (Roff, chap. 3). Quantitative genetic models also have the attractive property that their dynamic variables are not gene frequencies, but the mean values of one or more phenotypic traits in a population; hence their application to empirical data is relatively straightforward. In addition, most quantitative genetic models submit readily to mathematical analysis; this makes their results much easier to interpret than those of models that must be studied by computer simulation.

The Issue of the Evolution of Mate Choice

The evolution of female mating preferences for male courtship displays is by no means the only issue in sexual selection theory, but it has been a dominant one for most of the century since the theory was first articulated by Darwin (1871). Although the existence of female "choice" as a cause of sexual selection on males is now widely accepted, the mechanisms responsible for the evolution of mating preferences remain in dispute. Most biologists make the reasonable presumption that mating preferences have evolved largely as a result of natural selection. If so, then there are two general ways in which this could occur. The first is that the mating preferences are or were in the past the objects of *direct selection.* Direct selection is the probable cause of mating preference when females choose mates on the basis of resources or protection offered by males. However, when males make no obvious reproductive contribution beyond their gametes, or when their attractive traits appear to be unrelated to resources males provide, direct selection provides no obvious explanation, and a second mechanism, often called *indirect selection,* is proposed. In these cases mating preferences are seen as evolving in response to selection acting on other, genetically correlated traits. Most of the current controversy in sexual selection theory focuses on two issues: whether direct or indirect selection is more important in the evolution of mating preferences; and if indirect selection occurs whether or not it increases a population's level of adaptation.

The idea of indirect selection for mating preferences originated with Fisher (1915, 1930). He posited four phases of evolution of female mating preferences and attractive male displays. At first, a novel mating preference begins to evolve because it causes females to discriminate a male trait that indicates overall genetic fitness. As a consequence of preferential mating, genes for the preference become correlated with genes for survival, and the preference thus increases in frequency. In the second phase, the preference has become sufficiently common that it creates an additional force of sexual selection promoting further elaboration of the male trait. By the third phase, the male trait is so elaborated by sexual selection that it now reduces viability. An equilibrium is eventually reached in which sexual selection favoring increased ornamentation is offset by viability selection against too highly ornamented males. The first three phases of "Fisherian" evolution have been modeled using both quantitative and population genetic approaches, and their feasibility is established (O'Donald 1967, 1980; Lande 1981b; Kirkpatrick 1982; Heisler 1984b). The fourth phase Fisher proposed has yet to be substantiated theoretically; in it an established preference is displaced by preferences directed at new, fitness-indicating traits (Fisher 1915).

The major alternatives to the Fisherian model are sometimes called "good genes" models because they propose that attractive male traits continue to act as indicators of male genetic quality; thus the eventual opposition of sexual and natural selection that Fisher proposed does not occur. These models assume various mechanisms, such as conditional expression of the male display or variation in attractiveness owing to parasite load, that cause a sustained positive correlation between the attractive trait and genetic differences in viability (Andersson 1986; Hamilton and Zuk 1982). Genetic models have substantiated some of these mechanisms (e.g., Andersson 1986; Pomiankowski 1987b, 1988; Heywood

1989), but the general feasibility of the process as well as the interpretation of existing models remains in dispute (e.g., Kirkpatrick 1986a,b).

In many cases, both direct and indirect selection may occur at the same time. One interesting class of models assumes that there is direct selection against female mating preferences because attractive males have reduced fertility (caused for example by sperm or resource depletion or by inferior parental care provided by highly polygynous males). There has been considerable debate about these "sexy son" models, centered on the question of whether indirect selection favoring mating with attractive males can effectively oppose the direct selection against such matings (Weatherhead and Robertson 1979; Kirkpatrick 1985, 1988; Curtsinger and Heisler 1988, 1989). All three of the above types of models—Fisherian, "good genes," and "sexy son"—have been studied using the quantitative genetic approach. The results, some of which are described below, have resolved many questions but raised yet others.

Structure and Assumptions of Quantitative Genetic Models

Any model of evolution by natural selection has three necessary components: (1) trait variation, (2) a causal relationship between traits and individual fitness, and (3) inheritance of trait variations. The first steps in constructing such a model are thus to define, symbolically, the traits of interest, their distribution in a population, and the set of relationships that describe how these traits influence survival, mating success, and/or fertility. This constitutes the model of phenotypic selection per se, which then can be used to deduce the selection differentials that summarize how selection affects the average phenotypic values of the population. To complete the model's specification, one must further invoke some rules of inheritance, and it is at this point that quantitative genetic models use the time-honored equations of quantitative genetics. In the following sections, I will outline and discuss the various assumptions made by quantitative genetic models at each of these steps in model building.

Genetic and Phenotypic Distributions

If there is a single key to the success of quantitative genetic models, it is their assumption that phenotypic traits are normally distributed within populations. The manifold effects of this assumption provide the foundation for the biological realism, measurability, and mathematical tractability that make quantitative genetic models such an attractive approach to modeling evolution. Assuming normality affords at least four major advantages.

1. Actual quantitative characters are often normally or nearly normally distributed within populations. The assumption therefore approximates natural patterns of trait variation.
2. Normality provides a logical connection to the empirical and theoretical foundations of quantitative genetics. The normal distribution is by definition that expected when a variable is the sum of a large number of independent random effects; hence, it is the natural consequence of additive, polygenic inheritance combined with developmental "noise" and other environmental effects on trait expression. Many actual quantitative traits show substantial additive genetic variance, and even when non-additive genetic variance or environmental effects are present, it is the additive genetic component of the variance that determines the response to selection on a character (Falconer 1989).
3. Empirical testing of quantitative genetic models is greatly aided by the normality assumption. The means and variances of characters in a population are highly accessible quantities. When traits are scored on a scale that yields approximate normality—for example, by using a logarithmic transformation (Wright 1968)—the powerful methods of parametric statistics can be used to estimate model parameters and test hypothesized effects. However, not all characters are easy to score in a way that allows for parametric statistical tests, and decision-making behaviors such as mate choice are especially problematic (see below).
4. Normality provides ease of mathematical analysis. Quantitative genetic models typically use simple exponential-type functions to characterize the fitness effects of trait variation. When combined with the normality assumption, this allows for dynamic equations that are relatively easy to solve. Furthermore, by using the multivariate form of the normal distribution, it is possible to model selection acting simultaneously on multiple traits. This is a significant theoretical innovation, because it is very difficult to model complex interactions between traits using classical population genetic approaches.

Because quantitative genetic models assume *exact* normality (for analytic purposes) they technically make some unrealistic assumptions: characters are determined either by an infinite number of genetic loci or by a finite number of loci each of which has an infinite number of alleles; allelic effects are themselves normally distributed; and environmental effects are symmetric and normally distributed. Fortunately, moderate violations of these assumptions do not substantially affect the overall tendency of traits to be distributed normally. For example, normality is reasonably approximated by as few as ten additive, diallelic loci or sixteen loci with complete dominance (Lush 1945, 84).

Fitness Functions

Historically, quantitative genetic theory was developed primarily for animal and plant breeding, and models therefore assumed the same form of selection as that practiced by breeders: truncation selection in which all individuals whose scores fell to one side of a certain threshold value were "culled" from the breeding population. Unlike artificial selection, natural selection on continuous characters rarely involves truncation; rather, survival probabilities, mating success, and fertility vary continuously with phenotypic value. Quantitative genetic models of natural selection therefore use continuous, typically exponential-type, functions to model components of fitness. These fitness functions mimic realistic features of selection in nature yet keep the models mathematically tractable.

The most commonly used function for modeling the fitness of quantitative traits is the bell-shaped "Gaussian" curve defined by the equation

$$W(z) = c \cdot \exp\left(\frac{-(z - \theta)^2}{2\omega^2} \right).$$

This function has the property that absolute fitness, denoted $W(z)$, takes on a maximum value of c only when an individual's phenotypic value equals an optimum value, defined by the parameter θ. An example is illustrated in figure 5.1, where Gaussian curves are used to represent two components of male fitness—viability, $W(z)$, and mating success, $M(z)$—as functions of male phenotypic value for an attractive trait.

Gaussian fitness functions have an additional parameter, ω, that

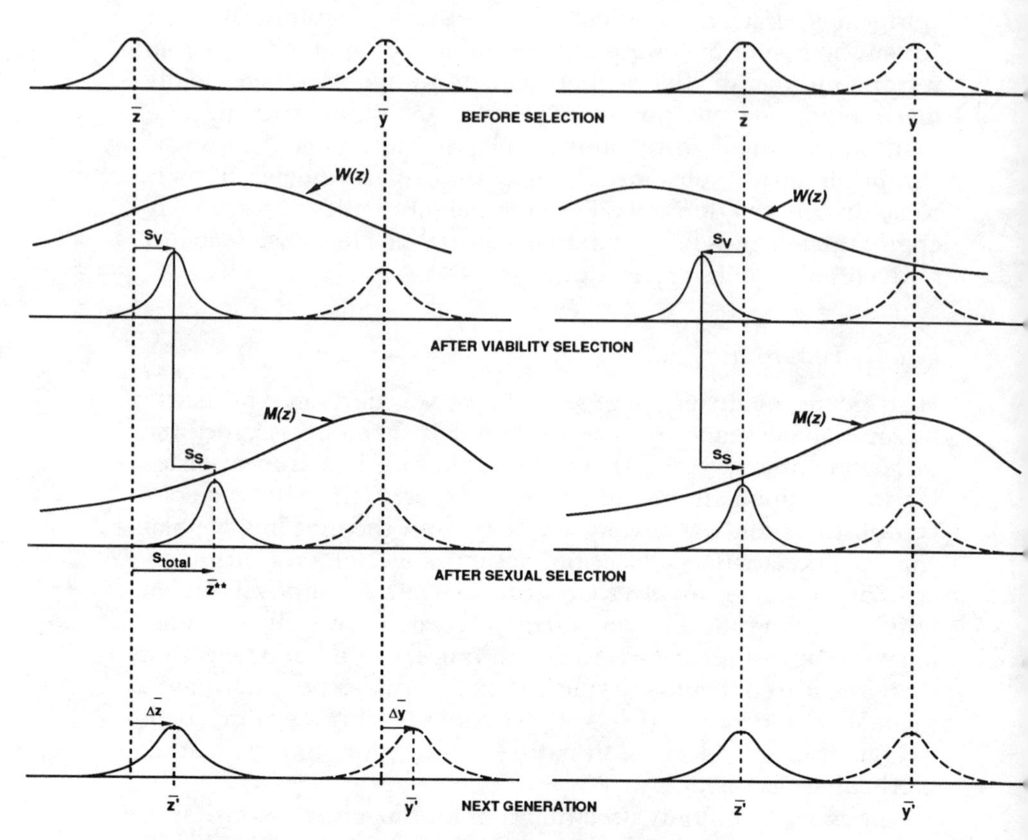

Fig. 5.1 Two phases of evolution under the Fisherian model. (*a*) The sequence of events in one generation for a population that is undergoing directional evolution. Both viability selection, defined by the fitness function $W(z)$, and sexual selection, defined by the mating success function $M(z)$, favor larger male phenotypic values. The total selection differential on the male trait is the sum of the selection differentials caused by viability and sexual selection ($S_v + S_s = S_{total}$). Both trait and preference show directional responses to selection, given by $\Delta\bar{z}$ and $\Delta\bar{y}$, respectively. (*b*) The sequence of events in an equilibrium population. Viability and sexual selection are here equal and opposite in their effect on the mean of the male trait ($S_v + S_s = 0$); hence, the trait has equilibrated at a mean value that lies between the two optima for survival and mating success.

determines the intensity of selection. Smaller values of ω cause the curve to drop steeply on either side of the optimum, implying strong selection on the trait. Larger values define broader, flatter curves, for which a wide range of phenotypes have similar, relatively high fitnesses. The two parameters ω and θ together define the strength and direction of optimizing selection. The parameter c determines only the scaling of fitness in absolute units (such as absolute survival probabilities or numbers of offspring) and has no effect on the magnitude of selective effects, as these depend only on fitness relative to the population mean.

Optimizing selection is the typical model used to consider viability selection on attractive male displays (e.g., Lande 1980b, 1981b). It has also been used to model variation in male fertility in "sexy son" models (Kirkpatrick 1985). Other exponential-type functions can be used to model different types of selection. For example, the function $W(z) = c \cdot e^{\alpha z}$ simulates sustained directional selection on a character. This is the type of selection expected for major components of fitness such as survivorship or fecundity that, when other traits are held constant, are always positively related to total fitness. Alternatively, one can model disruptive selection by changing the sign of the exponent in the Gaussian function above; then θ becomes the phenotype with the lowest fitness. Other more complex functions have been used to model selection when there is more than one optimum phenotype (Lande and Kirkpatrick 1988).

The simple fitness functions above gain interest when they are extended and combined to model complex types of natural selection. There are three general ways in which this is done. First, the functions can be generalized to model selection acting simultaneously on more than one character (e.g., Lande 1979; Lande and Arnold 1983). The multivariate Gaussian model, for example, includes a phenotypic optimum for each of the selected traits and a covariance-like matrix whose entries define both the strength of selection on each character and fitness interactions between characters. Such models have been used to consider female choice based on multiple male traits (Heisler 1985) and to model handicap-like interactions between a male display and a second, fitness-related character (Kirkpatrick 1986a).

Second, frequency-dependent selection, such as selection acting on competitive or social behaviors, can be modeled by allowing the fitness value of a character to depend on the distribution of the same or of different traits within the population. Such frequency-dependent selection occurs in all the models of female mating preference described below.

Third, multiple fitness functions can be combined to consider different episodes of selection during the life cycle. Total fitness then becomes the product of the different fitness components. For example, total male fitness in the basic Fisherian model is the product of viability and mating success, and the total selection differential on the male trait can be expressed as a sum of two component selection differentials associated with viability and sexual selection, respectively (fig. 5.1). Quantitative genetic models that consider all three major components of male fitness—viability, mating success, and male fertility—have been investigated by Kirkpatrick (1985).

Representations of Mating Behavior

Mate choice is a form of decision making, and decision making differs from other, simpler traits in a fundamental way. Specifically, the characters that evolve are the rules by which animals make decisions—their "preferences"—whereas fitness is determined by the actual "choices" that are made. Modeling the evolution of a preference thus requires that one specify the link between an individual's "preference phenotype" and the distribution of its choices. Lande (1981b) has termed these theoretical links "preference functions."

A preference function is a weighting system that shifts the distribution of a female's mates from what would be expected if she mated at random. Preference functions ignore the complex behavioral interactions that actually occur during courtship, and consider only how these interactions ultimately affect mating probabilities. Symbolically, a preference function can be written as $\Psi(z|y)$, which denotes the attractiveness of a male with trait z to a female with preference y. The probability of such a mating, that is, the actual distribution of a female's "choices," depends on both the preference function and the distribution of available males. In the case of a polygynous mating system in which all females mate, the probability that the mate of a type y female is a type z male can be defined as the product $p(z)\Psi^*(z|y)$, where $p(z)$ is the frequency distribution of phenotypic values among adult males and $\Psi^*(z|y)$ is the female's "relative preference" (her absolute preference divided by her average preference for males in the population; see Lande 1981b). The simplest biological interpretation of this model is that females encounter courting males at random but differ in their receptivity to different male phenotypes. Then $p(z)$ defines the rate at which females encounter type z males and $\Psi(z|y)$ is proportional to the probability that mating occurs, given an encounter. However, the same model can mimic cases in which females do not assess males but rather behave in a way that biases their encounter rates; then the preference function acts as a weighting system that determines the nonrandom distribution of encounters.

Lande (1981b) introduced three specific types of preference functions that have been used widely in quantitative genetic models of sexual selection. These are illustrated in figure 5.2. Each model represents a different class of decision-making rules. The "psychophysical preference" model assumes that attractiveness increases (or decreases) strictly with male phenotypic value (fig. 2a). In contrast, unimodal preferences assume that every female has an "ideal" male type (figs. 2b and 2c). The unimodal models include a parameter ν that determines female "choosiness;" when ν is small, females are unlikely to mate with males whose value deviates far from their ideal; when ν is large, females are likely to accept a wide range of males. If preferences are absolute, a female's preference phenotype, y, is equal to the male phenotype that she most prefers (fig. 2b). If preferences are relative to the mean, y determines the deviation of the ideal mate from the average of available males (fig. 2c).

These preference functions are readily extended to consider preferences based on more than one male character. Multivariate preference functions can be envisioned as a surface whose height defines the preference of a female with preferences $y_1, y_2 \ldots, y_n$, for a male with trait values $z_1, z_2 \ldots, z_n$. Multivariate preference functions can assume independent assessment by females of different male traits, or they can allow trade-offs in which the value attached to one trait depends on other aspects of a male's phenotype.

How preference influences fitness depends on additional assumptions. In the simplest models, only male mating success is affected by preference (Lande 1981b, 1982b; Engen and Sæther 1985; Lande and Arnold 1985; Heisler 1985; Kirkpatrick 1986a). When there are costs to females in searching for mates or differences in male fertility, then female fitness is also a function of preference (Kirkpatrick 1985, 1987a). An even more complex model is that of Kirkpatrick, Price, and Arnold (1990), which treats male reproductive success as a function not only of female preference, but also of breeding time and female nutritional state.

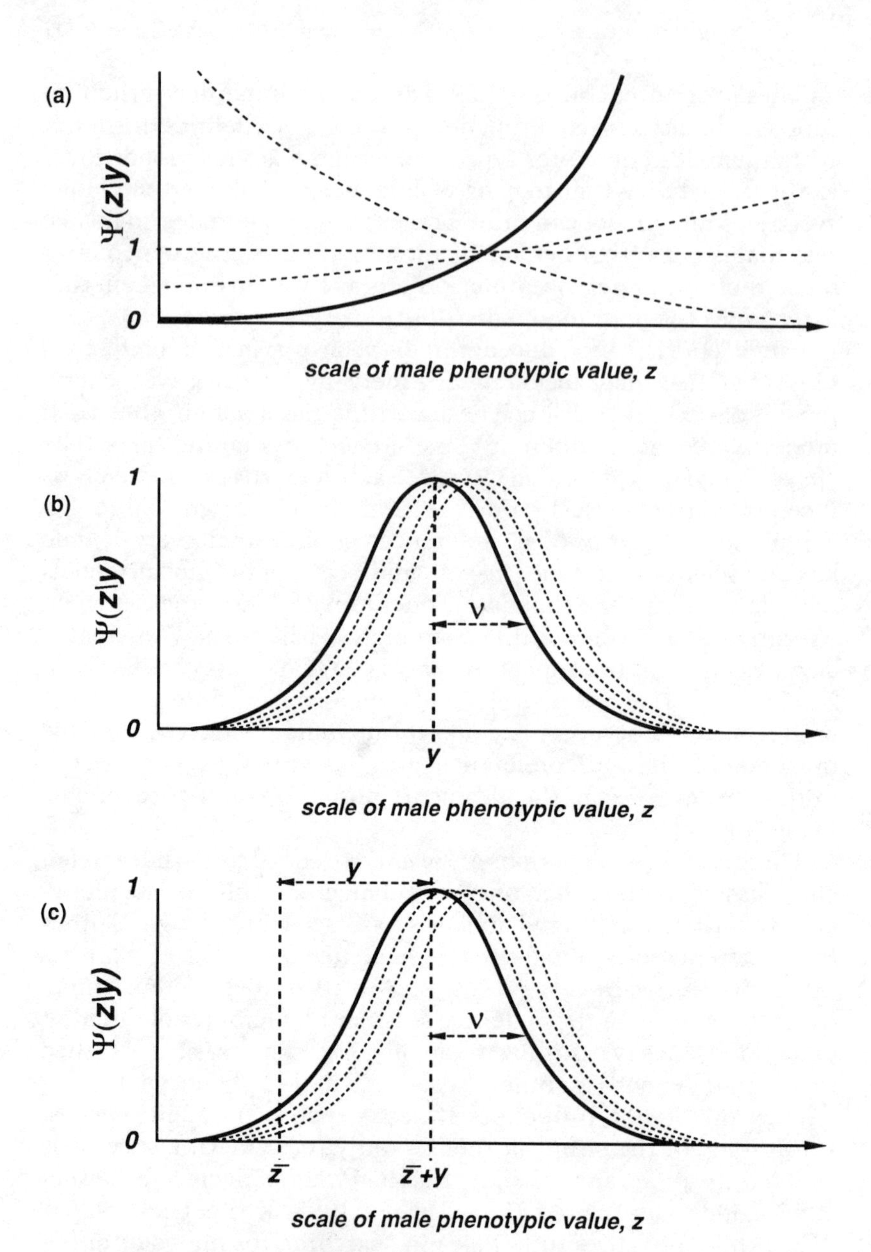

Fig. 5.2 Three preference functions used in sexual selection models. Each curve defines the preference of a female having preference phenotype y for a male having trait phenotype z. Different curves represent preferences of females with different values of y. (*a*) Psychophysical preference: $\Psi(z) = \exp(yz)$. (*b*) Absolute unimodal preference: $\Psi(z) = \exp[(z - y)^2/2\nu^2]$. (*c*) Relative unimodal preference: $\Psi(z) = \exp\{[z - (\bar{z} + y)]^2/2\nu^2\}$.

There are some obvious limitations to the preference functions used in quantitative genetic models. When females have unimodal preferences, the function has two parameters, y and ν. Modeling the evolution of the mean of y is straightforward, but it is not simple to analyze how female "choosiness," ν, evolves. Also, the preference functions do not readily lend themselves to modeling the evolution of the degree of polygyny itself, or to considering situations in which females have only a small sample of males from which to choose. Most important, the very generality that allows preference functions to mimic many different types of mate choice makes it difficult to apply them to specific situations: preference functions are difficult to measure empirically because mapping the responses of individual females to a wide range of male phenotypes is not feasible in most species.

The Dynamic Model

The assumptions discussed up to now are those used by quantitative genetic models to characterize the selective forces acting within any generation in a population. With these, one can deduce the selection differentials on the set of characters considered in any model. In order to go from a model of strictly phenotypic selection to a dynamic model of evolution, quantitative genetic models use the general equation $\Delta\bar{\mathbf{z}} = \mathbf{G}\mathbf{P}^{-1}\mathbf{S}$ for the response to selection on a set of quantitative characters. This equation, its uses, and its limitations are discussed by Arnold (chap. 2). In our present case, the dynamic model for the joint evolution of a female mating preference and an attractive male trait reduces to a slightly modified system of two equations:

$$\Delta\bar{z} = \frac{G_{zz}}{P_z}\left(\frac{S_z}{2}\right) + \frac{G_{zy}}{P_y}\left(\frac{S_y}{2}\right);$$

$$\Delta\bar{y} = \frac{G_{zy}}{P_z}\left(\frac{S_z}{2}\right) + \frac{G_{yy}}{P_y}\left(\frac{S_y}{2}\right).$$

This is the basic model originally studied by Lande (1981b) and Kirkpatrick (1985), and it forms the core of more complex quantitative genetic models that consider additional characters or types of selection (e.g., Lande 1982b; Engen and Sæther 1985; Lande and Arnold 1985; Heisler 1985; Kirkpatrick 1986a, 1987a; Lande and Kirkpatrick 1988; Kirkpatrick, Price, and Arnold 1990). In the above equations, the terms S_z and S_y denote the total selection differentials on the trait and preference, respectively. Each selection differential is halved because the trait and preference are assumed to be sex-limited; hence, one-half of all the population's alleles for each character are not exposed to direct selection in each generation. The ratios, G_{zz}/P_z and G_{yy}/P_y, are the respective heritabilities of the trait and preference; these determine the response to direct selection on the two characters. G_{zy}/P_z and G_{zy}/P_y are analogous quantities that predict offspring phenotypes for one character on the basis of parental values for the other character. These determine the magnitudes of the correlated responses in each trait to selection on the other trait. In Fisherian sexual selection models, the genetic covariance, G_{zy}, is considered to arise primarily from the nonrandom mating between attractive males and preferring females, rather than from pleiotropy of allelic effects.

As written above, the model assumes some direct selection on the female preference, hence the inclusion of the selection differential S_y (e.g., Kirkpatrick 1985, 1987a). If the preference is not subject to direct selection, then S_y is zero, and one obtains the "pure" Fisherian model in which the preference evolves only as a result of selection on the male trait and the genetic covariance between trait and preference (Lande 1981b). Figure 5.1 provides a graphical view of evolution under this model.

In order to apply the predictions of quantitative genetic models to empirical data, it is important to understand how the assumptions required by the above model influence the results that are obtained. Probably the most important assumption is that the genetic and phenotypic variances and covariances are constant over time. A considerable amount of theory now exists that defines the conditions under which this assumption is valid (e.g., Bulmer 1972; Lande 1976a; Turelli 1984; Barton 1986); however, the dynamics and equilibrium magnitude of genetic variance for characters undergoing selection remains an open theoretical question. The assumption has several implications for the empirical application of quantitative genetic models. First, the model is strictly valid only if selection is assumed to be relatively weak. Weak selection is necessary because it allows the genetic variation lost due to selection to be replenished each generation by new mutation. Second, so long as the genetic variances are constant and genetic correlations between characters are less than perfect, a necessary consequence of the above equations is that the equilibria are determined only by the pattern of selection. Regardless of how little genetic variation may be available, populations will eventually evolve to a state in which the net selection differentials on every character are zero. If the assumption of weak selection is violated in an actual population, genetic variance for characters may be lost and the resulting equilibrium would be one in which one or more characters is genetically fixed. Such equilibria cannot be predicted by quantitative genetic models; this means that empirical tests for heritability of traits are essential to determining whether the long-term predictions of quantitative genetic models might apply to a given population.

Results and Predictions of Quantitative Genetic Models

A frequent but avoidable error in applying evolutionary models is the failure to distinguish between results that bear on different phases of a population's evolution. Models can be analyzed to answer questions about the evolutionary origin of traits or about the period of their directional evolution, as well as about the long-term equilibria that occur. For the basic model outlined above, I will describe some results bearing on each of these phases of evolution, placing emphasis on their importance in identifying relevant tests of sexual selection theory.

Equilibria

The first goal in analyzing an evolutionary model is usually to determine the properties of populations at evolutionary equilibrium. For the "Fisherian" model in which there is no direct selection on females, equilibrium occurs when the forces of sexual and viability selection on males, measured by their respective selection differentials, are equal and opposite (Lande 1981b). Figure 5.1b illustrates the pattern of selection that would occur during a generation at such an equilibrium. More generally a line of equilibria occurs in which the means of the preference and trait can take on any pair of values (fig. 5.3a). The particular equilibrium that is obtained depends on the initial conditions assumed. If the genetic covariance between trait and preference is sufficiently large, the line will be unstable and populations will evolve indefinitely in what has been called a "runaway" process (Fisher 1930). This unstable situation is illustrated in figure 5.3b. When the line of equilibria is unstable, quantitative genetic models cannot predict the long-term results of sexual selection. The most likely outcome is that genetic vari-

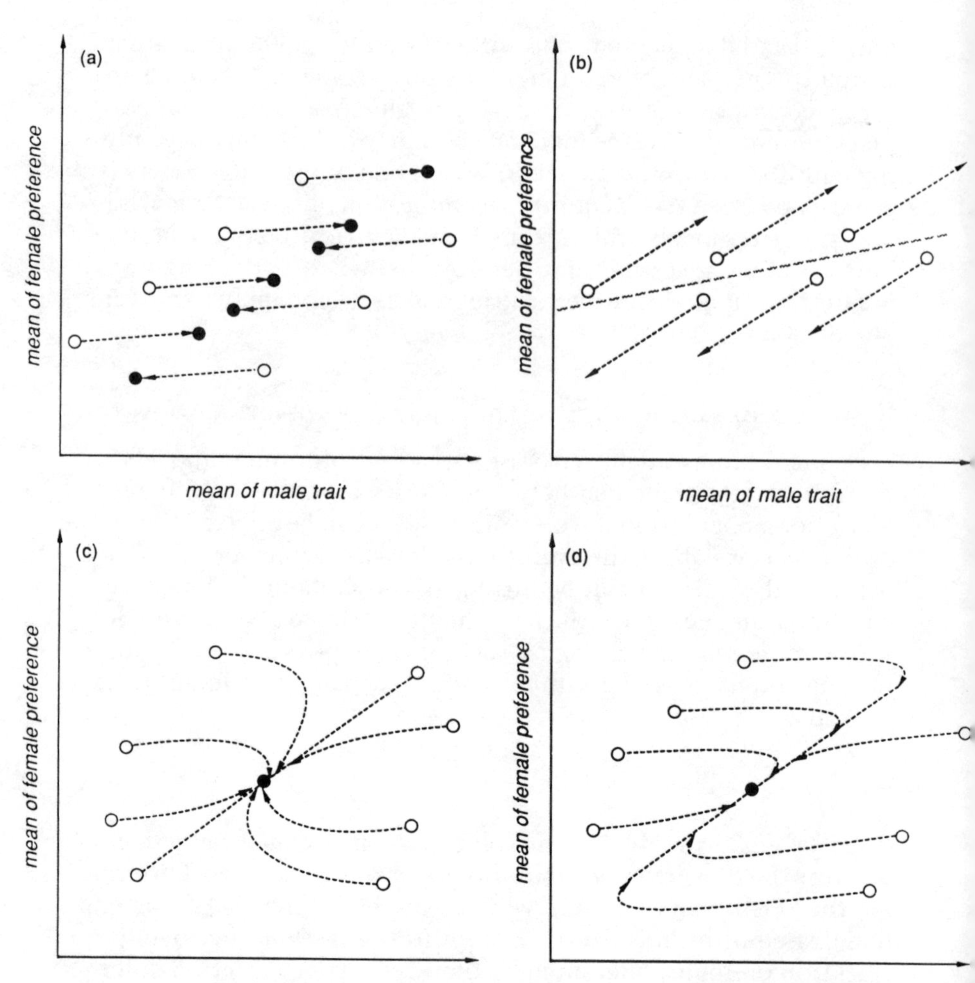

Fig. 5.3 Evolutionary trajectories in some sexual selection models. Open circles represent initial means for the trait and preference; arrows are evolutionary trajectories; solid circles represent equilibria that are eventually obtained. (*a*) The Fisherian model with multiple stable equilibria along a line defined by the row of equilibrium points. (*b*) The Fisherian model with "runaway" sexual selection; populations evolve away from a line of unstable equilibria. (*c*) The "sexy son" model with strong fertility variation among males; a unique equilibrium occurs at the mean values for trait and preference that maximize average female fertility. (*d*) The "sexy son" model with weak fertility selection; populations converge on a trajectory very similar to the line of equilibria in (*a*) and then evolve slowly along it toward the unique equilibrium.

ance for the male trait would be lost as a result of sustained intense selection.

Unstable runaway processes are probably unlikely to occur in nature, because the magnitude of genetic covariance required to destabilize the line of equilibria is quite large. In actual populations, the expression of trait and preference genotypes is probably imperfect, and nonrandom mating may be unable to generate sufficient covariance at the genetic level. Sampling error in finite populations also causes random fluctuations in the genetic covariance; this too may reduce the chances of a runaway (Nichols and Butlin 1989). An exception should be cases in which the trait and preference are influenced by common, pleiotropic genes; genetic correlations large enough to cause an unstable runaway process might well occur under these conditions. It should be emphasized, however, that the evolution of dramatic courtship displays is not dependent on unstable runaway evolution. Random genetic drift along neutrally stable lines of equilibria (e.g., fig. 5.3a) might eventually lead to the evolution of exaggerated traits and preferences. This would occur at an evolutionary rate much slower than that of the runaway process.

In contrast to the Fisherian model, direct selection on the female preference leads to a single equilibrium in which the preference on average maximizes female fitness. This is illustrated in figures 5.3c and 5.3d for the "sexy son" model of Kirkpatrick (1985).

Multiple equilibria along lines (actually, along curves) also occur in two-locus population genetic models of Fisherian selection (e.g., Taylor and Williams 1981; Kirkpatrick 1982; Seger 1985; Seger and Trivers 1986). This observation led to the idea that neutral curves of equilibria were a robust theoretical result that occurred in all Fisherian models regardless of their specific genetic assumptions. As such, they would form a basis for testing the Fisherian model in actual populations (Arnold 1983b, 1985; Heisler et al. 1987). However, recent models that make different genetic assumptions, such as mutation at a preference locus or diploid inheritance, do not predict lines of equilibria (Bulmer 1989; Heisler and Curtsinger 1990; Barton and Turelli 1991). The empirical significance of this difference in equilibria is not known; hence it is probably premature to consider the line of equilibria as a firm prediction that can be used to test the Fisherian model. This uncertainty is highlighted by considering the dynamic properties of the models, discussed below.

Initiation of Evolution

Because quantitative genetic models assume that populations always contain some additive genetic variance for any trait, they are not particularly well suited for constructing models for the evolutionary origin of qualitatively new characters. It is nonetheless possible to define conditions under which a novel trait, introduced to a population by mutation or migration, will increase when it is very rare. The method used to do this is not unique to quantitative genetic models but is similar to those used both in evolutionary game theory to determine when a novel strategy can invade a population (Maynard Smith and Parker 1976), and in population genetics to identify conditions for a "protected polymorphism" (Prout 1968). When this approach is used to examine the conditions that would favor the increase in frequency of a rare, novel mating preference in an otherwise randomly mating population, the results confirm that preferences for males having high viability may arise (Heisler 1984b). However, the force promoting such preferences is likely to be strong only in populations in which the viability indicator is undergoing directional selection toward some new optimum. This result demonstrates the importance of distinguishing between different phases of a population's evolution. Under Fisherian selection, positive correlations between male attractiveness and viability may occur during the early phase of evolution of a preference, but at equilibrium these correlations are expected to be negative. This contrast between dynamic and equilibrium states can be seen by comparing figure 5.1a with figure 5.1b.

Dynamic Properties

The dynamics of sexual selection models have not been studied as thoroughly as have their equilibria. Partly this is because we are more interested in predicting the results of long-term evolution. It is also mathematically more difficult to analyze the dynamics of a model than to solve for its equilibria. However, predictions about rates and directions of evolution in nonequilibrium populations can be very important in identifying the empirical implications of a model. This is especially true of models of sexual selection, because the rate of evolution can vary by orders of magnitude depending on the specific process that is occurring. For processes that occur very rapidly we would expect to find natural populations at or near their equilibria. However, if very slow evolutionary processes are involved, testing a model on the assumption that equilibrium has been reached may not be valid. For example, runaway Fisherian selection may occur at an extremely rapid rate, which implies that we may be unlikely ever to find a runaway in progress in a natural population (Fisher 1930; Lande 1981b). In contrast, selection for a sexual dimorphism can proceed very slowly, because pleiotropic gene effects limit the amount of genetic variation available for the independent evolution of the character in males and females. Although quantitative genetic models predict that if appropriate genetic variance persists, sexually dimorphic traits will ultimately reach their optimal expression in both sexes, many generations may be required before this occurs (Lande 1980b; Lande and Arnold 1985). We therefore can expect to find natural populations that are not at equilibrium for sexually dimorphic traits such as body size or breeding plumage. Testing equilibrium predictions in this situation might yield negative results, not because the model was incorrect but because the assumption that populations had reached equilibrium was wrong.

A second reason why dynamic properties are important is that nonequilibrium populations under one model may closely resemble equilibrium populations under another. This is the case with the Fisherian and "sexy son" models discussed above (Lande 1981b; Kirkpatrick 1985). Although the equilibria predicted by the two models are qualitatively different—an infinity of equilibria along a line in Fisherian models versus a single equilibrium in "sexy son" models—the situation is less clear-cut when deviations from equilibrium are considered. This can be seen by comparing the different cells in figure 5.3. Figure 5.3c illustrates a "sexy son" model that assumes large fertility differences among males. These differences generate strong direct selection on the female preference, which leads to rapid evolution of both trait and preference to the unique equilibrium at which average female fertility is maximized (Kirkpatrick 1985). In contrast, figure 5.3d illustrates a case of the same model in which fertility effects are assumed to be weak relative to viability and sexual selection. The preference, which is subject only to fertility selection, evolves much more slowly than the male trait, which is also subject to viability and sexual selection. Populations that deviate from equilibrium under weak "sexy son" effects spend many generations evolving along a trajectory that is very similar to the pure Fisherian model's line of equilibria, illustrated in figure 5.3a. Hence if the possibility of nonequilibrium is allowed, the predictions of Fisherian and "sexy son" models are less easily distinguished.

Testing Quantitative Genetic Models

There has been considerable discussion about whether the appropriate way to test evolutionary models is to test their predictions or their assumptions (e.g., Andersson 1987). Tests of predictions can involve either testing for an equilibrium pattern of variation within a single species, or comparative tests for predicted relationships between traits across a range of species. Tests of assumptions involve determining whether a group of organisms has, at present, the pattern of genetic, phenotypic, and/or fitness variation assumed by a particular model. In principle, tests of long-term predictions are preferable to tests of assumptions, because if they succeed a particular model can be soundly rejected as a historical explanation. Tests of assumptions only allow one to reject the hypothesized model as operating at the present time. However, it should be evident from the previous discussion that discriminating predictions of sexual selection theory are hard to come by. Tests of assumptions, although limited, are almost always informative. There are several reasons for this.

First, the long-term predictions of sexual selection models are not yet fully understood. The theory itself is relatively young, and new genetic models continue to undermine what have previously been seen as clear distinctions between the predictions of earlier models. Important features of natural populations, such as age structure and finite population size, have yet to be addressed thoroughly by quantitative genetic models. It is virtually certain that our current collection of long-term predictions will be altered by future theoretical work.

Second, even if the ultimate aim is to test long-term predictions, certain assumptions must be tested first, simply to verify that a general class of models might apply to a particular species. For example, quantitative genetic models assume that traits are heritable at equilibrium. Without testing the validity of this assumption, one cannot be certain that the equilibrium results of quantitative genetic models are in fact the appropriate predictions to test.

Third, whatever one's preferred approach to testing models may be, most evolutionary biologists share the ultimate aim of making inferences about historical mechanisms—inferences that rely fundamentally on combining estimates of the rates and magnitudes of current evolutionary processes with an assumption of evolutionary uniformitarianism. For example, if we wish to test a model's predictions, we must assume that past processes have operated consistently enough over time that their long-term results have been obtained. Similarly, if we wish to gain historical insight by testing a model's assumptions, we must assume that current processes approximate historical ones. In either case, only studies of current processes can identify probable mechanisms of past evolution and provide direct estimates of their relative strength.

The results of quantitative genetic models of sexual selection hinge on four classes of assumptions that are in principle testable in a variety of species. These are (1) the existence of genetic variance for traits and preferences; (2) the pattern of selection on traits and preferences; (3) the form of female mating preferences; and (4) the existence of genetic covariance between traits and preferences.

Heritability. Genetic variance for traits involved in mate choice is a central assumption of quantitative genetic models of mating behavior. If a male trait subject to sexual selection is not heritable within a natural population, then there can be no indirect selection on female mating preferences. Studies like those of Hedrick (chap. 11) are needed to determine whether male courtship displays typically show genetic variation. Likewise, if there is no genetic variance for mating preferences, then their maintenance requires no explanation in terms of ongoing forces of direct or indirect selection. At present only a handful of studies have identified genetic variation for mating preferences (Heisler 1984a; Breden and Stoner 1987; Engelhard, Foster, and Day 1989) and only one study (Majerus, O'Donald, and Weir 1982) has shown such variation to exist at the within-population level, where it is needed if it is to provide a basis for evolution in response to sexual selection. Moreover, if populations do not possess genetic variance for traits and preferences, the equilibria predicted by quantitative genetic models may not apply; hence, it may be foolish to test the equilibrium predictions of quantitative genetic models without first determining whether genetic variance is present.

Selection. The principal differences between alternative models for the evolution of mating behavior are in the types of selection they assume to be operating. In order to identify the general class of models that might apply to a given population, we need to know such things as whether female mating preferences are subject to direct selection, whether attractive traits are expressed and selected in one or both sexes, and whether fitness interactions like those proposed in viability indicator models occur. It is not easy to collect the large amount of data needed to determine the form of selection in natural populations, but when this is possible there are several methods that can be used to estimate the parameters relevant to quantitative genetic models (Lande and Arnold 1983; Schluter 1988; Heisler and Damuth 1987; Mitchell-Olds and Shaw 1987).

Preference functions. Some results of sexual selection models, such as the presence or absence of lines of equilibria, do not depend on the exact form of female mating preferences. A few do. For example, the likelihood that a population has undergone runaway sexual selection is much greater if females possess directional (psychophysical) preferences or preferences relative to the mean of the male trait (e.g., figs. 5.2a,c). Absolute preferences are unlikely to induce a runaway process because the genetic covariance required to initiate it would have to be very large, probably too large to occur in actual populations (Lande 1981b). Unfortunately, estimating the form of individual females' preference functions is virtually impossible, because in most species it is not feasible to test a female a sufficient number of times to estimate her receptivity to a large range of male phenotypes. By using females from inbred lines or families, it may be possible to estimate preference functions for particular genotypic classes, but under most circumstances the most we can hope to obtain is the average preference function for an entire population of females (e.g., Andersson 1982a; Gerhardt 1991).

Genetic covariance between traits and preferences. Genetic correlations are central to all models that assume indirect selection of female mating preferences (Partridge, chap. 6). These include not only the classic Fisherian model but also the various viability indicator and handicap models (e.g., Zahavi 1975; Hamilton and Zuk 1982; Nur and Hasson 1984; Andersson 1986; Pomiankowski 1987b). The ability of preferential mating to generate genetic covariance has never been empirically established. In theory, this could be done by using parent-offspring regression. However, because the genetic correlation is generated by the mating system rather than by pleiotropic gene effects, it is no trivial task to devise an experimental breeding scheme that will not itself disrupt the correlation one wishes to estimate. Because these genetic correlations are likely to be small, detecting them statistically will be difficult, and there is a hazard of misinterpreting genetic correlations due to other causes, such as random linkage disequilibrium, natural selection, or artifacts of the past breeding history of the experimental stocks. It may be that our best estimates of the magnitude of genetic covariances between attractive traits and mating preferences will be obtained, not from empirical studies, but from detailed, realistic computer simulations.

Conclusion

This chapter has focused on the relationship between the structure of quantitative genetic models and their empirical application. Although testability is a desirable quality in any model, it is important to recognize that quantitative genetic models of sexual selection have proved to be valuable apart from any attempts to test them. For example, quantitative genetic models have substantiated the logic of Fisher's early, verbal model of sexual selection. They have also clarified several issues concerning the "good genes" models: verifying that genetic variation in male viability can promote the origin of female mating preferences (Heisler 1984a); identifying problems in maintaining male traits as viability indicators (Lande 1981b; Kirkpatrick 1986a); and pointing out conceptual difficulties with the "sexy son" hypothesis (Kirkpatrick 1985). Formal symbolic analysis is a necessary safeguard against errors in biological reasoning. It is especially helpful when dealing with very complex evolutionary processes such as those involved in the evolution of communication systems.

The quantitative genetic theory of sexual selection is by no means complete, and a number of models that will be important to empirical workers await development. The evolution of mating behavior in species with parental care is only beginning to be explored (Cheverud and Moore, chap. 4). Age structure, especially female choice of traits that change with male age, has not yet been addressed by genetic models of mate choice. The effects of stochastic variation in behavior, of experience, and of conditional behaviors such as female copying or alternative male "strategies" remain to be investigated using the quantitative genetic modeling approach. All of these phenomena are encountered by behavioral biologists, some frequently, and their effects on the results of quantitative genetic models are yet to be determined. In addition, some difficulties in measuring mating preferences would be reduced if theoretical links could be created between detailed behavioral models and the generalized "preference functions" used in quantitative genetic models. In the interim, more quantitative genetic models of behavioral evolution will appear, and a major task for empiricists will be to identify the natural processes that it would be most fruitful for these future models to address.

Summary

This chapter introduces quantitative genetic models of phenotypic evolution using examples from recent models of sexual selection. Quantitative genetic models provide perhaps the best compromise yet achieved between the conflicting needs for realism, applicability, and ease of interpretation in mathematical models of evolution. The virtues of quantitative genetic models derive largely from three major assumptions: (1) that phenotypic traits are normally (or multivariate normally) distributed within populations; (2) that exponential-type fitness functions provide reasonably realistic models of stabilizing, directional, or more complex forms of phenotypic selection; and (3) that genetic variance is purely additive and of constant magnitude over time. A major advantage of quantitative genetic models is that they employ variables that can be measured in actual populations and that are amenable to established methods of statistical analysis.

The application of quantitative genetic models to sexual selection has largely been focused on a debate about the kind of selection—"direct" or "indirect;" "Fisherian" or "good genes"—that is most important in the evolution of female mating preferences for attractive male displays. A significant feature of quantitative genetic models of sexual selection is the use of "preference functions"—formulations that provide an explicit link between female "preference phenotypes" and the pattern of actual choices that females make in mating. Preference functions provide a weighting system that shifts the distribution of mating males from that expected if mating were random, thereby providing a model of female-determined, frequency-dependent sexual selection on males.

The results of quantitative genetic models of sexual selection have had an important influence on the conception and interpretation of empirical studies of sexual selection, and there has been a great deal of interest in testing these models in the field. The long-term predictions of different quantitative genetic models are not clearly distinguishable, especially when one allows the possibility that actual populations may not have reached evolutionary equilibrium. For this reason and others, tests of the validity of the assumptions of quantitative genetic models are more likely to provide lasting, informative results than are tests of the models' long-term predictions. There are four classes of assumptions that bear importantly on model results and hence on the implications of sexual selection theory: (1) the existence of genetic variance for traits and preferences; (2) the pattern of selection on traits and preferences; (3) the form of female mating preferences; and (4) the existence of genetic covariance between traits and preferences. Of these, the first two classes of assumptions are quite amenable to empirical analysis, whereas the last two classes will be difficult, and in practice may be impossible, to establish empirically.

Acknowledgments

I thank David Cox, Russ Lande, and Don Price for helpful comments on earlier drafts of this chapter. Chris Boake deserves bundles of credit, as well as special thanks for her patience and diplomacy.

6

Genetic and Nongenetic Approaches to Questions about Sexual Selection

Linda Partridge

Evolution by sexual selection occurs when additive genetic variance in a trait causes variation in the number or quality of the fertilizations obtained by different individuals. Sexual selection mainly affects males, because their potential mating rate is usually higher than that of females. Traits that cause increased mating success may do so by enhancing the ability of their bearer to compete with members of the same sex for access to matings, and by increasing the likelihood of acceptance as a mating partner by members of the other sex. Sexual selection has been the subject of intense theoretical and empirical scrutiny (e.g., Lande 1981b; Bradbury and Andersson 1987; Kirkpatrick and Ryan 1991; Heisler, chap. 5; Hedrick, chap. 11). Many of the characters known to cause variation in mating success are behavioral (e.g., McComb 1991), although much research has been directed toward morphological characters, especially body size.

As with any other form of evolution by natural selection, hereditary variation in a character affecting a fitness component is the fuel for sexual selection. Is it then the case that all empirical studies of sexual selection should involve the measurement of genetic parameters? That many do not is revealed by a glance through any journal or book concerned with behavioral ecology (e.g., Pitnick 1991; Weatherhead 1990; Krebs and Davies 1987), and even evolution (e.g., Hews 1990) and genetics (e.g., Pitafi et al. 1990) journals carry purely phenotypic studies. Behavioral ecologists working in the field typically do not use genetics as a tool, in part because it would not be practical to do so; artificial selection on elephant seals would be undertaken only by the foolhardy, and heritability studies on most insects would be a nightmare. On the other hand, behavioral ecologists have usefully tackled questions that would make any rational geneticist blanch, such as the effect of male roaring on mating preferences of female red deer (McComb 1991) and the costs of long tails in male European swallows (Møller 1989).

It is the aim of this chapter to consider how genetic and non-genetic approaches to sexual selection can complement one another. Most of the data will be drawn from observations and experiments on *Drosophila melanogaster.* A set of specific topics has been selected for discussion; some are directly relevant to studies of other forms of natural selection (identifying targets of selection; costs of selected characters) while others are peculiar to sexual selection (distinguishing competition from choice) or at least to the evolution of sex differences (genetic correlations between the sexes).

Identifying Targets of Sexual Selection

The tacit assumption underlying most empirical work on sexual selection is that if a character has evolved to its present form by sexual selection, then it should be possible to detect the effects of the character on male mating success in extant populations. This assumption is probably reasonable because, as originally argued by Darwin (1871), sexually selected characters look costly and have sometimes been demonstrated to be so (e.g., Møller 1989). Their continued presence therefore requires an explanation in terms of some form of balance between selection for mating success and selection for viability. Full understanding of this balance requires a quantitative approach that measures not only whether sexual selection is acting but how strong it is.

A common method of identifying sexually selected characters is to measure the correlations between candidate characters and mating success in the field. For instance, in *Drosophila melanogaster* males found in copula have been shown to have a higher mean wing length than those found unpaired (Partridge, Hoffmann, and Jones 1987; Taylor and Kekic 1988). Studies of this kind abound, and give a useful preliminary indication of candidate characters. However, age and cohort effects are both possible confounding variables, if longevity differs between size classes (e.g., Partridge and Farquhar 1981), or if size varies between cohorts in correlation with some other variable that is the true determinant of male mating success. It is therefore necessary to examine the mating success of males in a single size-variable cohort. For *Drosophila,* this confines work to the laboratory, where several studies have shown that predominantly environmental variation in wing length (Wilkinson 1987) and in thorax length (Partridge, Ewing, and Chandler 1987) were correlated with the mating success of males within a single cohort.

Both wing and thorax length are heritable traits in captive outbred populations of *D. melanogaster* (Robertson and Reeve 1952; Wilkinson 1987) and, possibly, in field populations (Coyne and Beecham 1987; Riska, Prout, and Turelli 1989). Either or both of these traits might therefore be expected to show an evolutionary increase under sexual selection unless some form of balancing selection is occurring through an effect on another fitness component. However, there are objections to this line of reasoning. First, this kind of study of mating success suffers from the general objection to correlational studies, namely, that the true target of sexual selection may be some unidentified variable correlated with those measured. Second, the correlation between the measured character and the true target of selection may be environmental, so that the measured character would not be expected to evolve with the true target of selection. Purely environmental correlations between characters are perhaps especially likely to occur for suites of characters that are sensitive to early nutrition. For instance, in collared flycatchers (*Ficedula albicollis*), survival from fledging to first breeding showed a positive correlation with tarsus length, which was itself heritable (0.5–0.6) in the study population. However, when the effect of body weight at fledging was included in the analysis, the effect of tarsus length on survival became nonsignificant, suggesting that body weight was the real target of selection. Body weight at fledging was not itself heritable, nor were the two characters genetically correlated, so no response to selection would be predicted in either character (Alatalo, Gustaffson, and Lundberg 1990).

Two methods can be used to circumvent the difficulties inherent in correlational studies. One is to make an experimental alteration of the suspect character, a method strongly recommended by Grafen (1988). Experimental reductions of the wing area of male *D. melanogaster* (Ewing 1964) and alterations of the tail length of widowbirds (Andersson 1982a) and swallows (*Hirundo rustica*) (Møller 1988) have revealed that these characters are determinants of male mating success. However, experimental alterations are not a universal panacea. It is necessary to manipulate only the character of interest, which is not always feasible. Surgery, with appropriate controls, is a useful approach, but it cannot be successfully applied to all characters. Thorax length of *D. melanogaster* is an example; it can be changed by altering the level of nutrition (Robertson 1960, 1963) or the temperature during development (Alpatov 1930), but these manipulations are likely to introduce the kinds of environmental correlations that the technique is intended to break. A second possible problem with environmental manipulation is that it may not give a quantitatively accurate picture of the effects of selection on genetic variation for the character. For instance, male birds with long tails have had to grow their tails themselves, which may to some extent be achieved by allocating resources to growing tails at the expense of other activities. If growing a long tail lowers nutritional reserves, then there could be a negative effect on mating success through less vigorous courtship. Trade-offs between different characters important for mating success have been documented in phenotypic studies (e.g., Hasselquist and Bensch 1991). Last, different characters involved in producing high mating success, such as long tails and vigorous courtship, could be to some extent under the control of the same genes. The characters would then tend to evolve together, so that the evolutionary response to selection on one of them would be for the other to change as well (a correlated response to selection). An experimental manipulation could alter each character independently, with potentially misleading consequences for predicting the effects of selection.

A second method of avoiding environmental correlations between characters is to examine the effects of purely genetic variation, using either resemblance between relatives or artificial selection. For instance, Fowler and Partridge (unpublished data) used artificial selection to examine the effects of thorax length on male mating success in *D. melanogaster.* Four replicate lines of flies were selected for large and small thorax length, and four control lines were maintained at the same effective population size, with parents selected randomly. There was a significant response to selection in both directions, and other characters, including wing size, were also affected (Wilkinson, Fowler and Partridge 1990). Lifetime mating success of the males of the lines was examined, using F_1 hybrids between the lines (line 1 crossed with line 2 and line 3 with line 4) to avoid inbreeding depression. F_1 hybrid males were then each allowed to compete with a *scarlet*-eyed male for matings with two *scarlet* females; because *scarlet* is recessive, the resulting progeny could be scored for paternity. The measure of mating success therefore included not only copulations, but also any postmating effects. The results (fig. 6.1) suggested that small line males sired significantly fewer progeny than control or large line males, which did not differ significantly from one another. These results are similar to those of Ewing (1961), who examined behavioral measures of mating success.

These findings could be taken to suggest that thorax length, or some character genetically correlated with it, shows a genetic correlation with mating success over the normal size range (i.e., up to control size), but that the correlation breaks down once the normal size range is exceeded. There are alternative interpretations. First, body size in the small lines could have been reduced in part by an increase in the frequency of deleterious recessive alleles, so that the flies were generally unfit. This is perhaps unlikely, because the larvae of the small lines were as viable as the controls (Partridge and Fowler 1993). As a second interpretation, some third character, genetically correlated with thorax length over the small-control size range but not over the control-large range, could have been the true target of sexual selection. The true target would then have been successfully manipulated downward but not upward. For instance, growth could have been stopped early in the small lines, so that all parts of the fly were small, whereas in the large lines selection on thorax length could have selectively increased the size of the external structures. Third, the correlated responses to selection on body size may not have been those that would have occurred in nature, because it is clear that the nature of responses to selection depends upon the environment in which selection occurs, and the laboratory can provide a peculiar and benign environment. For instance, Ewing (1961) showed that in his selected large lines, the amount of wing vibration during courtship dropped relative to that of controls, whereas in the small lines it increased. These changes were not an inevitable effect of body size, because they disappeared in the F_2 progeny of crosses between different size lines. When Ewing selected on body size, removing the opportunity for sexual selection on courtship by pairing flies for mating, the amount of wing vibration in the small lines did not change relative to that of controls, while the decrease in the large lines again occurred. These results suggested that there had been selection for increased wing vibration to compensate for the deficiencies in courtship consequent on small body size, but that the drop in wing vibration in the large lines may have been attributable to a trade-off between wing area and amount of wing vibration. Whatever the correct interpretation, these experiments imply, un-

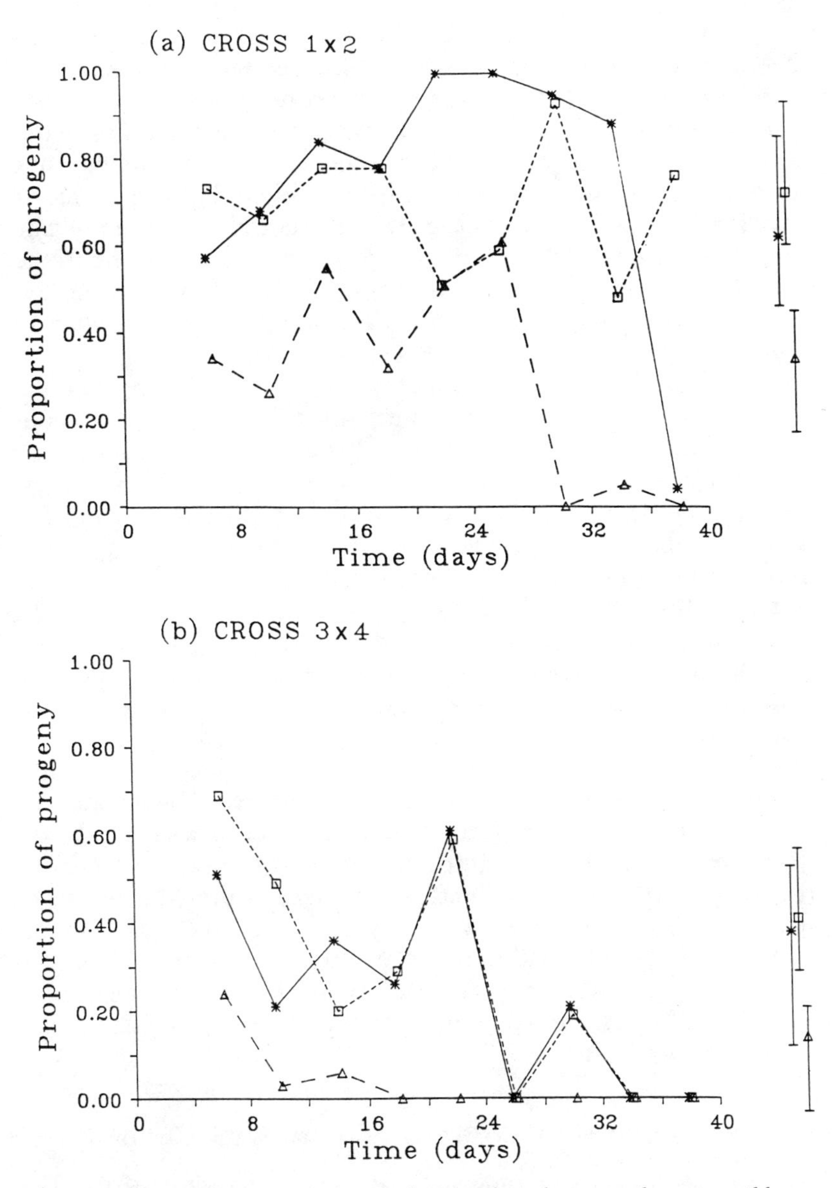

Fig. 6.1 Median (and interquartile range) proportion of progeny that were wild type plotted against age for males of the large (asterisks), control (squares), and small (triangles) lines, and their lifetime medians. Two F_1 crosses between the four parental size lines were used. Both age-specific and lifetime progeny production was lower in the small than in the large or control line males.

surprisingly, that a variety of correlated responses can occur, depending upon the circumstances in which selection is done.

The difficulties with selection experiments could be circumvented only partially by the use of a two-generation study of an unselected population. Any effects of altered frequency of deleterious recessive alleles would then not be a problem, but the character studied might still merely be genetically correlated with the true target of sexual selection, and the results would be specific to the environment in which they were obtained. In addition, selection experiments do have the advantage that they can take the suspect character outside the natural range of variation.

Identifying sexually selected characters and measuring their fitness functions are both important aims. It would be easy to conclude that these goals can be accomplished only by genetic studies if the results are to have any evolutionary relevance (e.g., Wills, Coyne, and Kirkpatrick 1991). However, genetic studies can have logical difficulties at least as serious as those of nongenetic ones, and short-term genetic correlations may be a poor guide to long-term evolutionary responses (e.g., Partridge and Sibly 1991). To date, few checks have been made on whether an experimental manipulation and a genetic one of the same magnitude produce the same effect on male mating success. Some data for *D. melanogaster* come close to achieving this aim: surgical reduction in wing area lowered male mating success (Ewing 1964), and genetically small flies had lower mating success (Ewing 1961; Fowler and Partridge, unpublished data). However, it is not clear whether the effects were of identical magnitude. They might not be expected to be, because the surgical reduction of wing area altered wing structure and shape as well as size. Only by a better biological understanding of the roles of different characters in determining mating success can progress be made, and it seems likely that this kind of insight will come from a combination of experimental and quantitative genetic studies.

Distinguishing Male Competition from Female Choice

Assuming that targets of sexual selection can be identified, the mechanisms by which they affect mating success are of interest. Darwin originally distinguished between the effects of male competition and female choice. Competition between males can take the form of various types of interference, for instance, through fighting or gamete destruction, and also various forms of scramble competition, such as searching for mates, production of large numbers of sperm, or manipulation of maternal investment. Mating preferences are more common in females than in males, and include any female characters that affect the probability that males with particular characteristics will succeed in mating (Halliday 1983; Maynard Smith 1987). Female discrimination between males is not necessary for a mating preference. For instance, if females tend to move up a sound gradient, then they will also tend to approach males that call loudly (Arak 1983). Similarly, if females produce eggs for fertilization at the top of a long reproductive tract and do not store sperm, then males that produce many mobile sperm are likely to be favored. This type of cryptic mating preference could be hard to detect.

At first glance, a quantitative genetic approach to detecting male competition and female choice does not seem likely to be productive; it would be laborious, and purely phenotypic studies should suffice. However, despite the notable successes that phenotypic studies have achieved (e.g., Davies and Halliday 1978; Andersson 1982a; Møller 1988; McComb 1991), a quantitative genetic approach may sometimes be useful. First, many of the problems inherent in identifying targets of sexual selection apply with equal force to studying their mechanisms of action. Manipulation of the sexual character is needed to break correlations with confounding variables. However, environmental manipulations are not always feasible. Second, females often do not appear to discriminate between different classes of males, but nonetheless show a cryptic mating preference in the form of behavior, morphology, or physiology that could favor males with particular traits, as in the examples mentioned above. The female trait value could then have evolved at least in part because it had the effect of causing particular kinds of males to have a mating advantage. Thus, one way to examine the idea that some form of female preference is at work would be to look for a genetic correlation between the male and female traits, for instance, between male call loudness and female approach tendency. Linkage disequilibrium as a consequence of assortative mating could produce such a genetic correlation if both male and female traits are heritable.

Male body size in *D. melanogaster* can be used to illustrate the issue, if we accept for the moment that this character is influenced by sexual selection and that phenotypic variation in body size is adequate material for a preliminary investigation of mechanisms. Success in fighting between males is correlated with increased mating success (Dow and von Schilcher 1975; Hoffman, chap. 9), and large males are more likely to win fights and have higher lifetime mating success (Partridge and Farquhar 1983; Hoffmann, chap. 9). However, success in fighting cannot be the whole explanation, because when the effects of male interference are eliminated by allowing single males to court single females, larger males mate more quickly (Ewing 1961; Partridge and Farquhar 1983). This result could be taken to show that females prefer larger males, but there are alternative explanations in terms of male scramble competition. Larger males have been shown to move more, which may have the effect of bringing them into contact with females more frequently, and they also court more and deliver more and louder courtship song (Partridge, Ewing, and Chandler 1987; Partridge, Hoffmann, and Jones 1987). Thus females could have a constant probability of accepting any courting male, but larger males would still be more successful.

Females may nonetheless discriminate between size classes of males. One method of doing so would be to behave so as to terminate courtships by nonpreferred males. Most courtships of virgin females end when the female moves away, breaking contact with the male. However, females are equally likely to move away from large and small males, and are, paradoxically, more likely to be moving when courted by large males. The probable explanation for this last result is that small males are more likely to lose the female if she moves; in running wheel tests small males run more slowly than large males (Partridge, Ewing, and Chandler 1987).

There is at present no evidence that female *D. melanogaster* discriminate between large and small males, although the issue is by no means settled. The results leave open the possibility that females could be exerting choice in at least two ways. First, the amount or rate of courtship they require before mating will affect the success of different male size classes. Second, the amount and speed of female movement during courtship will have a similar effect. One way of examining this idea would be to measure the genetic correlation between female courtship requirement or female movement and male size. A positive correlation would indicate that the genotype of the female affected the mating success of different male genotypes; in other words, that there was female choice.

This ambiguous situation is typical of species in which females do not overtly discriminate between males with different phenotypes (e.g., McCauley 1981). Although measurement of genetic correlations is likely to be laborious, and would have to be supplemented with other evidence about selection on female traits, it is important to understand the subtle means by which mates could be chosen. They may be the commonest.

Genetic Correlations between the Sexes

Genetic correlations can be caused by two different mechanisms: linkage disequilibrium and pleiotropy. Both mechanisms can lead to genetic correlations between the sexes. For instance, sexual selection by mate choice may cause a genetic correlation, through linkage disequilibrium, between genes for the mating preference expressed in females and genes for the preferred character expressed in males. These genes may become associated in the same individuals because they caused their parents to mate with one another at a higher frequency than random (Lande 1981b). Usually the mating preference and the preferred character will not be homologous between the sexes and will be controlled by different genes, unless both, for instance, are a preference for mating in a particular kind of habitat.

A second type of genetic correlation is brought about by pleiotropy, in which different characters are controlled by the same genes. Pleiotropy can also lead to a genetic correlation between the sexes for characters that are homologous. If selection acts differently on a homologous trait in males and in females, then a pleiotropic genetic correlation between the sexes for that character will slow their divergence under selection and keep one or both sexes away from an evolutionary optimum for a time (Lande 1980b; Slatkin 1984). A correlated response in one sex could occur even in the presence of direct selection on the character in that sex, provided that selection on the homologous character in the other sex was strong enough and of a different form. Differences in selection pressures on the sexes are perhaps especially likely in the context of reproduction and sexual selection (but see Slatkin 1984).

One suggested example of maladaptation due to conflicting selection on the sexes involves mating rate. In many animals males are sexually selected to mate frequently, while females, at first glance, gain little from multiple mating. Yet females of many species do mate more than once. In addition to a long list of possible adaptive explanations, such as male investment in the female or her offspring, prevention of harassment by males, a safeguard against male infertility, or the production of genetically diverse progeny (Halliday and Arnold 1987; Petersson 1991), females may mate frequently as a correlated response to stronger selection on males to do so (Halliday and Arnold, 1987; Arnold and Halliday 1988; but see Sherman and Westneat 1988). Remating in females could be selectively neutral; however, this is unlikely in view of the possible ecological and physiological costs of mating (e.g., Lewis 1987; Fowler and Partridge 1989).

Although often used as evidence (Sherman and Westneat 1988), comparative data are not suitable for testing the hypothesis of a genetic correlation between the sexes, because males and females of different species may covary in a trait for reasons other than pleiotropic genetic correlation. For instance, selection may favor high (or low) trait values in both males and females, leading to selective covariation. Teasing apart the effects of constraint and those of selection is therefore tricky. There is an additional difficulty in using comparative data on remating rates, depending upon how they are collected. Since every mating involves one male and one female, high *average* male remating rate must be accompanied by high average female remating rate unless the operational sex ratio is biased. Some measure of the potential mating rates of the two sexes or of the variance in male mating rate is required. Behavioral evidence that mating is entered into unwillingly by females (Sherman and Westneat, 1988) may also be misleading; as mentioned above, movement of female *D. melanogaster* away from males during courtship could be an evolutionary mechanism to enhance the likelihood of mating with larger males.

To measure genetic correlations between the sexes, quantitative genetic studies are required. These have not been frequent, and their results have been mixed. So far, data on remating frequencies do not support the idea of a genetic correlation between the sexes; in *D. melanogaster*, poultry (*Gallus domesticus*), and Japanese quail (*Coturnix japonica*), artificial selection for remating rate on one sex did not produce a correlated response in the other (Gromko and Newport 1988; Cheng and Siegel 1990). This is perhaps not surprising in view of the different behavioral and physiological mechanisms determining the mating rates of males and females.

Other reproductive characters do appear to be genetically correlated between the sexes. One of the first such characters examined was mating speed in *D. melanogaster.* Artificial selection on males and females separately for both fast and slow mating produced a direct response only in the two replicate lines of males selected for slow mating. There was a correlated response in the females of these lines in the same direction, suggesting a genetic correlation between the sexes for this trait (Manning 1963). A subsequent study produced both a direct response and a correlated response to selection for slow mating in both sexes, confirming Manning's result for the effects of direct selection on males and extending it to direct selection on females (Stamenkovic-Radak, Partridge, and Andjelkovic 1992, 1993; Butlin 1993). In zebra finches (*Taeniopygia guttata*), there appears to be conflicting selection on the sexes for bill color, with redder bills favored in males through increased reproductive success, but less red bills favored in females because of an increase in longevity. Bill color shows a significant genetic correlation between the sexes, so this seems to be clear example of evolutionary conflict (Price and Burley, pers. comm.).

There is a need for much more information on genetic correlations between the sexes. They could reveal much about mechanisms of developmental and physiological control, as well as about evolutionary conflict. In order to have long-term consequences, genetic correlations between the sexes would need to be very high, and in at least one group of *Drosophila*, evolution of sexual dimorphism does not appear to have been impeded by them (Schwartz and Boake 1992).

Costs of Sexually Selected Characters

One of the reasons for Darwin's interest in sexually selected characters was that they appeared to be detrimental to survival. Their negative effects on other fitness components are presumably often responsible for halting the evolutionary increase of sexually selected traits, although it is possible that they are occasionally subject to stabilizing selection. There has been considerable recent interest in measuring the costs of sexually selected characters, and in the kinds of methods that should be used. Many of the issues that arise are very similar to those involved in studying trade-offs between life history traits (Reznick 1985; Partridge and Endler 1987; Partridge and Sibly 1991; Partridge and Barton 1993; Roff, chap. 3).

Costs of sexually selected characters arise because of functional constraints on what an organism can achieve: activities such as growth, reproduction, and repair are costly, and they cannot be simultaneously maximized by natural selection. In genetic terms, there will therefore be a pleiotropic, negative, genetic correlation between at least some of these costly activities (Pease and Bull 1988). Organisms are therefore forced to make trade-offs between competing costly traits. The characters used to compete for mates are part of this set of life history trade-offs (Partridge and Endler 1987), and are suspected to incur two kinds of costs. Ecological costs occur if the sexually selected trait makes the organism more vulnerable to an ecological hazard such as predation (e.g., Endler 1983, 1988) or parasitism (e.g., Cade 1979), while costs are physiological if some mechanism intrinsic to the organism results in a negative effect of the sexually selected character on some other aspect of performance (e.g., Clutton-Brock, Albon, and Guinness 1985; Partridge and Fowler 1993).

Naturally occurring correlations between fitness components across different individuals are unsuitable material for examining the direction or magnitude of evolutionary correlations between fitness components, because natural correlations can arise due to the presence of confounding variables. For instance, an individual in good condition or one with a high-quality territory would be able to invest more in all costly activities, which would tend to reduce or even reverse the effects of costs (see Partridge and Sibly 1991). Some form of character manipulation is required, and genetic and environmental manipulations have both been used.

The genetic mechanism underlying costs is pleiotropy, so one obvious approach to studying them is to measure genetic correlations between the sexually selected trait and the value of other fitness components. For instance, the lines of *D. melanogaster* artificially selected for thorax length, described above, were also scored for their development times and competitive viability as larvae at different larval densities. The large lines took significantly longer to develop than the control or small lines. Larvae from the large lines had significantly lower viability than larvae from the control or small lines at higher densities, presumably because their slow development resulted in some failure to pupate before the food became unsuitable for larval feeding (Partridge and Fowler 1993). The data suggest that an evolutionary increase in body size would be accompanied by a drop in larval viability. The study has some limitations; artificial selection may not have manipulated the true target of sexual selection and may have inadvertently manipulated characters not involved in determining mating success. An earlier study by Wilkinson (1987) of the relationship between wing length and male mating success required the heritability of wing length to be measured outside the context of the measurement of sexual selection, so there may have been an environmental correlation between wing size and mating success in the study of sexual selection. Furthermore, both studies were conducted in the laboratory in an environment different from that in which the characters had evolved. Last, the ranges of the character values that were studied were small, smaller for instance than the difference in size between many *Drosophila* species. These limitations are not atypical of genetic studies, and mean that the results should be supplemented with other information.

If costs of the sexually selected character are incurred as a consequence of growth, as apparently is the case for body size in *D. melanogaster,* it is hard to see how phenotypic studies could have been used to demonstrate this, at least using the methods currently available. We do not have the means to manipulate the growth rate or target size of the whole organism or of its parts by nongenetic means, except ones that would themselves induce confounding correlations.

Comparative studies can be misleading indicators of costs of reproduction, in part because reproductive traits can affect demography, which can in turn have direct effects on the fitness components whose trade-off is being studied (Partridge and Sibly 1991). However, comparative studies can be useful indicators of developmental costs of sexually selected characters. It is the sex putting more parental investment into each offspring that determines the reproductive rate of the population, and this is usually the female. Males are merely competing to gain access to as much of this female reproduction as possible; they do not alter the total quantity of it. In this sense variation in the effects of sexual selection on males does not cause variation in the vital rates of populations. A comparative approach has been used to demonstrate that, across species of mammals and birds, the degree of sex bias in juvenile mortality is correlated with the degree of sexual dimorphism in adult body size (Clutton-Brock, Albon, and Guinness 1985). The high growth rates necessary to produce large adults may increase the likelihood of starvation during the juvenile period.

Costs of sexually selected characters can occur before the character is used in acquiring mates if there are developmental costs, during mate acquisition if there are contemporary costs, and after reproduction is over if there are delayed costs. It is in the second and third contexts that environmental manipulations have proved particularly effective, although comparative work can again be useful, because of the minimal effect of sexual selection on males on demography. The association across guppy populations between bright male colors and high predation risk is an excellent case of a contemporary ecological cost (Endler 1983, 1988). Contemporary costs can be physiological as well as ecological (e.g., Partridge and Andrews 1985). So far, the data on delayed costs suggest that they are mainly physiological. For instance, Møller (1989) showed that male swallows that had had their tails artificially lengthened attracted better mates, but went on to have poorer subsequent breeding success, apparently because the longer tail induced mechanical problems with food capture and hence with subsequent feather growth. Tail elongation experiments cannot explore the full costs of a long tail because any cost of growth, including delayed cost, is excluded, but they can provide information about costs incurred by the character after the time of its manipulation.

It seems likely that quantitative genetic information will have to be supplemented with data from other approaches if we are to understand the costs of sexually selected characters. Much of the enterprise boils down to a need to understand how organisms work and what intrinsic limitations there are on survival and reproductive success. Although genetic methods have been very revealing, unless we understand the physiological mechanisms producing pleiotropy we are unlikely to be able to make logical extensions of genetic findings on one species to others less amenable to genetic study.

Summary

The chapter discusses how genetic and nongenetic studies can complement one another in understanding sexual selection. Illustrations are drawn mainly from work on *Drosophila melanogaster.* In identifying targets of sexual selection, experimental manipulations of suspect characters can be useful for breaking correlations with potentially confounding characters. However, such manipulations are not always a practical possibility, and may not give a quantitatively accurate picture of the effects of selection. The effects of genetic variation on the suspect character can be used to circumvent these problems, but may misidentify the true target of sexual selection because of a genetic correlation with it. A genetic approach, based on genetic correlation through linkage disequilibrium, could be useful in detecting cryptic female mating preferences, which occur not because of female discrimination among males but instead because female behavior or morphology has the effect of making matings by particular phenotypic classes of males more likely. Genetic studies are essential for detecting genetic correlations between the sexes. Both genetic and nongenetic approaches are useful for studying the costs of sexually selected characters, and which to use depends in part upon the timing of the costs.

2

Applications of Quantitative Genetics to Studies of Behavioral Evolution in Natural Populations

7

Genetic Analyses of Animal Migration

Hugh Dingle

Milkweed bugs (*Oncopeltus fasciatus*). (*a*) Nonmigratory bugs on *Asclepias curassavica* in Puerto Rico. (*b*) A migratory milkweed bug in tethered flight.

The Variability of Migration

If motivation were necessary to prompt genetic and ecological analysis of behavioral variation, one would need to look no further than migration. Variation in migratory behavior that begs for explanation is present at all taxonomic levels. Among birds, for example, the globe-traversing movements of the arctic tern have become legendary, while at the opposite extreme the North American scrub jay may move so little that populations become geographically isolated and evolve unique behavioral (Florida) or morphological (Santa Cruz Island, California) traits (Atwood 1980; Woolfenden and Fitzpatrick 1984). Similar variation is present in insects, as represented by the long-distance flights of monarch butterflies and African locusts on the one hand and the sedentary wingless beetles of alpine zones and the water striders of large permanent lakes on the other (Dingle 1980; Roff 1990a). Of even greater interest from the perspective of evolution as a process are variations in migratory behavior within species. These may occur both among populations and among individuals in the same population. To the extent that this variation is genetic, it provides potential for these populations to evolve under the influence of natural selection. Indeed, behavioral variation may be one of the more important routes to new adaptive radiations (Mayr 1974).

Added impetus for studying the genetics of variation in migratory behavior comes from evidence that new migration patterns are continuing to evolve, most recently because of environmental changes of human origin. In the southeastern United States, for example, large areas of former forest have been converted to agriculture, resulting in expanded habitat for the fall armyworm moth, *Spodoptera frugiperda,* an insect of tropical origin. By expanding northward via migration, *S. frugiperda* invades an area extending well beyond its original range and has considerable economic impact (Mitchell 1979). Increased migration capacity has undoubtedly played a role in this range expansion, although this has not yet been fully investigated. More direct evidence for the evolution of a new migration system comes from the European blackcap warbler, *Sylvia atricapilla* (Berthold 1988b; Berthold and Terrill 1988). In the last few decades overwintering populations of this bird in Britain and Ireland have increased dramatically, at least in part due to an increase in the number of both fruiting garden shrubs and bird feeders. This population breeds in central Europe and generally migrates southwest to Spain in the autumn, but a portion has evidently evolved a new migration route, northwest in the autumn and southeast in the spring. As we shall see below, a great deal of genetic variation for migratory behavior exists in this species (Berthold 1988b), resulting, apparently, in a ready response to a changing selection regime.

Animal migration is thus a dynamic and diverse process that allows organisms to adapt to changing habitats. From the perspective of the animal, it renders the environment much more equable (Leggett 1985) and so is an important homeostatic behavior. It also allows the rapid exploitation of new habitats as they become available. Because migration is a complex behavior that is probably influenced by many genes, the methods of quantitative genetics provide an excellent means for analyzing the genetic and environmental contributions to observed variation (Dingle 1991). Ultimately, analyses of this important contributor to life history flexibility should lead to greater understanding of evolutionary processes, including speciation and adaptation.

Measuring Migratory Capability

In order to successfully carry out genetic analyses of quantitative traits, one must be able to make reliable and repeatable measurements. Because migration per se does not take place in a laboratory setting, this has proved difficult for migratory behavior. What is required is a behavior that occurs in the laboratory and can be relatively easily quantified, and at the same time provides an index of the migration that actually takes place in the field. Three examples illustrate how laboratory-based migration indices can be successfully validated by comparisons with field data. (See Partridge, chap. 6, for a discussion of validating sexual selection with field data).

Several investigators have used various types of tethered flight to index migratory tendency or capability in insects (reviewed in Dingle 1985). Gatehouse and his colleagues have used an automated electronic system to study tethered flight performance in the African armyworm moth, *Spodoptera exempta*, with particular reference to the genetics of migration (Gatehouse 1986). The duration of flight on the tether was used as a measure of migratory capability. The longer-flying moths concentrated their flight in the earlier hours of the night, the time when migratory flight takes place in the field (Gatehouse and Hackett 1980), thus suggesting that tethered laboratory flight indeed indexed field migratory performance. Tethered flight indices of migration have been successfully developed for other insects as well, including some used to study the genetics of migratory behavior (Dingle 1985; Rankin 1985).

Useful indices of migration have also been developed for vertebrates, two examples of which I discuss briefly here. Berthold (1988a,b) has used *Zugenruhe,* the nocturnal migratory restlessness of caged birds, to indicate migratory activity in the blackcap warbler. The amount and duration of restlessness correlates well with the amount and duration of migratory flights in natural populations varying in the extent of migratory behavior. More recently (Berthold 1988a), the filming of wing whirring during restlessness has allowed calculation of estimated migration distances based on duration of wing whirring and estimates of flight speed, and these estimated distances correspond remarkably well to observed distances traveled by wild birds. In the final example, Rasmuson, Rasmuson, and Nygren (1977) summed transits through runways between cage compartments to index migratory activity in meadow voles (*Microtus agrestis*). The magnitudes of these measures differed between migratory and nonmigratory populations in ways that reflected behavior in the field. In all the above cases the indices developed have allowed quantitative genetic analysis of differences among populations and individuals in the expression of the migratory trait.

The Genetic Basis for Migratory Behavior

Population differences in a trait have both genetic and environmental causes. The most direct way to sort out these two sources of variation is to hold one of them constant, as discussed by Arnold (chap. 2). The genes can be held constant by using highly inbred lines (e.g., mice: Lynch, chap. 13) or by using clonal organisms (e.g., aphids, *Daphnia*). Usually, however, it is the environment that is controlled (a "common garden" experiment). Two examples of common garden experiments involving migratory behavior are illustrated in figure 7.1. The first case is that of the cowpea weevil, *Callosobruchus maculatus* (Messina 1987), a pest of stored legumes. High population densities induce a migratory form, while low densities produce a sedentary morph; populations of the weevil vary in their tendency to produce the migrant morph. The different propensities to produce migrants are retained when beetles from two populations are reared in a "common garden" under sim-

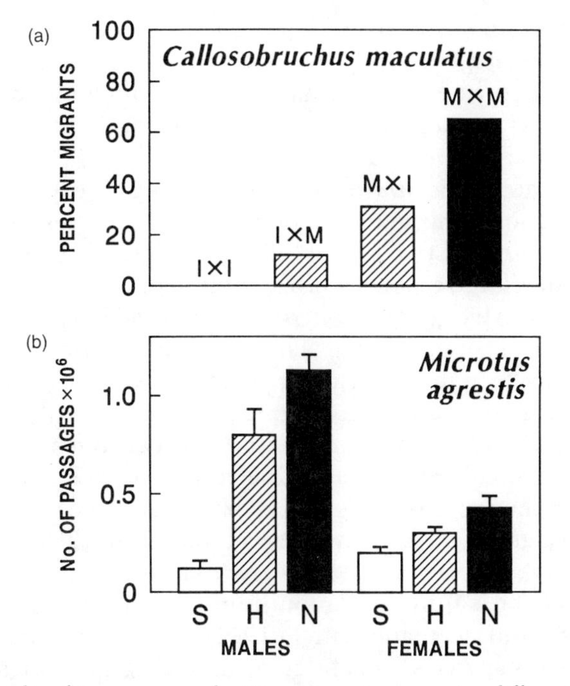

Fig. 7.1 Results of "common garden" experiments comparing different populations of cowpea weevils (a) and voles (b). In the weevils a nonmigrant population (I × I) and a migrant population (M × M) displayed differences when reared under identical conditions in the laboratory. Crosses (hatched bars) were intermediate between parental populations, but resembled the maternal parent more (first letter of cross). (Adapted from Messina 1987). In the voles, northern migratory (N) animals showed more movement between cage compartments than southern nonmigratory animals (S). Crosses (H) were intermediate, with some "dominance deviation" toward the migratory parent especially noticeable in males. (Adapted from Rasmuson, Rasmuson, and Nygren 1977.)

ilar abiotic conditions (fig. 7.1a), demonstrating that the variation has a genetic component. The nonmigrant morph displays higher reproductive potential than the migrant, and may be selected to reproduce in large and rich supplies of seeds, as would be found in granaries, while the migrant morph can locate new resources when crowding signals overexploitation (Messina 1987). The source areas of the two experimental populations presumably reflect different balances of selection for stasis or movement. The second example (fig. 7.1b) involves the vole, discussed above, *Microtus agrestis* (Rasmuson, Rasmuson, and Nygren 1977). Populations of

M. agrestis from northern and southern Sweden differ in migratory tendency (and degree of population cycling), and these differences are maintained when the animals are reared under identical laboratory conditions, again demonstrating genetic contributions to the behavioral differences.

In both the *C. maculatus* and the *M. agrestis* experiments the migratory and nonmigratory populations were crossed, and the performances of the two sets of hybrids are also given in figure 7.1. The intermediate responses of the hybrids in both cases suggest that the respective genetic contributions are additive, in all probability the result of polygenes. The additivity of the traits implies that they have a high potential for evolution, at least over the short term, because additive genes contribute specifically to resemblance among relatives (e.g., parents and their offspring) and can therefore be acted upon by natural selection (see below and Arnold, chap. 2). In the voles there is a bias or "dominance deviation" toward the migrant parent as well as additivity. In the cowpea weevils there is a deviation towards the maternal line. This sort of maternal effect is quite common in migratory and other life history traits (Mousseau and Dingle 1991) and has important consequences for responses to selection, including lag times and the continuation of response after selection has ceased (Kirkpatrick and Lande 1989). Maternal effects will be discussed further below and are discussed in detail by Cheverud and Moore (chap. 4).

The examples of population crosses above both involve univariate traits. A multivariate analysis of life history traits (e.g., clutch size, egg production rate) in several populations of the milkweed bug, *Oncopeltus fasciatus,* was carried out by Leslie (1990) to investigate relations between life history syndromes and migration. Twenty-six crosses were made among populations from California (C), Iowa (I), Maryland (M), Texas (T), Florida (F), and Puerto Rico (P). The Iowa and Maryland populations are migrants that invade their respective areas each year, the California and probably the Texas and Florida populations migrate locally, and the Puerto Rico population is nonmigratory. The effects of combinations of genomes from the different populations were examined using Mahalanobis distances to compare each interpopulation cross and the two potential lines contributing to it (e.g., PC was compared with PP and CC). (Mahalanobis distances are multivariate measures of the "distance" or degree of difference between sets of measures in multidimensional "space;" see, e.g., Williams 1983 for some uses in ecology. In Leslie's study the measures were of the various life his-

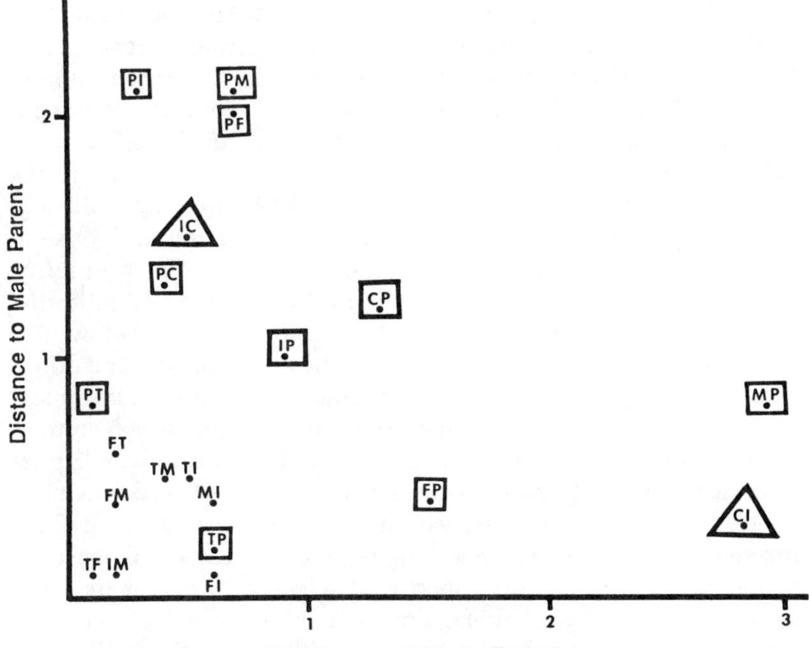

Fig. 7.2 Mahalanobis distances for hybrid offspring from 20 between-population crosses in milkweed bugs. Populations were Iowa (I), Maryland (M), Texas (T), Florida (F), California (C), and Puerto Rico (P) (female given first in crosses). Note especially the "dominance deviations" toward Puerto Rico (all crosses involving Puerto Rico are enclosed in squares), as indicated by shorter distances to the Puerto Rico parent, and the similarity of central and eastern United States populations, as indicated by clustering of their hybrids near the origin. Strong dominance toward Iowa parents occurred in Iowa × California crosses (enclosed in triangles). (Adapted from Leslie 1990).

tory traits such as clutch size, egg production rate, and so forth.) The results are displayed in figure 7.2. If genomic effects were strictly additive, hybrids would be expected to be exactly intermediate between parental populations. For all the mainland crosses except those involving California, this seems to be approximately the case, as indicated by the cluster of points near the origin of the graph; the clustering near the origin also indicates that there is very little difference between hybrids and parentals of these populations. In contrast, having a Puerto Rico parent (e.g., PI, FP, etc.) resulted in multivariate coordinates strongly displaced toward Puerto Rico (a dominance deviation). Two exceptions were CP and IP, which were approximately equidistant between their parentals, probably as a consequence of maternal effects, which have been demonstrated for life history traits in *O. fasciatus* (Groeters and Dingle 1987, 1988). Strong dominance deviations toward Iowa were displayed by IC and CI crosses. The multivariate analysis of life history traits in these interpopulation crosses indicates the evolution of a quite distinct genome in Puerto Rico and suggests a separate genome in California as well. The relative isolation of these two populations from those in eastern North America has thus apparently resulted in the evolution of genetically distinct life history syndromes. I shall return to this issue below.

To this point I have discussed only the existence of gene differences contributing to phenotypic differences among populations. It may also be desirable to estimate the contribution of each component of genetic and environmental variance to the overall phenotypic variance of a population. The most interesting component is often the additive genetic variance, i.e., the variance contributing specifically to variance among relatives, because this best describes the potential for response to selection, at least over the short term (selection changes gene frequencies, which can complicate response over the long term; see, e.g., Barton and Turelli 1989). The ratio of additive genetic variance to total phenotypic variance is the heritability, a parameter that can be estimated in several ways. Two of the more accurate ways are estimates based on half sibs and on offspring-on-parent regression, because they are not confounded by variance due to dominance components or to the common environment in which full sibs are reared (Falconer 1989; Arnold, chap. 2).

Because the experimental design is more straightforward and usually less labor-intensive, offspring-on-parent regression is the means more frequently used. Two examples of its use for estimating the heritability of migratory behavior are found in studies of the small milkweed bug, *Lygaeus kalmii* (Caldwell and Hegmann 1969) and the African armyworm moth (Gatehouse 1986). (For a half-sib analysis of life history traits in a migratory population of *O. fasciatus,* see Hegmann and Dingle 1982). In both cases the behavior measured was tethered flight in the laboratory, and regressions were computed for male and female parents separately. In *L. kalmii,* heritabilities were .20 $\pm$ 0.06 SE from regression on males and .41 $\pm$ 0.05 from regression on females; in the armyworm moth, values were .88 and .71 for males and females respectively

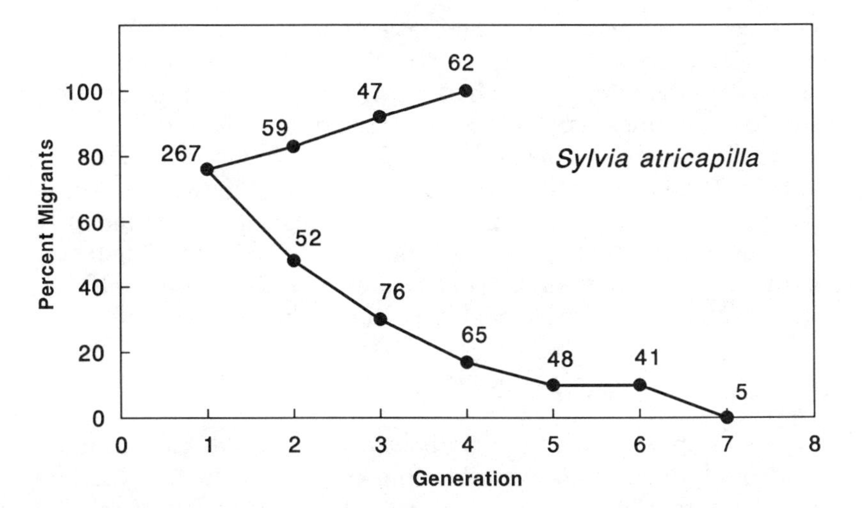

Fig. 7.3 Results of selection for migratory restlessness in the European blackcap warbler. Note that both lines go to fixation for the presence or absence of the trait. Numbers on graph indicate sample sizes measured each generation. (Adapted from Berthold, Mohr, and Querner 1990.)

on male parents and .54 and .50 on female parents (SE's not given, but all heritability estimates significantly different from zero). Differences among male parent- and female parent-derived estimates presumably reflect maternal effects (curiously, these seem to be negative here), but these do not obscure the fact that there is substantial additive genetic variance for flight duration in both cases, leading to the prediction of rapid responses to selection on flight.

Another way to estimate heritability is via responses to artificial selection, the so-called "realized heritability," where heritability is the ratio of the response to selection over the selection differential ($h^2 = R/S$) (see also Arnold, chap. 2 and Partridge, chap. 6 for other uses of artificial selection). The underlying assumption is that the trait shows a Gaussian distribution, but for most studies of natural populations this is unlikely to be very restrictive. Using artificial selection, Berthold (1988b; Berthold, Mohr, and Querner 1990) obtained heritability estimates of .58–.87, depending on the generation used, for nocturnal migratory restlessness in the European blackcap warbler. The responses to selection were quite dramatic and produced lines that went to 0% and 100% migrants in six and three generations of selection, respectively (fig. 7.3).

Because the selected lines were not replicated (for reasons of cost and labor), the selection experiment does not control for possible drift or founder effects, which can occur in selected lines if initial samples sizes are small. Nevertheless, there is the strong implication that the genetic variance for migratory behavior would permit a rapid response to a changed selective regime. There is also some evidence that variation in orientation is at least in part genetically based (P. Berthold, pers. comm.), and is therefore a possible factor contributing to the evolution of the new migration route for blackcaps to and from the British Isles (Berthold and Terrill 1988).

Migration Syndromes

Migration and other important behavioral traits do not occur in isolation, but often show correlations with traits such as life history characters that are expressed in the form of syndromes or "strategies" (Dingle 1986). The causes of these correlations can be either environmental or genetic, and if we wish to determine how natural selection acts on syndromes, it is necessary once again to distinguish among sources of variation. The various methods discussed above for estimating heritability, i.e., offspring-on-parent regression, full-sib and half-sib analysis, and artificial selection, can also be used to estimate genetic correlations. These methods are summarized in detail by Falconer (1989), Arnold (chap. 2), and numerous other sources. It should be noted that genetic correlations, especially if estimated by offspring-on-parent regression or sib analysis, are subject to large sampling errors that considerably reduce precision (Falconer 1989); therefore large sample sizes are required, a considerable inconvenience with most organisms and any behavior whose measurement is not easy and rapid. To some extent these difficulties can be eliminated by using selection, especially if one is primarily interested in the genetic correlations involving a particular trait of interest. If selection on one trait, such as flight, results in statistically significant differences among selected lines in other traits, such as size or fecundity, then these other traits are genetically correlated with the character selected. Because one is selecting on only one trait, however, it is possible to construct only a vector of genetic correlations, not a complete matrix of all traits examined (e.g., selection on flight does not give information about correlations between size and fecundity).

In my laboratory we have examined migratory syndromes in a comparative study of milkweed bugs (Palmer and Dingle 1986;

Dingle and Evans 1987; Dingle, Evans, and Palmer 1988; Dingle 1988; Palmer and Dingle 1989). We compared two populations. The first was a highly migratory one from Iowa that migrates into the upper Midwest in early summer and returns south in the fall to overwinter in Texas and along the Gulf Coast. The second was a nonmigratory population from Puerto Rico that reproduces year round and moves only among local milkweed patches. We first examined these populations using offspring-on-parent regression. Heritabilities estimated from the regressions indicated high levels of additive genetic variation in two morphological traits, wing length and head capsule width, in both populations; there were also statistically significant genetic correlations between these traits. A difference between the populations occurred in fecundity, measured as the number of eggs produced in the first 5 days of reproductive life, an important period for a migrant colonizer. In the Iowa population the heritability of fecundity was high (.50 ± 0.17) and significant, and there was a significant genetic correlation between body size (as measured by head capsule width) and fecundity. Neither the heritability of fecundity nor its genetic correlation with body size were statistically significantly different from zero in the Puerto Rico population, suggesting the possibility of an interesting difference in life history syndromes between migratory and nonmigratory bugs.

This possibility was investigated further using selection. For two reasons, the trait initially chosen for selection was wing length. First, it has a high heritability in both populations, and we therefore anticipated strong response to selection. Second, studies of several species have suggested a positive relationship between wing length and migration. Selection for several generations on wing length revealed, as expected, a strong response in the selected trait in both populations and a positive genetic correlation between wing length and fecundity in the Iowa population, but not in the bugs from Puerto Rico. In addition, there was a positive correlation between long-duration tethered flight and wing length in Iowa bugs, but not in Puerto Rico bugs. In neither population was there a correlation between wing length and age at first reproduction. These selection experiments were then extended to selection on flight length itself in the Iowa population, with the results illustrated in figure 7.4. These results show significant correlated responses in wing length and fecundity to selection on flight, and once again no correlated response in age at first reproduction. What all these experiments demonstrate is that in the migratory

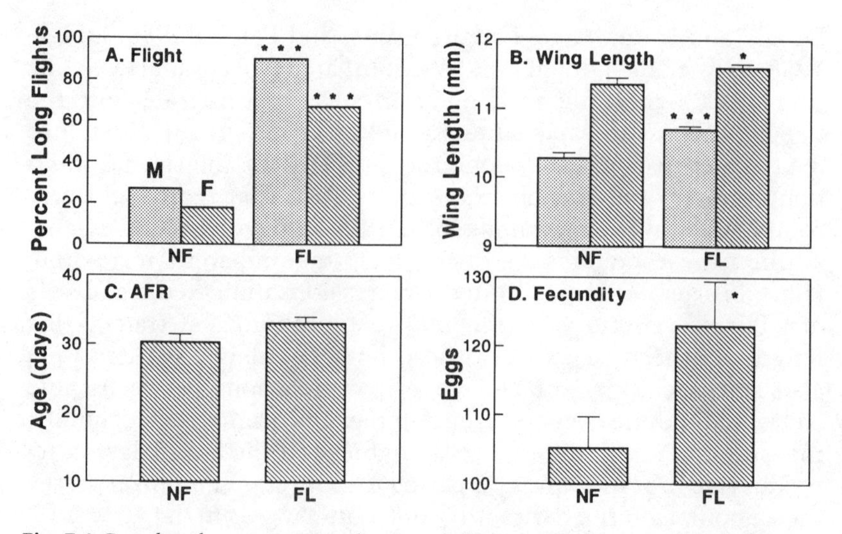

Fig. 7.4 Correlated responses to selection on flight in migratory milkweed bugs from Iowa. Bugs were selected for both longer (FL or "flying") and shorter (NF or "nonflying") flights, using tethered flight as an index of migration, for two generations. The selected lines were then compared with respect to (*A*) flights over 30 minutes duration, (*B*) wing length, (*C*) age at first reproduction (AFR), and (*D*) fecundity during the first 5 days of reproduction (when migrant colonists would be newly arrived). Standard error bars are shown, and asterisks give significance levels for comparisons within males (M) and females (F) separately (***$p < .001$, *$p < .05$, so that, for example, females differ in wing length at $p < .05$ and males at $p < .001$). Note correlated responses in wing length and fecundity but not age at first reproduction. (Adapted from Palmer and Dingle 1989).

population there is a migration syndrome defined by positive genetic correlations among flight, wing length (and body size), and fecundity, but that this syndrome is absent in the nonmigratory population. A consequence of the absence of correlations with age at first reproduction is flexibility of reproductive timing independent of migration. The two populations thus not only differ genetically in terms of gene frequencies, but also in the way their genomes are organized. Natural selection has evidently acted both on particular traits and on the genetic correlations among the traits (Bradshaw 1986). Genetic correlations are, of course, population phenomena and are most likely influenced indirectly by individual selection acting on pleiotropic effects or by changing gene frequencies.

A migratory syndrome involving genetic correlations among traits that determine migratory tendency has also been demonstrated in the sand cricket, *Gryllus firmus*, by Fairbairn and Roff (1990). This species is wing dimorphic: macropterous individuals have long hind wings and the ability to fly, while micropters have both the hind wings and the dorsal longitudinal flight muscles (DLM) reduced, rendering them incapable of flight. In addition, macropterous females vary in their tendency to histolyze the DLM within the first few days following adult eclosion. Fairbairn and Roff selected lines for increased and decreased proportions of macropters, as well as maintaining unselected control lines, and examined macropters from all lines for flight propensity (tethered flight) and extent of DLM histolysis. An increased proportion of macropters was positively correlated, both genetically and phenotypically, with increased flight and retention of functional DLM. Thus in lines with a higher proportion of macropters, the macropters were also more likely to display long flights. Fairbairn and Roff propose that these correlations form the basis for a migratory syndrome coordinated by juvenile hormone, which is known to influence all the traits involved (Dingle 1985; Rankin 1991). Levels and timing of juvenile hormone production influence migration in at least three insects (Rankin, McAnelly, and Bodenhamer 1986), and it is likely that genes influence migration at least in part through their control of hormone production. Migration in such cases could be a threshold trait expressed when hormone levels are sufficient. Threshold characters require some modifications of methods when subjected to quantitative genetic analysis (Falconer 1989).

Migratory syndromes are not confined to insects. Differences between migratory and nonmigratory populations in genetic correlation structures have also been found in the stickleback *Gasterosteus aculeatus* (Snyder 1988; Snyder and Dingle 1989). Populations of these fish from the Navarro River on the northern coast of California display two life history patterns. Estuary populations migrate from salt water into fresh water to breed, while upstream populations are confined to restricted areas of the stream and so are largely sedentary. When reared in a common environment, the migratory estuary fish display later age at first reproduction (AFR), are larger at reproductive maturity, and have higher fecundities than the upstream fish, all suggesting gene differences and adaptation for migration in the estuary individuals. To determine genetic correlations, full-sib families of both populations were divided and reared in both fresh and salt water. One significant genetic correlation was demonstrated in the estuary fish, between AFR and size at reproduction in fish reared in salt water. In contrast, fish from the upstream source displayed three statistically significant genetic correlations: AFR with size, size with clutch size in fresh water, and size with clutch size in salt water. The results thus reveal some differences in genome organization for life histories in migratory versus sedentary sticklebacks, although it should be noted that the correlation estimates were obtained from an analysis of full sibs and so contain unknown biases from dominance and maternal effects.

Environmental Complications

Different phenotypes may be produced when a given genotype experiences different environments. Such differential responses are known as reaction norms and yield a great deal of information about potential for response to selection when the selective regime encompasses more than one environment (Via and Lande 1985; Groeters and Dingle 1987, 1988; Scharloo 1987; Stearns 1989; van Noordwijk 1990). In many cases genotypes may perform well in one environment but poorly in another. If one considers performance in the two environments as two characters, then there is negative genetic correlation across the environments (e.g., Via and Lande 1985). In such cases selection for improved performance under one set of conditions may result in reduced performance in the other (a genotype by environment interaction resulting from the fact that the selected and unselected genotypes differ in their response to the two environments). The evolution of a trait optimal in both environments may thus be constrained (Clark 1987). In contrast, if there is positive genetic correlation across environments, that is, if the genotypes perform well in both conditions, evolution toward an optimum can proceed. The genetic variation may thus lead to plasticity, resulting in flexibility in traits when environments are spatially or temporally variable (Groeters and Dingle 1987; van Noordwijk and Gebhardt 1987).

Two examples of genotype by environment interactions involving insect migratory genomes are illustrated in fig. 7.5. The African armyworm moth is one of several species of insect that display a density-dependent polyphenism, originally described as "phase polymorphism" in African migratory locusts (Uvarov 1921). In the armyworm moth, crowded populations have dark, active larvae, develop rapidly (2–3 days faster than uncrowded larvae), and display much more long-distance migratory flight as adults. Experiments have revealed that long-duration tethered flights are more frequent

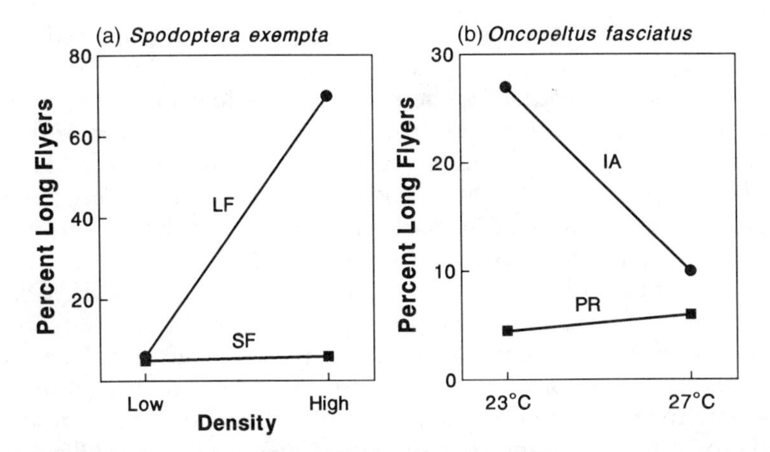

Fig. 7.5 Interaction plots for flight behavior in (*a*) African armyworm and (*b*) milkweed bug. The long flyer (LF) genotype in *Spodoptera* is expressed only at high density, while the short flyer (SF) genotype displays little flight at either density. In *Oncopeltus*, the migratory Iowa (IA) population flies at 23° C but not at 27° C, while nonmigratory Puerto Rico (PR) bugs show little flight at either temperature.

in offspring of long-flying parental pairs (Gatehouse 1986). Offspring of these long-flying pairs were divided between crowded and uncrowded rearing conditions; the crowded or gregaria sibs produced 70% long flyers, compared with 6% for the uncrowded or solitaria sibs (fig. 7.5a). The offspring of short flyers displayed very little long-duration flight under either rearing condition. Parker and Gatehouse (1985a,b) suggest that the phase polyphenism is part of a migration syndrome that includes genes for long-distance migration, which are expressed when populations reach high densities following onset of the rainy season. Migration allows the moths to escape from crowding and to colonize new habitats made available by the flush of vegetation following the rains.

The second example in fig. 7.5 displays the tethered flight responses of a migratory Iowa milkweed bug population and a nonmigratory Puerto Rico population, each tested at two temperatures. As expected, the nonmigratory bugs showed little long-distance flight under either set of conditions. The migratory Iowa population showed differential expression of migratory genes at the two temperatures. At 23° C some 27% of the population expressed long-duration flight (defined here as flights of over 30 minutes duration), while at 27° C only 10% of the population displayed long flights. Under cool conditions in spring and fall migratory flight would thus take place, but when warm conditions prevail, as when the bugs arrive at northern latitudes during the midsummer flowering and fruiting of the milkweed host plant, migrants would tend to be relatively sedentary. Across-environment genetic correlations that could have an important influence on the evolution of this migration system remain to be investigated.

Another type of environmental influence on behavior and other traits results from maternal effects, or the nongenetic transmission of environmental variation from mother to offspring (see also Cheverud and Moore, chap. 4). A form of maternal effect directly influencing migration occurs in aphids, where the degree of maternal crowding determines the wing form displayed by the offspring. Typically, crowded mothers produce winged migratory offspring, while uncrowded ones give birth to wingless sedentary adults in the next generation (reviewed in Mousseau and Dingle 1991). The extensive telescoping of aphid life cycles, in which the older embryos already contain the germaria of future offspring, i.e., embryo daughters contain the germaria of grandaughters, creates a situation highly favorable for the evolution not only of maternal effects, but also of effects lasting for several generations (Blackman 1975). Grandmaternal effects have indeed been described in aphid clones. An example (fig. 7.6) from the pea aphid, *Acyrthosiphon pisum*, demonstrates that the age of the grandmother when future mothers are born influences whether those mothers will produce alate or apterous daughters (MacKay and Wellington 1977). Mothers born to young grandmothers produce predominantly wingless offspring; those born to older grandmothers, winged ones.

There may also be genetic variance for maternal effects among clones or populations of aphids. Groeters (1989) examined differences in production of winged offspring under crowding in Iowa, California, and Puerto Rico populations of the milkweed-oleander aphid, *Aphis nerii*. Iowa aphids were expected to have a higher proportion of winged offspring since populations in the upper Midwest cannot overwinter and must therefore be migratory. Puerto Rico aphids were expected to be the most likely to remain apterous, because host plants are continuously available there. California aphids were expected to be intermediate to these extremes. This prediction was only partially realized, with Iowa mothers producing 37.7% winged offspring, Puerto Rico mothers 31.6%, and California mothers 25.7%. The results do, however, demonstrate gene

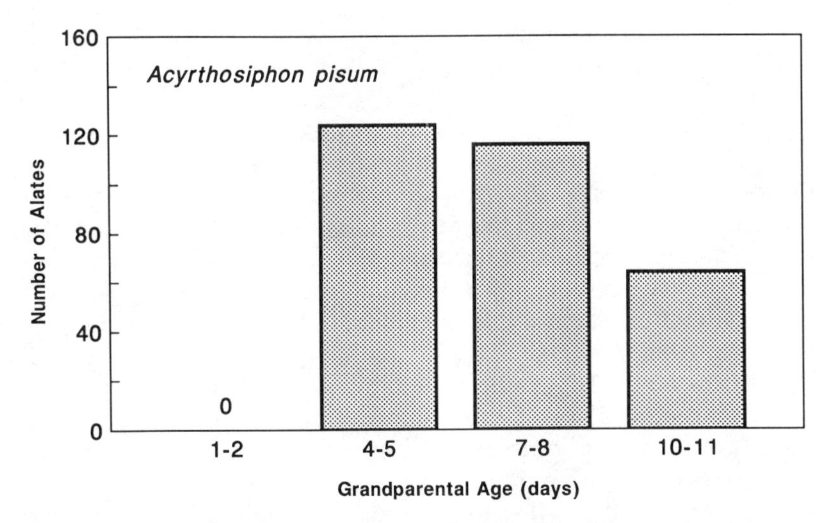

Fig. 7.6 "Grandmother effects" in the pea aphid. The age of the grandmother when the mother is born determines the number of offspring with wings (alates) the mother produces. (Adapted from MacKay and Wellington 1977).

differences influencing the trait. Variation was also found among clones within populations, further indicating gene influences. Interestingly, there seems to be no syndrome involving migration, morphology, or life history, as was observed in milkweed bugs, for any population of *A. nerii.* Where host plants can be exploited through migration and colonization, this is accomplished through the production of greater numbers of colonists, not of colonists with a distinct syndrome of life history traits.

Finally, to make our picture of environmental influences complete, it is worth noting that parental effects need not be confined solely to the maternal line. If a father influences the size of his daughters, for example, and maternal size in turn influences the performance of offspring, those offspring will express "grandfather effects" (Reznick 1981). Such grandpaternal effects are straightforward once it is realized that the expression of gene differences may require more than one generation. Much more difficult to explain are data that imply nuclear transmission of paternal environmental effects to offspring, for example, the transgenerational effect of paternal photoperiod on development in *Drosophila* (Giesel 1988; Giesel, Lanciani, and Anderson 1989). The surprising nature of these results clearly calls for further investigation. The point to be made from parental effects, however, is that environmental influences and genotype-environment interactions are not confined to single generations.

Assessment

Barton and Turelli (1989) subtitled their review of evolutionary quantitative genetics "How little do we know?" and concluded that although there is descriptive value in quantitative genetic models, their usefulness for making evolutionary predictions is uncertain. Unfortunately, with respect to behavior in natural populations, even the descriptive value of quantitative genetics is largely unknown. For example, in their review of heritabilities estimated for natural or outbred populations, Mousseau and Roff (1987) could cite only 21 cases of heritability estimates for behavioral traits (out of a total of 270 estimates) in only 14 species. The situation with the well-studied genus *Drosophila* is little better: only 38 studies of behavior were found and 17 of those involved phototaxis or geotaxis (Roff and Mousseau 1987). Recent studies have improved matters somewhat, but do not change the general picture. There are, of course, reasons for this dearth of behavior genetic analyses. It is a distinctly nontrivial task to keep enough animals healthy and behaving normally to accumulate enough data to make estimates of genetic parameters meaningful.

To fully understand the development or evolution of behavior, however, it seems to me that we need a far greater understanding of its genetics. The entire sad history of the nature-nurture controversy is a reflection of the failure of both sides to understand that the important question was not whether a behavior is due to genes or to environment, but rather, what are the relative contributions of genes, environment, and, I stress, genotype by environment interactions. This second question is easily recognized as the verbal form of the basic equation of quantitative genetics, **P**(henotype) = **G**(enotype) + **E**(nvironment) + cov **G,E**, and should be reason enough to explain why quantitative genetic studies of behavior do matter. Since the transmission of gene differences is a fundamental Darwinian precept, quantitative genetic studies are important to behavioral ecology as well.

In this chapter I have tried to show how quantitative genetics has increased our understanding of migratory behavior. Much of this increased understanding has come from experiments which, by the standards of genetics, are really very simple and unrefined. There is much debate among evolutionary quantitative geneticists about the assumptions underlying various methods (e.g., Barton and Turelli 1989), and among applied quantitative geneticists about the best methods for parameter estimation. It would be regrettable, I think, if behaviorists failed to use quantitative genetics because they were unable to satisfy assumptions or reach sample sizes allowing refined computations. Our knowledge is at present so limited that even approximations of genetic and environmental contributions have value in adumbrating the issues we need to address. As one suggestion, I propose that if behaviorists would keep pedigrees of their sample populations, an enormous amount of information would be made available at the cost of only some extra bookkeeping.

To return to migration, studies of the genetics of migratory behavior have revealed gene differences within and among populations, and thus ample variation upon which natural selection can act. In the case of the European blackcap warbler, changes in selection acting on available genetic variation have evidently produced new migration routes within just a few decades. Our studies of milkweed bugs have revealed a distinct migration life history syndrome defined by the nature of genetic correlations among traits. Such syndromes suggest that selection can act both on traits and at least indirectly on the correlations among traits (Bradshaw 1986), and so have profound implications for the evolution of adaptations. Thus quantitative genetics not only has great potential for helping us to understand behavior, especially if combined with appropriate optimality modeling (see Roff, chap. 3), but also can magnify contributions made by behavioral studies to the analysis of developmental and evolutionary processes. This should be reason enough to pursue quantitative genetics despite the sometimes formidable barriers the methods present.

Summary

In this chapter I have discussed several examples of how the methods of quantitative genetics can be used to study the genetic and environmental variation present in migratory behavior. These applications require a measure of migratory capability, such as tethered flight in insects or migratory restlessness in caged birds. Genetic methods have ranged from the production of hybrid crosses to estimates of heritability or genetic correlations based on analysis of siblings, offspring-on-parent regression, or artificial selection experiments. All of these analyses suggest that there is much genetic variance for migratory behavior in a variety of migrants from insects to birds. Some of this variation may be contributing to the evolution of new migratory pathways in nature, as seen in the fall armyworm in the southeastern United States or the blackcap warbler in Europe. Genetic studies also reveal the presence of migratory syndromes involving genetic correlations among migratory behaviors such as flight duration and life history and morphological traits such as fecundity and size. Genotype by environment interactions are also evident in the evolution of migration syndromes, and in some cases the environmental influences may result from maternal effects. I conclude that even though there are difficulties in applying quantitative genetic methods, because of the need for large sample sizes and labor-intensive experiments, they are a valuable tool for studying the way behavior is organized and the way it evolves.

Acknowledgments

I thank Tim Mousseau, Peter Berthold, Frank Messina, Jerry Waldvogel, and Chris Boake for comments on the manuscript. My research and the preparation of this chapter were supported by grants from the National Science Foundation.

8

Size-Dependent Behavioral Variation and Its Genetic Control within and among Populations

Joseph Travis

Male sailfin molly, *Poecilia latipinna*.

One of the most striking discoveries of the last two decades is the ubiquity of extensive behavioral variation among individuals within natural populations, especially in the mating behaviors of males. In the majority of reported cases, this variation is discrete and encompasses a small number of alternative behaviors. However, in many cases individuals exhibit continuous variation in the rates at which they express a variety of behaviors (Beani and Turillazi 1988; Zimmerer and Kallman 1989). Regardless of the form of behavioral variation, in nearly all cases it is associated with other phenotypic attributes of individuals, such as age, size, or health (reviewed by Ryan, Pease, and Morris 1992). Most of the attention devoted to these observations has focused on two issues. First, do such associations reflect behavioral tactics that are independent of any genetic variation among individuals, or do they represent alternative strategies based on genetic variation? Second, is the behavioral variation associated with variation in absolute fitness (Austad 1984; Dominey 1984; Clutton-Brock 1988; Ryan, Pease, and Morris 1992)?

Two interwoven problems that arise from these observations have not been as well explored. First, does the expression of size- or state-dependent behaviors vary among populations, and if so, is such variation associated with specific ecological variation such that it might be adaptive? Second, if behavioral variation within a population is associated with a trait like body size, and both variables are demonstrably related to genetic variation among individuals, what is the precise nature of the genetic influence on behavior and its associated trait?

These questions are facets of a single problem; how does conditional behavioral expression evolve? I use the phrase "conditional behavioral expression" to denote the statistical relationship between the rates at which one or more behaviors are exhibited and the value of the phenotypic trait with which the behaviors covary. This meaning is distinct from other uses of "condition-dependent behavior." A broad use of this term might refer to the observation that nearly all behaviors are expressed under specific circumstances. For example, territorial behavior might be expressed only under specific conditions of population density and resource levels (Rubenstein 1981; Wilcox and Ruckdeschel 1982), or only in the presence of females (Ewing 1973; Otronen 1984), or only under very specific combinations of ecological and social conditions (Hoffmann and Cacoyianni 1990). In some usages "condition-dependent behavior" is taken to mean the expression of behavior as a direct function of an individual's health or vigor (Andersson 1982b; Kodric-Brown and Brown 1984; Nur and Hasson 1984).

In the present context, "conditional behavioral expression" encompasses the manner in which rates of behavior as functions of size or age vary within and among populations, and in this chapter I explore how statistical and quantitative genetic methods can be used to investigate those patterns. The thesis I offer is that knowledge of how variation in condition-dependent behavior is controlled can lead quickly and efficiently to testable hypotheses for the evolution of behavioral variation. In the next section I describe some patterns of conditional behavioral variation and how they can be quantified. In the subsequent section I describe how quantitative genetic methods offer insight into the control of behavioral variation. In the final section I illustrate how this information can be used to frame testable hypotheses for the evolution of behavioral variation in a specific case study.

The examples I have chosen to illustrate these points are drawn from studies of fishes and in particular from studies of male mating behaviors in the sailfin molly, *Poecilia latipinna.* The sailfin molly is a live-bearing fish of the same family as guppies and swordtails, the Poeciliidae. Sailfin mollies are native to the coastal marshes of the southeastern United States and northeastern Mexico. Fishes like the sailfin molly have the twin virtues of being observable in the field and sustainable in laboratory rearings, which means that behavioral variation within and among populations can be quantified and that feasible genetic studies can be calibrated by comparison with observations in natural populations. From an evolutionary point of view, the sailfin molly is especially appropriate for study because the patterns seen in this species are characteristic of the behavioral variation seen more generally within the family to which it belongs. Hence, testable hypotheses for the evolution of behavioral diversity in this species, derived from quantitative genetic studies, can offer insights into the evolution of a much larger scale of variation.

Size-Dependent Variation in Behavior

In most examples of conditional male mating behavior, specific behaviors are conditional on body size (e.g., Cade 1981; Thornhill 1981; Clutton-Brock, Guinness, and Albon 1982). Several well-known examples in fishes illustrate a general pattern in which larger males hold territories or court females while smaller males attempt to obtain matings through an assortment of surreptitious mechanisms (Warner, Robertson, and Leigh 1975; Dominey 1980; Gross 1982, 1985). This pattern is evident within populations of several species of poeciliid fish, especially in the tribe Poeciliini (reviewed by Farr 1989; see also Ryan and Causey 1989; Zimmerer and Kallman 1989).

Sailfin mollies have long been known to exhibit such a pattern.

Mollies, like all poeciliids, have internal fertilization: males transfer spermatophores through the modified anal fin, the gonopodium. The two principal behaviors that male mollies direct to females are a courtship display, which is an attempt to elicit female cooperation in mating, and a gonopodial thrust, which is an attempt at forced insemination without female cooperation. Although all males can be observed to exhibit both behaviors, larger males engage in courtship displays more often and resort to gonopodial thrusting less often than smaller males. Males do not defend territories or occupy specific mating sites; the fish move in loosely organized shoals. The initial quantitative studies of this species treated body size and its correlated behavioral variation as discrete (Baird 1968; Luckner 1979). Males grow only minimally after achieving maturity, and a male's behavioral profile is consistent throughout his adult life. Thus the initial discussions of the evolution of this pattern were predicated on the assumption that the variation represented a discrete genetic polymorphism at a single locus with very few alleles.

Subsequent work has shown that body size and behavioral variation within populations of mollies is unimodal and continuous. Body size distributions can vary widely among populations (Snelson 1982, 1984, 1985; Farr, Travis, and Trexler 1986; Travis and Trexler 1987; Travis 1989a; Trexler, Travis, and Trexler 1990); some populations contain males between 15 and 35 mm in length (average approximately 22 mm), whereas others contain males between 25 and 65 mm in length (average approximately 38 mm). Behavior also varies continuously among males of different sizes; the larger the male, the more often he will engage in courtship displays and the less often he will exhibit gonopodial thrusts (Farr, Travis, and Trexler 1986; Travis and Woodward 1989). These patterns are highly repeatable (Travis and Woodward 1989) and do not change with male age (Travis, Farr, and Trexler, in press). The absolute rate at which each behavior is expressed varies with temperature, social condition, and female receptivity (Farr, Travis, and Trexler 1986; Farr and Travis 1986; Travis and Woodward 1989; Sumner, Travis, and Wilson, in press), but the disparity in rates between the smallest and largest males can be enormous. In a 30-minute observation period at 25° C, a male approximately 55 mm in standard length can exhibit an average of 9 thrusts and 92 displays, whereas a male approximately 26 mm long can exhibit an average of 39 thrusts and 9 displays under the same conditions (Travis and Woodward 1989; see also Farr, Travis, and Trexler 1986).

Table 8.1 Rates of behavior exhibited by males of different size classes in a single natural population at Shepherd's Spring, Florida

	Male Size Class		
	Small	Medium	Large
Range of lengths (mm)	17–25	30–40	45–55
Number of observations	69	67	128
Gonoporal nibbles	3.6 ± 0.4	3.1 ± 0.5	1.4 ± 0.2
Gonopodial thrusts	2.0 ± 0.4	1.0 ± 0.2	0.2 ± 0.1
Courtship displays	0.1 ± 0.0	1.3 ± 0.4	2.8 ± 0.3
Total observed behaviors	5.6 ± 0.7	5.5 ± 0.7	4.4 ± 0.3

Note: Entry for each behavior is the average and standard error of the number of times that behavior was observed for a focal individual in a 1-minute observation period.

The variation among males in their rates of behavior under any single condition is not proportional to the variation in body length, as the numbers cited above indicate. The relationship of behavioral rates to body size is linear on a log-log scale. However, the rate of engaging in courtship displays increases at a faster rate on the log-log scale than does male body length; the slope of the best line through these points is greater than +1.0. The rate of engaging in gonopodial thrusting decreases more slowly on the log-log scale than does male body length; the slope of the best line through these points is greater than −1.0 but less than 0. In other words, display rates increase with a positive allometry to male body length, whereas thrusting rates decrease with body length in negatively allometric fashion.

These results were drawn from comparisons of males that were observed in a variety of controlled laboratory settings in which behavioral rates could be measured over an extended period and body sizes could be measured very precisely. Field observations are cruder: an individual can rarely be followed for more than 60 seconds, and observers can visually place males into only a few size classes with any precision. Despite these limitations, observations of males in the field reveal the very same patterns that can be seen in the laboratory (table 8.1). The rates of thrusting and displays differ significantly among male size classes ($F_{2,261} = 18.59$ and 23.07 respectively, both $P < .0001$), as does the rate of gonoporal nibbling (a direct inspection by a male of a female's gonoporal area; $F =$

24.18, $P < .0001$). However, male size classes did not differ significantly in the total number of sexual behaviors observed in an observation period ($F_{2,261} = 1.14$). The results indicate that males of different sizes differ more in the behavioral allocation of their investment in gaining female attention than in their total investment in activity rates.

Two aspects of these data are striking. First, the rates observed in the field are comparable to those elicited in laboratory studies. Although both sets of rates may seem high to the general reader, they are typical rates of sexual behavior for male poeciliids (Farr 1989). Second, even though males were placed in categories of body length to allow repeatable observations, the rates still appear to vary continuously but allometrically with male size. We can examine the display rate as a function of the midpoint of each male size class by first taking natural logarithms of the average rates and the midpoint for size. The difference between the logs of the averages for intermediate and large males is 0.77, and the difference in the logs of the midpoints for size is 0.36. The slope of a line that would connect these two points is 2.1. A similar calculation for the line that would connect the bivariate points of small and medium males yields a slope of 5.0. By any measure, display rates increase more rapidly than does body size. The observation of size-dependent behavior in the field and the similar rates of behavior observed in the field and in the laboratory indicate that the laboratory assays of the behaviors directed toward females are accurate indicators of natural patterns.

The more precise laboratory assays reveal a more subtle phenomenon: the relationships between male body size and the rates at which the different behaviors are expressed can vary dramatically among populations (Farr, Travis, and Trexler 1986). A male of a specific body length drawn from a population in which the average male size is large exhibits lower levels of courtship displays and higher levels of gonoporal nibbling and gonopodial thrusting than does a male of the same absolute body length drawn from a population in which the average male size is small (fig. 8.1). These differences are not trivial: at 20° C, a male of 33 mm length would exhibit 20 displays in a 25-minute period if he were from Melanie's Pond, where males are large, but would exhibit almost 67 displays in the same period if he were from Lighthouse Pond, where males are small (Farr, Travis, and Trexler 1986). The secondary sex characters dorsal-fin length and height, which also vary allometrically with male size, do not show a comparable pattern (Snelson 1985;

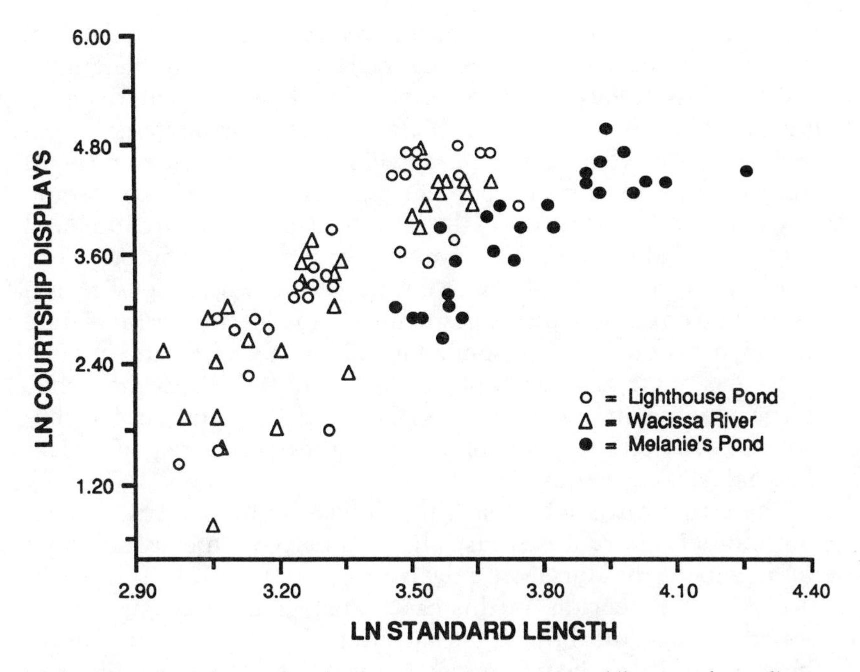

Fig. 8.1 Number of courtship displays in a 25-minute period (log transformed) as a function of standard body length (log transformed) of fish drawn from each of three populations of *Poecilia latipinna.* (Adapted from figure 4 of Farr, Travis, and Trexler 1986.)

Farr, Travis, and Trexler 1986; this is not true for all species, cf. Borowsky 1973). These analyses reveal that a male's relative body size in a population is as important a predictor of the rate at which he expresses various behaviors as his absolute body size (Farr, Travis, and Trexler 1986).

The quantitative work that yielded these conclusions about the allometry of behavior rates and the variation among populations in size-dependent behaviors used methods derived from morphometrics. In morphometrics researchers are often interested in both the absolute size of a structure and its size relative to body size in a number of populations. For morphological structures that covary with body size isometrically (i.e., those that increase in size at the same rate that the measure of body size increases so that the ratio of structure size to body size is constant), one can perform simple analyses of variance on the relative sizes of the structures. For example, if height of the dorsal fin in the sailfin molly varied isometrically with body length within any one population, we could test whether populations differed in the relative heights of their dorsal fins merely by testing the ratio of dorsal-fin heights to body lengths. If the populations differed in their body-length distributions, this test would still yield an unambiguous answer because the relative height of the dorsal fin would be the same at all body lengths in any one population. Some statisticians would frown on the direct testing of the ratios because they are compound random variables and may not be normally distributed. They would advocate an analysis of covariance in which body length would be a covariate and the germane test for differences in relative fin height would be the test of adjusted mean fin heights. In most cases of isometry, the mode of testing would make little difference to the conclusion. If populations differed greatly in their distributions of body size, the test of ratios might have more power than the analysis of covariance.

The situation is rarely so simple because many, if not most, morphometric structures vary allometrically with body size, and the height of the dorsal fin relative to body length in the sailfin molly is no exception. In this case, larger males have disproportionately high dorsal fins. This pattern makes a convincing test of the relative size of the fin more difficult to achieve, especially if the populations to be tested differ in their average body lengths. In such a case, the ratio method is unreliable, and its failure is easy to understand. If population A has only very small fish, the average ratio of fin height to body length will be low. If population B has only large fish, the average ratio will be high. For each population, because of the allometry, the value of the average ratio is largely determined by the actual distribution of body lengths. A test of these ratios would reveal significantly distinct relative fin sizes, but of course this result is unreliable. The two populations might have exactly the same relationship of fin height to body length, but the observed ratios would differ because of the different length distributions in the two populations. In this case, some form of analysis of covariance is the only reliable method.

An extensive literature exists on how best to describe allometric patterns quantitatively and test hypotheses about them (Cock 1966; Gould 1966; Sprent 1972; Mosimann 1975; Mosimann and James 1979; Humphries et al. 1981; Bookstein et al. 1985; Darroch and Mosimann 1985; Strauss 1985). Even though rates of specific behaviors are obviously not expressible in the same verbal terms as morphological structures, the statistical problems posed by the two types of data are identical because rates of behaviors can vary allometrically. Tests for population differences in size- or age-dependent behaviors must also take into account the potential differences in size or age structure among the populations and the allometric description of the size dependency.

This discussion has focused on the type of pattern seen in sailfin mollies, in which the slopes of various behaviors as functions of body size do not differ among populations and only the adjusted mean values differ. There is no reason not to expect slopes to differ in some cases, and the virtue of the more sophisticated statistical techniques is that they permit this situation to be diagnosed. When slopes differ among populations, we would state that the *level* of allometry varies.

Variation in the level of "behavioral allometry" can occur for a variety of reasons. For example, aggressive interactions among males during ontogeny or immediately after maturation might not influence the range of adult body size but might increase the range of behavioral rates. If the tendency of larger males to exhibit displays were exaggerated along with the tendency of smaller males *not* to display to females, then the level of allometry would be increased (a necessary consequence of expanding the dependent variable but not the independent variable and preserving the linear relationship between them). If this process operated in some populations but not in others, then the level of allometry would be observed to vary. Other processes might compress the variation in body size without affecting the ranges of behavior; the effect would be the same. Thus a considerable amount of interesting biology could be reflected in variation in the level of allometry.

In the case of the sailfin molly, statistical analyses have not revealed variation in the level of allometry. They have, however, revealed that populations differ in the expression of condition-dependent tactics (the adjusted mean values of behaviors differ) and that such differences are related to the *direction* of differences in body size, the variable on which the specific behaviors are conditioned. This is the answer to the first question I posed. The answer to the second question is not as easily obtained.

Genetic Analyses of Conditional Variation

Genetic analyses of behavior face two methodological problems. First, in what social context should the behavior be measured so that a valid genetic inference can be obtained (see also Dingle, chap. 7; Hoffmann, chap. 9)? Poeciliid males, like those of most animals, exhibit wide facultative variation in behavior rates. Males will vary their rates of sexual behavior in response to differences in the social milieu (Farr and Herrnkind 1974; Farr 1976; Travis and Woodward 1989), female receptivity (Liley 1966; Farr 1980; Farr and Travis 1986; Sumner, Travis, and Wilson, in press), light intensity (Winemiller, Leslie, and Roche 1991; Reynolds, in press), and perceived predation risk (Endler 1987; Magurran and Seghers 1990; Magurran and Nowak 1991). If the changes in rates of behavior are proportional across these distinct conditions for all behaviors and for males of all sizes, then the choice is merely a practical one, but this is rarely the case. In poeciliid species, altered conditions have been shown to induce greater facultative changes in some behaviors than in others (Liley 1966; Farr 1980; Farr and Travis 1986) or greater changes for some male phenotypes than for others (Sumner, Travis, and Wilson, in press). In our studies of the sailfin molly, a standard behavioral measure is made by observation of a single male's interaction with a single, nonreceptive female in a 38-liter aquarium. The male is kept in isolation for 24 hours before the introduction of a female, and an acclimation period of 10–60 minutes has been allowed to precede the observation period in various experiments. In the genetic studies reported in this section, the acclimation period was 30 minutes. This protocol appears to motivate males of all body sizes in similar fashion and yields the high rates of behaviors typically observed in natural populations.

The second problem is specific to the analysis of condition-dependent behaviors. When rates of behaviors and body size covary tightly, it will be difficult to disentangle the importance of genetic variation for the rates of behavioral expression and the importance of the phenotypic value of body size as an influential variable. This problem affects dissection of variation within and among populations and is the focus of this section.

Quantitative genetic analyses are predicated on a simple principle: if the phenotypic variation in a character is based on genetic variation, then any group of related individuals should resemble one another more in their values for that character than do a similarly sized group of individuals randomly chosen from the population. If one measures a genetically variable character in enough groups of related individuals (or families), those groups will differ significantly in their average values for the character because the resemblance of individuals within groups exceeds the resemblance among groups. Conversely, the variation among groups exceeds the variation within groups.

Specific quantitative genetic models translate a specific level of relatedness within groups and the magnitude of variation among group means into an estimate of the fraction of the phenotypic variation that is genetically determined. Consider the simplest case, that of measurements on several sets of identical twins where the twins were raised in different environments. If there is no variation in behavior within a set but statistically significant variation among sets, as diagnosed in an analysis of variance, then nearly all of the phenotypic variation we observed would appear to be genetically determined. More precise details of exactly how these inferences are made are provided by Arnold (chap. 2). Analogous models estimate the extent to which the phenotypic covariation between two characters is based on common genetic controls.

Quantitative genetic analyses are statistical summaries of the properties of variation and covariation in characters, and different genetic mechanisms can produce similar quantitative genetic results. This problem is especially critical when one examines the phenotypic covariation of two traits. Two phenotypic characters can vary concordantly among groups of related individuals through several distinct genetic mechanisms. First, variation in each character may be controlled by the variation at a common set of loci; this is denoted a pleiotropic effect of the alleles at each locus. Second, each character may be controlled by alleles at many completely distinct loci that are embedded in linkage groups that cannot be broken by recombination. This mechanism is an opposite extreme from the first one and is unlikely to account for the major part of any covariance except in unusual circumstances, such as when only a very few loci control the variation in each trait (this is not to deny that tight linkage contributes to some patterns of trait covariance: Davies 1971; Davies and Workman 1971). Third, the majority of the loci that control each character's variation are unique to that character, but the few shared loci may have large effects on both phenotypic traits while the majority of distinct loci have small effects on each trait (Carey 1988). Fourth, variation in each character is controlled by alleles at completely distinct loci, but the loci are temporarily in extreme linkage disequilibrium, which could occur following the admixture of two populations that are strongly differentiated for each trait. This mechanism is also an unlikely source of persistent covariances in natural populations, but it cannot be ruled out as a causal explanation in synthetic stocks of laboratory animals. Statistical estimates of genetic covariance cannot distinguish among these mechanisms.

The distinctions between the nontrivial mechanisms, the first and third, are important for probing the evolution of condition-dependent behaviors. An extreme form of the first mechanism, which I will denote "pure pleiotropy," constrains the evolution of different rates of behavior at the same value of the condition trait (e.g., body size) because the allelic effects on each trait are completely shared and therefore one trait cannot be altered without altering the other. The "limited pleiotropy" of the third mechanism provides greater latitude for each character in the short term because favorable mutants at the independent loci can be selected to alter either character without altering the mean value of the other. However, the changes that result will be small and will take a long time to accumulate because the independent loci each contribute only a small amount of variance to each character; the shared loci account for the majority of the variance.

Present knowledge of the genetic control of the correlated variation in size and behavior in the sailfin molly suggests that some control of behavior rates is indeed independent of the genetic variation for body size. Body size variation within a population appears to be inherited patriclinally through an allelic series on the Y chromosome of mollies (fig. 8.2; Travis, Trexler, and Farr, in press). The regression of a male's body size on that of his father has a slope of 1, which indicates Y linkage (Bulmer 1980). Pedigree analyses support this interpretation (Travis, Trexler, and Farr, in press), and nearly all of the phenotypic variation in body size within an experimentally reared cohort is accounted for by this mechanism. Variation in size is produced by variation in the onset of maturation; maturation, once initiated, slows growth continuously until all growth ceases upon attainment of sexual maturity. Large males initiate maturation much later than small males (Farr and Travis 1989; Travis, Trexler, and Farr, in press).

The males in our experimental study displayed typical patterns of size-dependent behavior rates (fig. 8.3), but the regressions of sons' behavior on that of their fathers had slopes that were significantly different from both 0 and 1 (table 8.2), which indicated that, although behavior rates were heritable, they were not completely controlled in the same manner as body size (Travis, Farr, and Trexler, in press; see Arnold, chap. 2, for details on how such inferences are made). Further support for this interpretation is available from an analysis of half-sib variance in courtship display rates with individual body size as a covariate (table 8.3). There were significant differences among half-sib families in the display rates beyond what would be predicted by accounting for body size (see Travis,

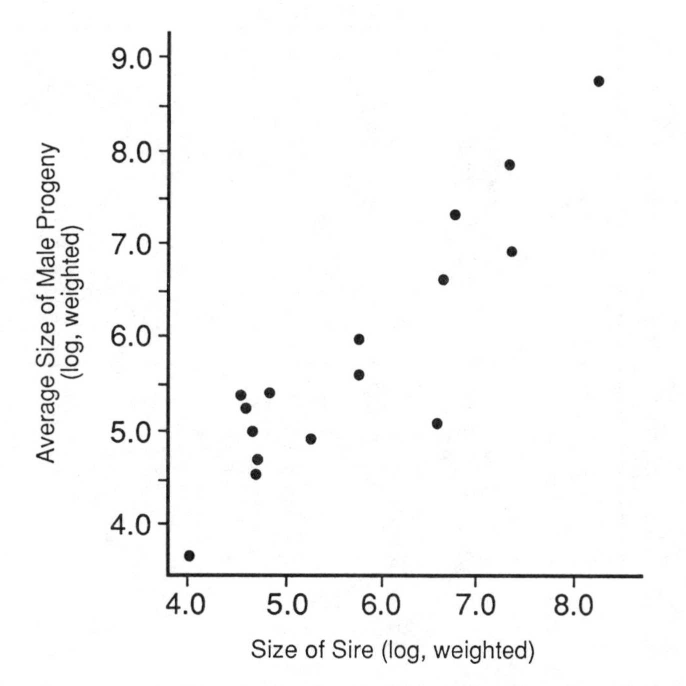

Fig. 8.2 Average standard length of male offspring as a function of standard length of sire (log transformed and weighted by family size) for *P. latipinna* raised in the laboratory.

Farr, and Trexler, in press, for fuller discussion). If the same precise mechanisms were controlling size and behavioral rates, the covariate, size, should account for all of the phenotypic variance in behavioral rates. The differences among families accounted for only a fraction of the variation in display rates, suggesting that any control of behavioral rates independent of body size is quite limited in scope.

Additional evidence on the control of the covariance between body size and display rate arises from comparison of the behavior rates of normal and stunted siblings (table 8.4; see Travis, Farr, and Trexler, in press, for further details). Some males were experimentally stunted, that is, raised at cold temperatures, at which they grow very slowly, for 120 days, then exposed to warmer, more normal temperatures. This procedure induces males that have not initiated maturation before being exposed to warmer temperatures to do so, and causes males that carry the alleles for a late initiation of maturation to mature at a much smaller size than normal. Stunted

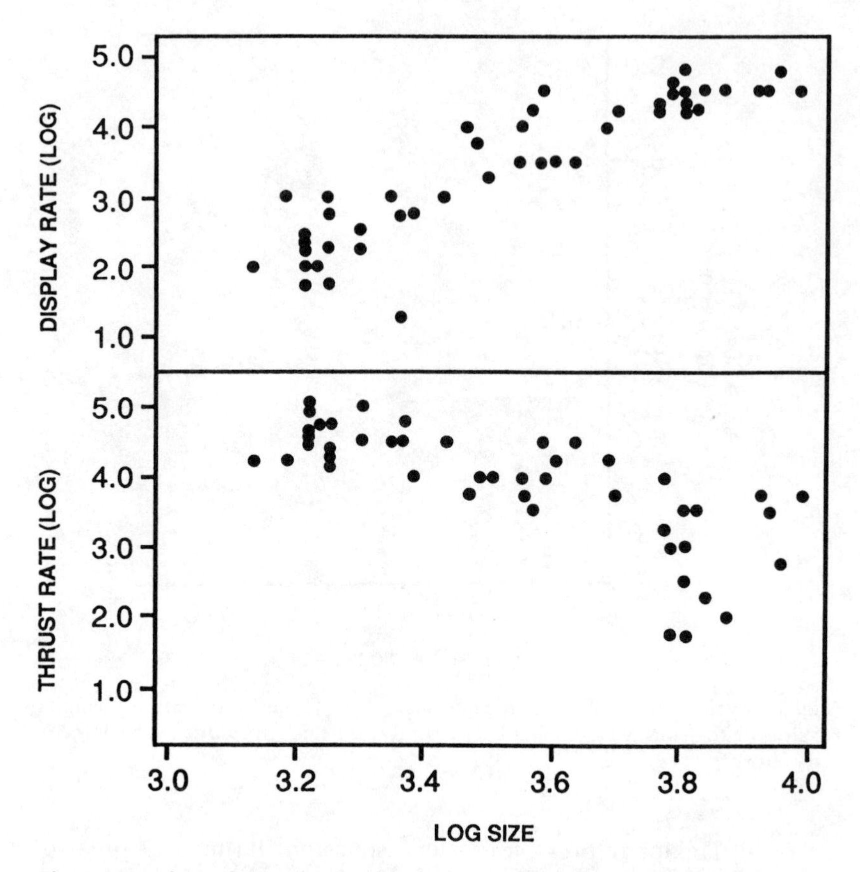

Fig. 8.3 Numbers of courtship displays and gonopodial thrusts in a 25-minute observation period as a function of individual body size for male *P. latipinna* from controlled crosses raised in the laboratory.

Table 8.2 Regressions of average phenotypic trait values for sons on their sires' phenotypic values

Trait	Slope ± SE	Intercept	r^2	*F* Value
Standard length	1.02 ± 0.09	−0.03	.90	132.46
Gonoporal nibbles	0.69 ± 0.10	1.87	.75	46.62
Gonopodial thrusts	0.65 ± 0.14	2.02	.59	22.39
Courtship displays	0.42 ± 0.17	3.49	.26	6.37

Note: Regressions were performed on log-transformed data and weighted by the size of each family (see Falconer 1964). Seventeen half-sib families are represented in each regression. All *F* values are significant at or below the .05 critical value.

Table 8.3 Analysis of covariance for courtship display rates

Source	df	Mean Square	F	r^2
Log standard length	1	36.57	457.11	.82
Family	15	0.37	4.34	.13
Residual	29	0.08	—	—

Note: Courtship display rates for 16 half-sib families were log-transformed; log standard length was used as the covariate. F values are significant at or below the .05 critical value.

males exhibited fewer displays, more nibbles, and more thrusts than did their larger sibs. Males exposed to the stunting treatment whose normal sibs were small were unaffected by the treatment, and their rates of behavior were comparable to those of their siblings. Overall, males of similar size exhibited similar rates of behaviors regardless of family origin, which suggests that most of the variation in rates of behavior is inherited together with body size. However, the small sample sizes of these data do not permit a precise estimation of the level of congruent control, and they must be taken only as qualitative results.

The second question I posed, how the variation in condition-dependent behavior is controlled genetically, has at least a partial answer through the preceding analyses. It is a partial answer because inferences about the variation within a population need not apply to the variation among populations. In principle, the Y-linked elements must be examined atop the different autosomal backgrounds represented in the different populations. Doing so would permit an estimate of the covariance between traits due to the Y-linked effects and the covariance due to other effects within or among populations, which can only be accomplished by development and use of several distinct "isomale" lines that differ in their Y-linked effects but that have had their autosomal complements randomized among lines. This randomization will preclude linkage disequilibrium between Y-linked effects and autosomal alleles that are passed on with the Y chromosome in experimental crosses. This work is presently under way.

In the specific case of the sailfin molly there is a final potential problem. In other species with similar systems for the inheritance of body size, different populations appear to have accumulated differences at autosomal loci that modify the Y-linked effects on male

Table 8.4 Average values of standard length and rates of sexual behavior for normal and stunted males

Treatment[a]	Family Designation	n	Standard Length (mm)	Gonoporal Nibbles	Gonopodial Thrusts	Courtship Displays
Normal	2	6	51.0	36	21	86
	3	4	48.0	38	32	60
	12	4	27.5	68	92	14
	13	2	35.0	67	74	25
Stunted	2	2	25.5	52	54	25
	3	2	30.2	79	79	26
	12	5	25.5	74	67	15
	13	2	24.8	71	76	14

[a] Males of four half-sib families were split between normal rearing and a stunting treatment.

size (Kallman 1984, 1989). The presence of such differences may preclude a true separation of the genetic control of each trait, although they do not impede the ability to distinguish an allelic series with strong effects on both traits (the Y-linked effects) from autosomal loci with smaller effects.

Hypotheses for the Evolution of Behavioral Variation

Some knowledge of the genetic control of variation within and among populations can help in framing testable hypotheses for the evolution of behavioral diversity in the sailfin molly. To illustrate this point, I begin with two hypothetical scenarios based on the presumption that most of the variation is controlled by a few loci of major effect along with many loci of small effect. Further, I assume that there is pleiotropy among the loci with large effects but not among those with small effects. It is critical to point out that this genetic system, although apparently common in poeciliids (Kallman 1984, 1989), is unlikely to be found more generally. This discussion is meant to illustrate how knowledge of a specific case can inform the framing of hypotheses for that case.

Consider two identical populations descended from a common stock but now subjected to different selective pressures for either increased or decreased rates of courtship display via sexual selection (body-size variation is under optimizing selection for the same mean in both populations). The initial genetic covariances in each population are determined by the joint effects of the alleles at the few shared loci with major effects. In one population mutations are favored at the many independent loci that increase courtship rates, whereas in the other population mutations that decrease courtship rates are favored. Mutations at the few shared loci that affect both courtship rate and body size are not favored because their detrimental effects on body size outweigh the selective value of changes in behavior rates. Over time the populations could accumulate fixed differences at the independent loci, if the shape of the fitness function for these coupled traits permits (Wagner 1988). Any such accumulation is likely to be a slow one. A graphical model of this process shows that evolution alters the vertical height of the regression line for courtship rate as a function of body size (fig. 8.4). Statistical comparisons like the ones described in the previous section will reveal different rates of behavior in the two populations, but the genetic covariance between size and courtship rate will be identical in the populations. The genetic co-

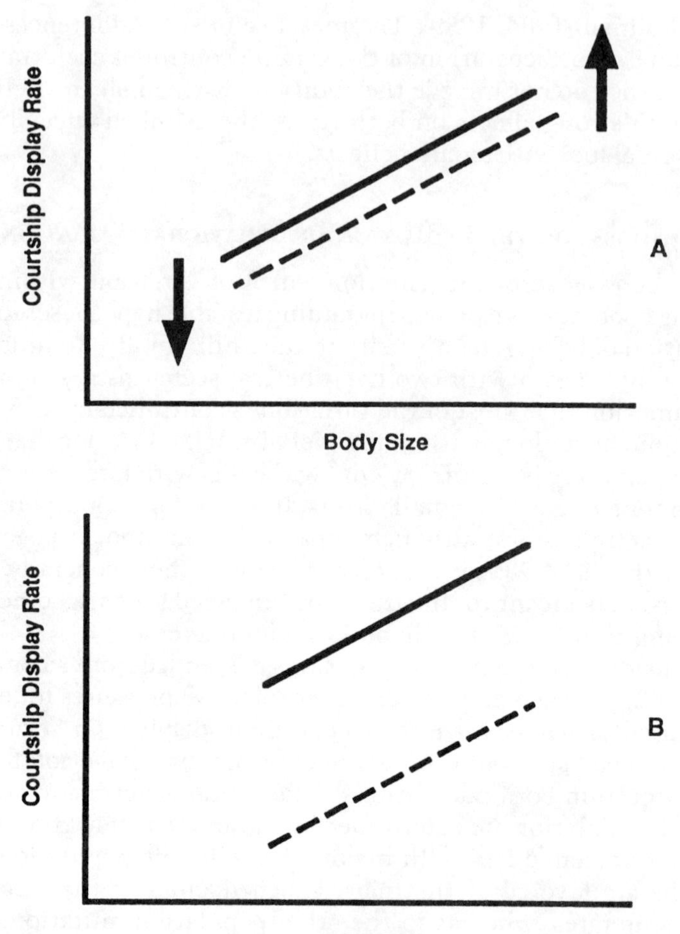

Fig. 8.4 Hypothetical course of bivariate evolution of the rate of courtship displays and body size. Two populations that are identical for bivariate distribution of average values experience opposite directional selection for courtship rates at loci that do not affect body size (*A*). Size is under identical optimizing selection in the two locations. The result (*B*) is differential elevation of the regression lines.

variances are unchanged because such covariances are determined only by the polymorphic loci, which have not had their allelic profiles altered by this process. The populations will each exhibit internal concordant genetic variation for body size and courtship rates, and will not be distinguishable genetically for body size, but will be readily distinguishable for courtship rates.

Imagine a second, slightly more complex scenario in which natural selection in one population favors smaller size while natural selection in the other population favors larger size. If the benefits of an allele's contribution to body size outweigh any detrimental effects of the allele on courtship rates, then the allelic profile at the few shared loci will change in each population. Over time the populations will diverge at the few loci that influence both traits because these loci produce the greatest effect on body size. For loci of large effect, like those seen in poeciliid males, this divergence can be rapid. When the disadvantages of changes in courtship rates outweigh the advantages of changes in body size, a new equilibrium for those loci will be attained, and further changes in the body-size distributions will accrue from changes at the independent loci. A graphical depiction of this process illustrates that evolution first slides the regression lines of the two populations apart diagonally, parallel to the initial slope, and then displaces the lines horizontally (fig. 8.5). The magnitude of the genetic covariances will have changed because of the altered allele frequency profiles within each population, but they will remain comparable (unless one population evolves farther from the original optimal body size through the major loci than the other). Yet the continued divergence in body size through changes at the independent loci will produce differences in the size specificity of behaviors.

This second scenario has an interesting feature. The horizontal displacements of the regression lines of the two populations mean that, when a trait that is positively correlated with body size, like courtship displays, is examined by means of an analysis of covariance, the population that has had its body-size distribution increased by selection (the line displaced to the right) will have an adjusted mean courtship display rate that is less than that of the other population. For a negatively correlated trait, such as thrusting rate, the horizontal displacement will make the population with the larger body size exhibit a higher adjusted mean value of the trait. Selection on the behaviors functions only to prevent further diagonal displacement of the populations' regressions; the major cause of the differences in size-dependent behaviors is the selection on body size. The patterns of adjusted means produced by this scenario are exactly those seen in the population comparisons of mollies that were discussed above. In that case, the patterns were discussed as concordant with an expectation derived from sexual selection on behaviors through a male's relative size. It should be clear that the patterns could also be produced by natural

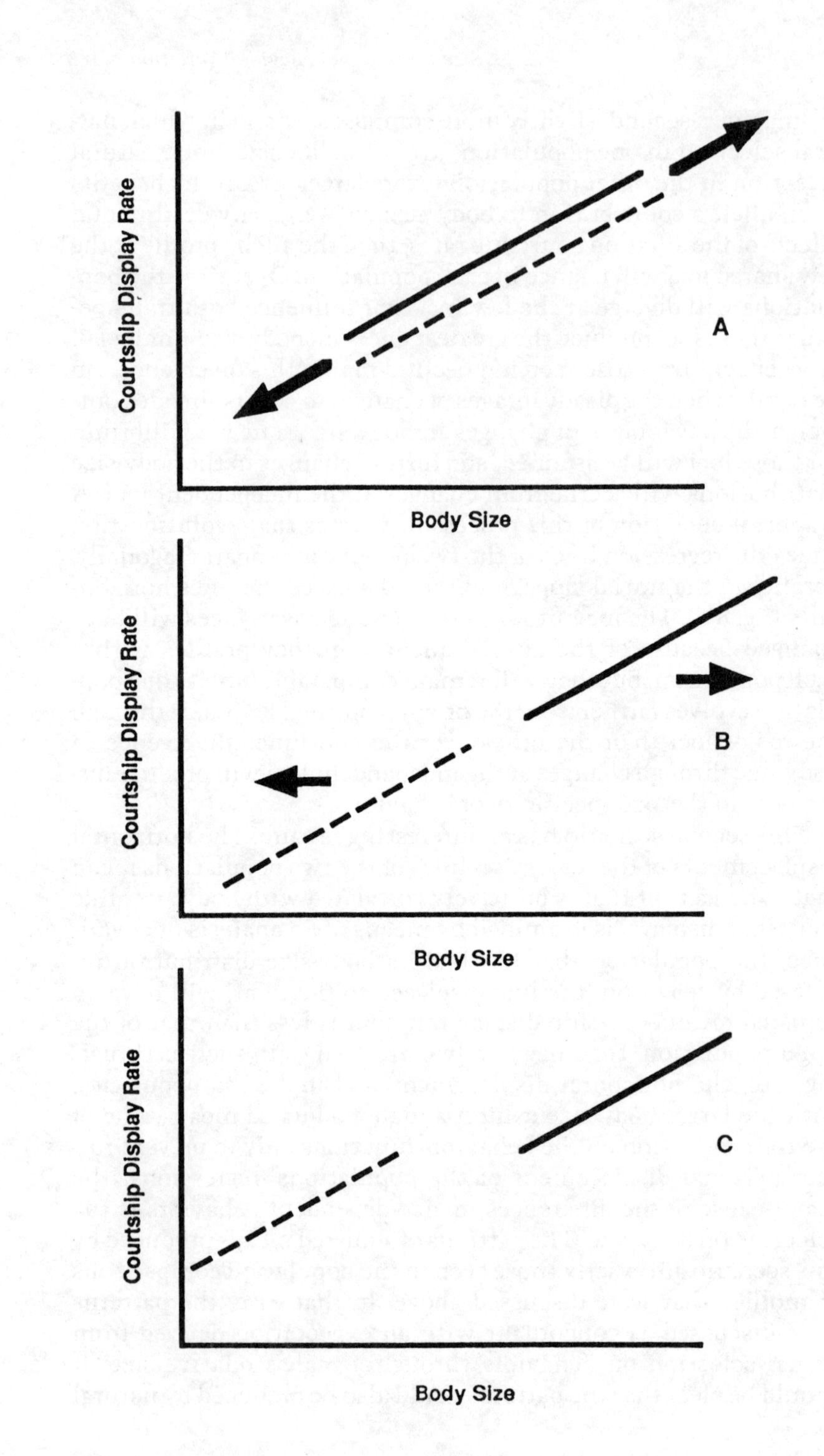
Courtship Display Rate
Body Size
A
Courtship Display Rate
Body Size
B
Courtship Display Rate
Body Size
C

selection on body size. The existence of a second, viable hypothesis for the pattern means that the reconstruction of the mechanism of population differentiation must involve tests for both sexual selection on behaviors and natural selection on body size.

The knowledge of how concordant variation is controlled brings into focus several aspects of devising hypotheses for the evolution of diversity in condition-dependent behavior. In the first scenario, the populations diverge slowly because of the small effects of each locus that would be acting. In the second scenario, natural selection produces the divergence quickly. Presumably, if the same pattern were to be generated largely by sexual selection, the time to achievement would be much longer. Estimates of divergence times through molecular methods and coalescence theory (e.g., Hudson 1990) could therefore further focus the efforts to dissect the causes of observable behavioral diversity.

For the sailfin molly, these are in fact the hypotheses currently under scrutiny. The variation among constituent populations in this species may be a useful model for the variation in the entire poeciliid family. Many species have no courtship at all, whereas others exhibit it sporadically and still others have elaborate, well-developed courtship displays (Farr 1989). The variation among species in behavioral diversity is not independent of the variation in morphological and secondary sex characters, and work on condition-dependent behaviors in the sailfin molly may contribute to understanding of the evolution of behavioral diversity and its correlates on a wider scale.

Conclusions

The ubiquity of alternative tactics or other forms of behavioral variation has inspired an enormous amount of research, but not enough on the magnitude, control, and evolution of variation in the expression of condition-dependent behaviors. In fact, natural

---

Fig. 8.5 Hypothetical course of bivariate evolution of the rate of courtship display and body size. As in figure 8.4, two populations begin at identical locations, but they experience opposite directions of selection for body size that produces indirect selection on courtship rates (*A*). After divergence of the regression lines (*B*), further selection on size can only progress through a response by loci that do not affect courtship display rates because further changes in courtship display rates oppose selection on body size. The result (*C*) is differential elevation of regression lines but in the direction opposite to that produced by the mechanism illustrated in figure 8.4.

populations can and do vary in these traits, often to a considerable extent. The detection and dissection of such variation can be a complex statistical problem, but a variety of methods exist for approaching it, requiring little more than careful thought and analyses of data that can be readily obtained. Variation among populations need not be genetically based, however, and the obvious possibility of facultative differences should be investigated first. The methods of quantitative genetics, even of the crude sort that is considered "minimal" by purists, can be used profitably to understand the control of such variation within and among populations. The information gathered from these methods allows development of more focused hypotheses for the evolution of behavioral diversity.

There are two subtexts within this theme. First, quantitative-genetic studies ought not to be ends in themselves for evolutionary biologists. I have shown how they can be useful tools for framing evolutionary hypotheses. However, they cannot provide definitive answers or tests of these hypotheses because quantitative-genetic parameter values have no one-to-one correspondence with particular modes of selection or evolutionary histories (Boake, chap. 14; Hoffmann, chap. 9; Roff, chap. 3). Second, realistic hypotheses for behavioral evolution will almost always involve many other aspects of the animal's ecology and evolution besides behavior per se, as shown elsewhere in this volume by Dingle (chap. 7) and by Hoffmann (chap. 9). This should be no surprise; evolution acts on entire phenotypes, and its processes are not nearly as specialized as are many of its students.

Summary

Although much attention has been devoted to the factors that might determine optimal patterns of sexual behavior in males, little empirical attention has been paid to the question of how such behavioral diversity evolves. This is especially true for behaviors whose expression varies as a function of body size or age. The correlation of behavior with other phenotypic attributes introduces several complications into the construction of evolutionary hypotheses and the empirical evaluation of those hypotheses. Quantitative-genetic methods can be used profitably to discern the control of behavioral variation and its correlates and thereby lead efficiently to testable hypotheses about behavioral evolution.

The study of male mating behavior in the sailfin molly, *Poecilia latipinna,* illustrates how this can be done. Male mollies engage in two discrete sexual behaviors, the rates of which vary with male body size. The rate at which males exhibit courtship displays increases as body size increases, and the rate at which males attempt forced insemination of females decreases as body size increases. These patterns can be seen in uncontrolled field observations as well as controlled laboratory studies and are unrelated to male age. However, populations differ in the rates at which males of the same size will exhibit each behavior.

Laboratory experiments indicate that this variation is not induced environmentally through differences in social context. Quantitative-genetic studies reveal that body size is inherited patriclinally such that fathers and sons are virtually the same size at maturity. However, although the behavioral profiles of sons resemble those of their fathers, the resemblance is not perfect. Moreover, manipulations of the developmental trajectory of males reveal that, although there is some shared genetic control of body size and rates of behavior (pleiotropic effects of genes on both traits), there is some amount of genetic control of behavior that is independent of the genetic control of body size.

With this knowledge, two different hypotheses for the evolution of behavioral diversity in sailfin mollies can be constructed. One hypothesis invokes direct selection on rates of behavior independently of body size, which generates population differences in size-dependent behavior patterns through genetic changes that influence only rates of behavior. This hypothesis predicts a limited range of differences in behavior that may or may not show a pattern of directional association with body size variation among populations. The competing hypothesis invokes direct selection on body size in different directions in different populations and only an indirect effect on rates of behavior through the pleiotropic effects of the genes controlling body size. This hypothesis predicts a specific pattern of directional association of size-dependent behavior rates with patterns of body size variation: males from populations in which the average body size is smaller should exhibit higher rates of courtship displays than males of the same body size from populations in which the average body size is larger. Current research is designed to distinguish between these hypotheses by making direct experimental tests of the extent to which body size and rates of behavior are selected for independently of each other.

9

Genetic Analysis of Territoriality in *Drosophila melanogaster*

Ary A. Hoffmann

Territorial interactions between two *Drosophila melanogaster* males. (*a*) The male on the right is defending the food surface of the cup and approaches an intruder entering his territory. (*b*) The territorial male has lunged onto an intruder, forcing it from the defended area.

Early studies in the behavioral genetics of *Drosophila* were usually concerned with demonstrating genetic differences among individuals. Much of this work involved the characterization of behaviors that could be easily measured under laboratory conditions (reviewed in Parsons 1973). Examples include geotactic and phototactic responses of flies in a maze, duration of copulation, and "mating speed," which is the time taken for males to mate with virgin females. Genetic analyses of such traits could often be taken to a relatively sophisticated level by crossing selected lines to strains with dominant markers and inversions that inhibit crossing over. This procedure enabled chromosomes from selected lines to be combined in all possible ways so that the effects of individual chromosomes and their interactions on the behavior could be studied (see Kearsey and Kojima 1967). For example, Hirsch and Erlenmeyer-Kimling (1962) showed that genes controlling geotactic behavior were found on all three of the major chromosomes of *Drosophila melanogaster.*

These studies served to illustrate that behavioral traits could be studied from a genetic perspective just like morphological and physiological traits. However, the relevance of such laboratory-defined behaviors to *Drosophila* behavior in the field was generally unknown. It is not clear whether geotactic and phototactic responses in a maze measure behaviors used by flies to locate natural resources. Even a trait such as mating speed, which would seem to provide a measure of mating success, may have little relevance to the field situation, where virgin females are rare and females are not confined to a small mating arena.

One way of determining the types of evolutionary forces that have acted on a laboratory behavior is to divide variation in a behavior into genetic and environmental components, and to further divide the genetic component into additive, dominance, and epistatic effects (see Arnold, chap. 2, for an explanation of these terms). This approach has been used extensively in behavioral genetics (reviewed by Broadhurst 1979). Genetic parameters are interpreted in terms of the history of selection acting on a trait. Traits that are closely related to fitness and that have therefore been under directional selection are expected to show a low level of additive genetic variance as well as dominance in the direction of increased fitness, while traits under stabilizing or weak selection are expected to show higher levels of genetic variation. For example, Fulker (1966) crossed six lines of *D. melanogaster* and found that the F_1 hybrid in crosses between fast and slow mating lines tended to have a mating speed more similar to that of the fast mating line than that of the slow mating line. This result indicated directional dominance for fast mating and suggested that rapid mating had been under directional selection in the past. Unfortunately, this approach is based on assumptions that may not be valid (as discussed in Kacser and Burns 1981).

More recently, *Drosophila* workers have realized that the genetic basis of behaviors important in the field needs to be studied. For example, Sokolowski and coinvestigators have conducted experiments on variation in the foraging behavior of *D. melanogaster* larvae which showed that the variation has a relatively simple genetic basis and can be related to pupation behavior in the field (e.g., Sokolowski et al. 1986). In mycetophagous *Drosophila,* Jaenike has shown that flies differ genetically in their responses to natural resources where they breed (e.g., Jaenike 1989). Genetic variation in the mating behavior of *Drosophila* has also been investigated in laboratory situations likely to be relevant to field conditions (e.g., Gromko and Newport 1988).

In this chapter, I describe a behavioral genetic analysis of territorial behavior in *D. melanogaster.* This behavior is ecologically important and is usually studied by behavioral ecologists using optimality approaches (see Roff, chap. 3) rather than from a genetic perspective. My review will attempt to illustrate the usefulness of a behavioral genetic approach for understanding territorial behavior. In particular, I will emphasize the limitations of a phenotypic approach in predicting evolutionary changes and the usefulness of a genetic approach in assessing costs and determining traits underlying territorial success.

TERRITORIALITY IN *DROSOPHILA*

It has been known for some time that aggressive interactions occur between males in several *Drosophila* species and that many of these involve the defense of a territory (Spieth 1968). In some Hawaiian *Drosophila,* males aggregate in a lek and defend leaves as mating territories within the lek (e.g., Shelly 1988). In *D. pseudoobscura, D. immigrans,* and other species utilizing rotting fruit as breeding sites, males defend areas of fruit used by females for oviposition (Partridge, Hoffmann, and Jones 1987; Hoffmann, unpublished observations).

Males of the widely studied cosmopolitan species *D. melanogaster* also exhibit territorial aggression. This was first described by Jacobs (1960), who noted that males in laboratory cages defended patches of food medium against other males by charging at them or tussling with them. This behavior fits classic definitions of territoriality entailing the defense of a limited resource because the medium represents a resource for breeding and feeding. The defended area was small, which is typical for resources defended by male insects (Thornhill and Alcock 1983). Defense of food in *D. melanogaster* was independently reported by Dow and von Schilcher (1975). They observed six males in a cage and found that the male successfully defending the food had a relatively higher mat-

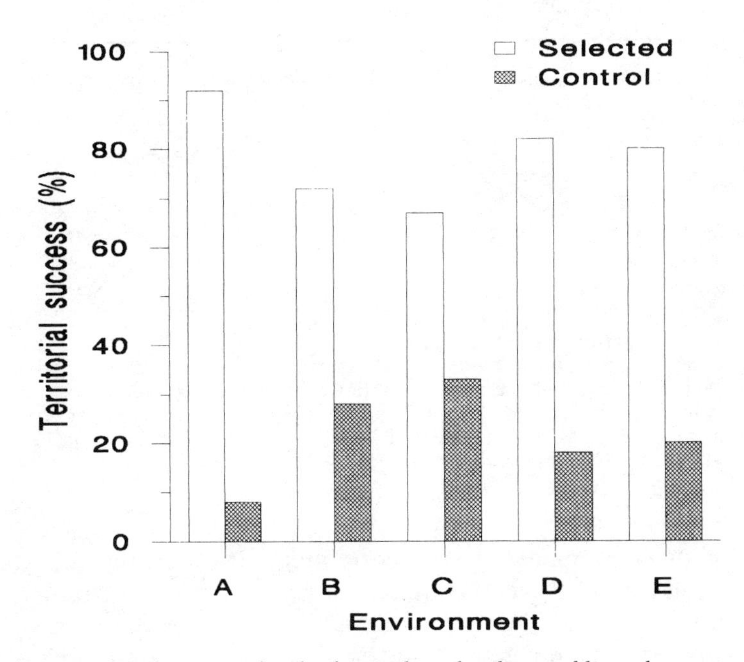

Fig. 9.1 Territorial success of males from selected and control lines that were cultured under different environmental conditions: A, cultured on laboratory medium at 20° C (standard conditions); B, cultured on laboratory medium at 29° C; C, cultured at high larval density; D, cultured on apple pulp; E, cultured on whole apples.

ing success. A mating advantage for territorial males was confirmed by Skrzipek, Kroner, and Hager (1979) using a much larger sample size. These authors also noted that most interactions between males were short and fighting rarely occurred.

In a quantitative analysis of territoriality in *D. melanogaster* (Hoffmann 1987a), I divided encounters into two categories: escalated and nonescalated. The nonescalated contests were usually brief; intruders were forced off the food after being lunged at by the territorial resident. Escalated encounters occurred when intruders turned to face residents and tussled with them; these encounters could last several minutes with intruders persistently tussling with residents. Residents inevitably won nonescalated encounters, while intruders were successful in displacing residents in some of the infrequent escalated encounters (table 9.1). Territorial success could be scored as the ability of residents to evict intruders over a time interval.

Table 9.1 Effect of size and resident status on the outcome of aggressive encounters in *D. melanogaster*

	All Encounters	Escalated Encounters
Won by resident	0.99	0.74
Probability	<.001	<.01
Won by heavier male	0.67	0.65
Probability	<.01	<.05

Source: Adapted from Hoffman 1987a.
Note: Encounters won represent the proportion of encounters won per trial. For each trial the behavior of six males was followed for 8 hours on videotape. Probabilities are from Sign tests on the number of trials having more or less than half the encounters won by resident or heavier males. Means are based on 23 (overall) or 18 (escalated) trials.

Body weight appeared to be correlated with fighting ability because heavier males tended to win escalated encounters (table 9.1). Weight differences were not a consequence of territory ownership because repeated measurements on males from the same cages indicated no significant weight changes with changes in territorial status. Heavier males had an overall advantage in territorial encounters, presumably because males usually held onto territories after winning an escalated encounter. Residents did not usually leave territories unless they were forced off after an escalated encounter or unless they mated with a female that had entered their territory. While residents usually won encounters when the size difference between opponents was small (Hoffmann 1987a), this changed when the size difference was greater because large intruders often escalated against much smaller residents, and fights between such opponents tended to be short (Hoffmann 1987b).

Similar findings were obtained with *D. simulans*, a sibling species of *D. melanogaster,* although there were some quantitative differences in the behavior of the two species (Hoffmann 1987a). *D. simulans* males tended to escalate more often against residents than *D. melanogaster. D. simulans* males also engaged in longer escalated encounters.

When the aggressive behavior of *D. melanogaster* males was studied in a number of situations, it soon became apparent that territories were defended only under some conditions. Optimality approaches predict that males should stop defending territories in situations when they no longer have a mating advantage and/or when the benefits of territorial defense are outweighed by other costs. This prediction was tested with respect to variation in size of the defended area, presence/absence of females, male density, and sex ratio (Hoffmann and Cacoyianni 1990). Males abandoned territorial defense when food cups were increased in size and when male density was increased; this coincided with conditions in which territorial males no longer had a mating advantage. We also found that males had a reduced tendency to defend territories when females were absent and when defensible resources consisted of food that was less attractive to females (Hoffmann and Cacoyianni 1990), suggesting that males were more likely to defend resources leading to encounters with potential mates. Finally, we found that males defended territories regardless of sex ratio, which was consistent with a mating advantage for territorial males even when females were twice as common as males.

Territorial behavior in *D. melanogaster* is therefore a conditional strategy, and switchpoints between territorial and nonterritorial behavior may have been influenced by natural selection. However, it should be emphasized that the experiments were carried out under laboratory conditions. It is much more difficult to study the effects of different environmental conditions on territorial behavior in the field because of the mobility and small size of *D. melanogaster.* Moreover, optimality predictions depend on an accurate assessment of the costs of territorial defense, and we have only been able to measure costs of defense in terms of reduced mating success. Other costs may also be important for *D. melanogaster,* such as an increased risk of predation as a consequence of territorial defense.

Information about the relative importance of territorial defense in obtaining mates is needed, and this will require observations on flies in different natural habitats. It is clear that in some situations *D. melanogaster* males do not attempt to defend territories in the field (e.g., Taylor and Kekic 1988; Partridge, Hoffmann, and Jones 1987), and these may represent situations in which territorial males do not have a higher mating success than nonterritorial males. Success in aggressive encounters between males has also been correlated with mating success in the fruit fly *Dacus dorsalis* (Poramarcom and Boake 1991), but not in *Drosophila silvestris* (Boake 1989a).

Genetic Variation in Territorial Success

From the data presented by Skrzipek, Kroner, and Hager (1979) and Hoffmann (1987a), it became obvious that males differed enormously in their tendency to behave aggressively. Some males never defended a food cup or escalated against a resident, whereas other males escalated in their first or second encounter with a resident. Males also appeared to have different tendencies to be territorial, as reflected by the quantitative differences in their tendency to escalate against a resident and the length of time they held territories. Even some small males showed some tendency to be territorial when faced with a much larger opponent despite having no chance of winning an escalated encounter with such an opponent (Hoffmann 1987b). The tendency to be territorial is probably a quantitative trait that is continuous, even though males fall into two discrete classes (territorial, nonterritorial) in some situations such as when the size difference between opponents is considerable.

Some of this phenotypic variation has a nongenetic basis. This includes behavioral variation associated with male age: young *D. melanogaster* males (< 30 hours posteclosion) rarely occupied territories in cages with older males unless they had territorial experience, and young males tended not to escalate against the older territory residents (Hoffmann 1990). Nongenetic variation may also arise because of experience with conspecifics. For example, *D. melanogaster* males showed a decreased tendency to be territorial when they were held in contact with other males rather than in isolation (Hoffmann 1990).

Genetic factors may also contribute to variation in territorial success. A number of techniques are available to evaluate the relative importance of genetic versus environmental factors, as discussed in texts such as Falconer's (1989) and in this volume by Arnold (chap. 2). Two widely used techniques have been applied to territorial behavior in *D. melanogaster:* parent-offspring comparisons and directional selection.

Parent-Offspring Comparisons

Parent-offspring comparisons were used to test for heritable variation in territorial success among field-collected males as well as among males reared under laboratory conditions. Culture conditions in the field differ from those in the laboratory and are likely to be more variable. This difference may influence heritability estimates because the environmental (i.e., nongenetic) component of the phenotypic variance is likely to be higher when culture conditions are variable. For example, fly size is influenced by larval density in fruit (Atkinson 1979), so variation in size due to environmental factors will be greater in the field than in the laboratory, and this will in turn affect territorial success. It is therefore important to obtain estimates of heritable variation under natural conditions as well as under artificial conditions.

Ideally, heritability studies should be carried out by following animals over more than one generation in the field, so that parents and progeny experience field conditions. This is not possible in *Drosophila,* but an indication of the heritable variation can be obtained by characterizing traits in field-collected flies and their F_1 progeny reared under laboratory conditions (Prout and Barker 1989; Riska, Prout, and Turelli 1989; Aspi and Hoikkala 1993; see Boake, chap. 14). Any similarity between the behavior of the field flies and their progeny will reflect heritable factors because the progeny of different parents are cultured in the same laboratory environment, although differences may also reflect maternal effects due to the preoviposition environment experienced by the F_1 flies or cytoplasmic factors passed on via the female parent.

I carried out parent-offspring comparisons for territorial success with males from two field populations and from a laboratory population (Hoffmann 1991). Males were collected from an orchard 35 km southeast of Melbourne and from a winery 100 km north of Melbourne, and their territorial success was characterized directly after bringing them into the laboratory. The laboratory population was founded by flies collected from an orchard.

The experimental design is complex and only a brief description is provided here. Males were marked with paint and introduced into cages at a density of five males per cage. The male in each cage that continuously defended the food cup over a 3-day period was characterised as "territorial." These males and two "nonterritorial" males from each cage were subsequently mated to females from an inbred stock. The F_1 male progeny were then tested to see if progeny of territorial fathers were more likely to hold territories than those of nonterritorial fathers. This was determined by setting up five cages with progeny for each parental cage and observing these cages over a 3-day period. Maternal effects are eliminated in the experimental design because all males were mated to females from the same stock. In any case, reciprocal crosses between lines with high and low levels of territorial aggression suggest that variation in territorial success is not influenced by maternal factors (Hoffmann 1988).

The results are summarized in table 9.2, which gives the num-

Table 9.2 Territorial success of offspring of territorial and nonterritorial males

More Successful[a]	Orchard	Winery	Lab Stock
Territorial offspring	39	24	33
Nonterritorial offspring	17	23	13
Ties	9	5	8
Probability	<.01	NS	<.01

Source: Adapted from Hoffmann 1991.
[a]The territorial success of offspring from each parental cage was determined by setting up five cages containing males from territorial and nonterritorial parents. Numbers represent five-cage groups in which males in each category were more successful. Probabilities are based on Sign tests.

ber of cases in which progeny of territorial or nonterritorial fathers were more successful. Ties arose when, for example, progeny from nonterritorial and territorial fathers each dominated in two of the five cages, and both progeny types defended the food cup equally frequently in the remaining cage. These results provide evidence for heritable variation in territorial success in males collected from the orchard but not in those from the winery. A significant difference between progeny of territorial and nonterritorial fathers was also evident in the laboratory males, and this difference was similar in magnitude to that obtained with the orchard flies. Heritable variation for territorial success could therefore be demonstrated in laboratory and field populations from an orchard.

Unfortunately, these data cannot be used to compute heritabilities for territorial success because scores were not assigned to individual males. Territorial behavior involves interactions between at least two individuals, and only the joint phenotype of these individuals can be measured (i.e., the behavior a male exhibits is not independent of its opponents' behavior). The only way to overcome this problem would be to test males against a standard stock, but this would be an onerous task for parent-offspring comparisons.

In tests with flies collected from the orchard, males holding territories tended to be heavier than nonterritorial males. However, there were no significant weight differences between progeny from territorial and nonterritorial males, indicating that body weight did not contribute to the heritable variation in territorial success (Hoffmann 1991). This means that a trait affecting territorial success at the phenotypic level did not contribute to variation in territorial success at the genetic level.

The reason for this finding became clear when heritabilities for body weight were calculated on the basis of parent-offspring regressions. When measurements are made on one parent (the male in this case) and its progeny, a regression of parental values on progeny values will give a coefficient that estimates half the narrow heritability, as discussed in Falconer (1989). Narrow heritability is a measure of the additive genetic variance that determines the similarity between parents and their offspring. The mean regression coefficients for three orchard collections were −0.04, −0.06, and 0.06, with standard errors of 0.05, 0.03, and 0.06 respectively. These values were not significantly different from zero. In contrast, parent-offspring regression on body weight in one of the experiments with the laboratory population gave a coefficient of 0.24, with a standard error of 0.06. This coefficient differed significantly from zero and suggested a heritability of about 50%.

These results mean that there is little heritable variation for weight in the field but some heritable variation in the laboratory. Thus the field environment generated variation in weight among the males that affected their territorial success, but factors unrelated to weight were responsible for the transmission of differences in territorial success to the next generation. Such findings highlight the difficulty of inferring a likely direction of evolutionary change from purely phenotypic data. The phenotypic results alone would predict a change in body size in response to selection for increased aggression, whereas no such change is predicted once the genetic data are taken into account.

Artificial Selection

Directional selection provides an alternative way of examining genetic variation within a population, and the above experiments indicate that territorial success should respond to selection in the laboratory. Males were selected for increased territorial success over 20 generations in two populations of *D. melanogaster* originating from Melbourne and Townsville, Australia (Hoffmann 1988). Cages were set up each with six males that could be individually distinguished because they had been marked on the thorax with different colors of paint. Males holding territories were identified and mated to females to initiate the next generation. Two lines were independently selected from each population in this

Table 9.3 Number of cages in which males from selected (S1, S2) or control (C1, C2) lines were territorial after 20 generations of selection

Lines Compared	Melbourne Lines		Townsville Lines	
	Selected	Control	Selected	Control
S1 vs. C1	14	1	16	4
S1 vs. C2	17	2	18	1
S2 vs. C1	13	1	18	0
S2 vs. C2	16	1	16	3

Source: Adapted from Hoffmann 1988.
Note: Each cage was set up with two males from a control line and two males from a selected line.

manner. Two control lines were also maintained for each population by setting up males and females in pairs so that mating success was not influenced by aggressive interactions between males. It is important to replicate selected and control lines in this type of experiments because divergence between lines might occur in the absence of selection as a consequence of genetic drift. Differences between replicate lines can be used to monitor the effects of genetic drift.

Selection led to a rapid divergence between selected lines and control lines, as is apparent from pairwise comparisons between lines after 20 generations of selection (table 9.3). The heritability of a trait can be estimated from the speed of the selection response, but this was not possible for territorial success for the reasons already discussed. Nevertheless, this experiment indicates that genetic factors had a large effect on the tendency of males to behave aggressively. Because few of the control males held territories in the pairwise comparisons, selection had produced a situation in which most males from the selected lines followed a different strategy than those from the control lines. This suggests that territorial behavior can evolve rapidly in populations.

The selection response in both populations was manifested as an increased tendency for intruders to escalate against territory residents, as well as an increased ability of males to win escalated encounters (Hoffmann 1988). Body weight (or wing length) did not increase as a consequence of selection because there were no detectable weight (or length) differences between males from the selected and control lines. This suggests that factors other than size contributed to variation in territorial success at the genetic level, which is consistent with the results of the parent-offspring comparisons. Size is therefore not the only trait influencing the fighting ability of males.

One question arising from the selection experiment is whether the selected genes also control territorial success when flies are cultured under different environmental conditions. Evidence from bristle number experiments in *Drosophila* suggests that different genes may be involved in different environments (Schnee and Thompson 1984), but equivalent data are not available for behavioral traits. I therefore compared the territorial success of selected and control lines under conditions different from those used during the selection process, by culturing the lines on fruit (apple) at a high larval density and at a high temperature. The results are summarized in figure 9.1 and indicate that differences between the lines were reduced somewhat by these treatments (environments B, C, D, E) compared with the conditions under which the lines were selected (A). However, the changes were relatively small, suggesting that the selected genes influence territorial success when flies are cultured under a range of conditions.

These experiments provide a starting point for understanding phenotypic variation in territoriality in *Drosophila,* but major gaps remain. Perhaps most important, no attempt has yet been made to test the behavior of strains with different levels of territorial success in a field situation. I have found that *D. melanogaster* will defend a range of resources in the field, such as damaged areas of fruit or areas of fruit not covered with mold (unpublished observations). It should be possible to release flies from laboratory strains in the field and monitor their behavior on such resources.

The genetic approaches described here are feasible in animals other than *Drosophila.* Artificial selection experiments or strain comparisons have been used to demonstrate genetic variation in agonistic behavior in a range of organisms, such as mice (van Oortmerssen and Bakker 1981), fish (Farr 1983; Francis 1984; Taylor 1988; Bakker 1986), quail (Bernon and Siegel 1983), and lobsters (Finley and Haley 1983). Regressions between individuals from the field and their laboratory-reared offspring have been used to demonstrate a heritable component to fighting ability in scorpionflies (Thornhill and Sauer 1992).

Assessing Costs of Territorial Aggression

Results from the above experiments indicate heritable variation for territorial success, implying that levels of territorial aggression can evolve rapidly within populations. A population's level of aggression should therefore reflect the costs and benefits of territorial defense in the environment it occupies. The presence of heritable variation also implies that individual flies in a population do not all follow the same "optimal" genetically based strategy. The persistence of more than one strategy in a population suggests that natural selection may not always favor the same phenotype, because if it did, a reduction in genetic variance would be expected. Different strategies may persist if costs and benefits vary so that one strategy is favored in one situation and an alternative strategy is favored in a different situation. This may occur via frequency-dependent selection when a strategy is favored when it is rare but not when it is common (e.g., Gadgil 1972), or via selection in heterogeneous environments when a strategy is favored in one environment but not in another environment (e.g., Hedrick 1986; Gillespie and Turelli 1989).

These costs and benefits need to be evaluated at the genetic level rather than at the phenotypic level if the above questions are to be addressed. Costs determined from phenotypic studies may reflect genetic costs, but this is not always the case. For example, consider the association between body size and territorial success discussed above. This association may form the basis of a phenotypic cost because large males have higher territorial success but longer development times. However, as discussed above, variation in size in the field may be largely due to environmental factors rather than genetic factors. This means that the phenotypic trade-off between development time and territorial success will not be evident at the genetic level. Thus, environments that select for shorter or longer development time will probably not influence levels of territorial aggression.

Mating Costs

Costs may occur when the effects of territorial defense on mating success depend on the environment. Territorial males could have lower mating success than nonterritorial males if the time spent defending a territory interfered with their courtship of females. Comparing the mating success of genotypes differing in territorial behavior under a range of situations provides a way of evaluating such costs. We therefore characterized the mating success of males selected for increased territorial success and males from control lines under a range of conditions (Hoffmann 1988; Hoffmann and

Cacoyianni 1989). Selection will have accumulated genes increasing territorial success, and comparisons between males from the selected and control lines can thus be used to investigate the effects of these genes on mating success. We were particularly interested in situations in which control males might have a mating advantage over selected males because these might help to explain the relatively low levels of aggression in stocks recently established from the field as well as the persistence of heritable variation.

Results from these mating comparisons can be found in Hoffmann (1988) and Hoffmann and Cacoyianni (1989), and are summarized in figure 9.2. In situation A, two males from a selected line and two males from a control line were held in cages with a single food cup and with two females that had already been inseminated. Receptivity of *D. melanogaster* females is drastically reduced after mating, so the inseminated females did not readily remate. Selected males had an enormous mating advantage in this situation, and most of the matings were obtained by territory residents. In situation B, 200 males and 200 females from a selected line and a control line were released in large population cages with eight large food cups, many of which were defended. The number of males per food cup (50) was therefore much higher than in A. Females in the cages had been previously inseminated to reduce receptivity, and most matings occurred on the food cups. A mating advantage for males from the selected lines was apparent in this situation, but it was smaller than in A.

We found several situations in which males from selected lines did not have a mating advantage (Hoffmann and Cacoyianni 1989). These included the situation in which receptive virgin females were introduced into cages with a food cup and two selected and two control males (C), as well as cages where there was no food cup or territorial defense (D). Selected males also did not have a mating advantage in cages set up with 20 males and 10 food cups even though females in these cages had been previously inseminated (E). This suggests that territorial males do not have a mating advantage when defensible resources are abundant or when females are highly receptive.

Control males seemed to have a small mating advantage in one case (F). These cages contained one food cup, and were set up with five selected males and one control male. Control males had a marginally significant mating advantage in two experiments carried out with the Melbourne lines and one of two experiments with the Townsville lines (table 9.4). These results suggest that males with a

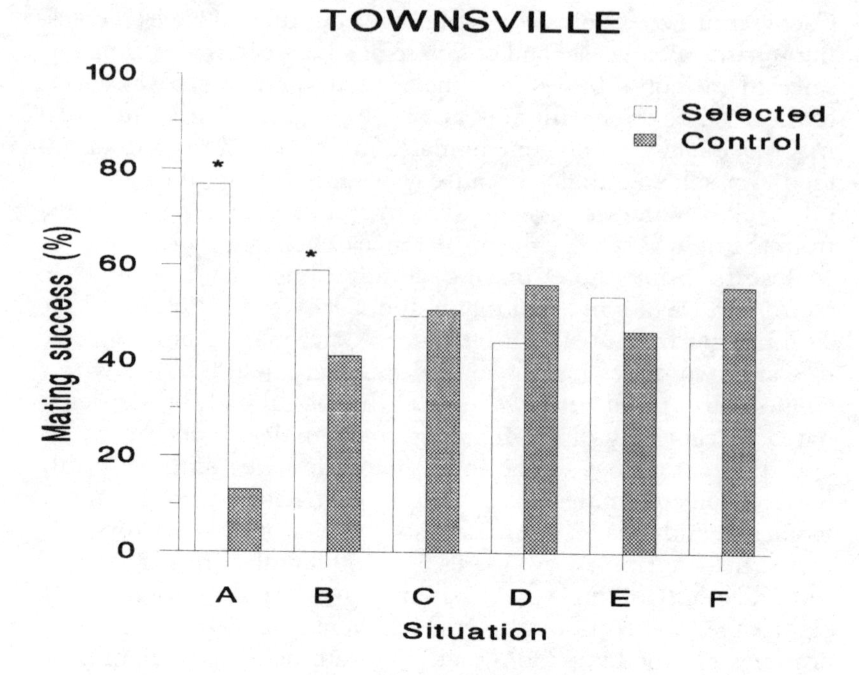
TOWNSVILLE
100
80
60
40
20
0
Mating success (%)
Selected
Control
*
*
A
B
C
D
E
F
Situation

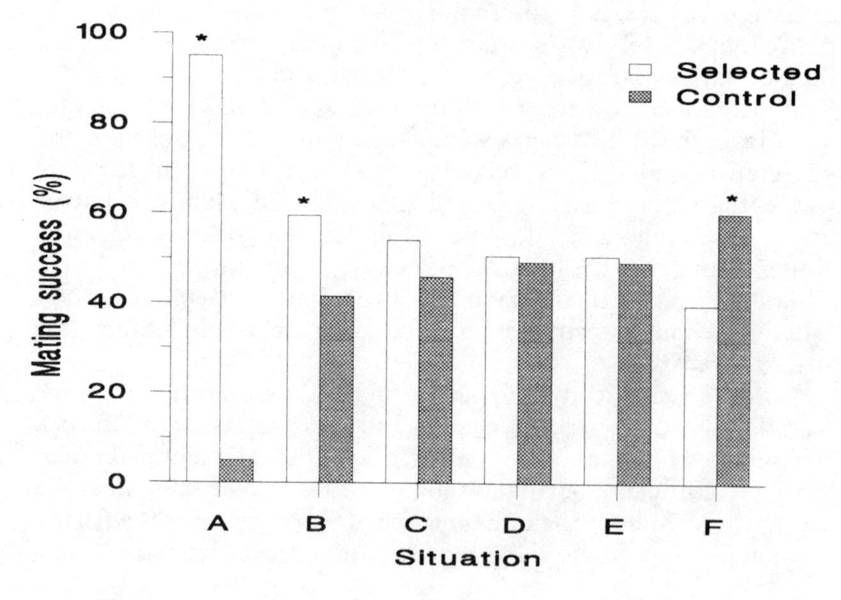
MELBOURNE
100
80
60
40
20
0
Mating success (%)
Selected
Control
*
*
*
A
B
C
D
E
F
Situation

Table 9.4 Number of matings obtained by males from selected and control lines when most males in a cage were from a selected line. (from Hoffmann and Cacoyianni 1989)

	Selected Lines		Control Lines		
	Observed	Expected	Observed	Expected	*P*
Townsville					
Experiment 1	107	115.0	31	23.0	<.10
Experiment 2	84	84.2	17	16.8	NS
Melbourne					
Experiment 1	93	100.8	28	20.2	<.10
Experiment 2	72	78.3	22	15.7	<.10

Source: A. A. Hoffmann and Z. Cacoyianni, 1989, Selection for territoriality in *Drosophila melanogaster*: Correlated responses in mating success and other fitness components, *Anim. Behav.* 38:23–34. Courtesy Academic Press.
Note: P values are for chi-square tests comparing the observed and expected number of matings in each experiment. Cages contained five males from a selected line and one male from a control line.

low level of territorial aggression are favored when they are rare, and that the relative fitness of males with different strategies is frequency-dependent.

These results can be interpreted in terms of genotype by environment interactions, which occur when phenotypic differences between genotypes depend on the environment (see Boake, chap. 14). In this case, genotypes are represented by the selected and control lines, mating success is the phenotype, and different environments are represented by the situations in which mating success was scored. The data in figure 9.2 indicate genotype by environment interactions because the mating success of males from the selected and control lines is situation dependent and switches in one instance. This means that the mating success of genotypes

Fig. 9.2 Mating success of males from selected and control lines in different situations, expressed as the percentage of matings males obtained: A, small cage with small food cup, females inseminated; B, large population cage with large food cups, females inseminated; C, small cage with small food cup, virgin females; D, virgin females in cage without food; E, large population cage with many small food cups, females inseminated; F, small cage with small food cup, selected males common (relative mating success determined by multiplying matings obtained by control males by 5). A significant difference between selected and control lines is indicated by an asterisk.

measured in one environment cannot be extrapolated to other environments.

Trade-Offs with Other Traits

Costs may arise because genes increasing levels of territorial aggression have deleterious effects on other fitness-related traits. These "trade-offs" can be examined by measuring genetic correlations between aggression and other traits (see Partridge, chap. 6; Roff, chap. 3). If negative correlations exist between traits, they may not evolve independently of one another, and any attempt to account for evolutionary changes in one trait becomes difficult without considering selection on the correlated trait (see Arnold, chap. 2).

We examined genetic correlations between terrritorial success and other traits by scoring correlated responses in the selected lines for several unselected traits (Hoffmann and Cacoyianni 1989). If all of the replicate selected lines have significantly higher or lower scores for an unselected trait than all the replicate control lines, genes controlling the selected character are also likely to influence the unselected character. However, weak genetic correlations may not be detected with this technique because of genetic drift within the lines (Henderson 1989).

To detect trade-offs, lines were scored for fitness-related traits other than mating success. There was no evidence for a trade-off between viability and territorial success: selected and control lines had the same viability regardless of whether larvae were cultured under high- or low-density conditions (Hoffmann and Cacoyianni 1989). This contrasts with the findings of other workers that genes increasing mating success are associated with reduced viability, at least in situations where mating success is scored using virgin females rather than previously inseminated females and where there is no territorial defense (Partridge 1980; Taylor, Pereda, and Ferrari 1987; Wilkinson 1987). Our finding that territorial success is closely related to mating success in some situations (figure 9.2) suggests that the detection of a trade-off between mating success and viability depends on the way that mating success is measured.

We did find suggestive evidence for a trade-off between territorial success and male longevity; mortality was significantly higher in cages with males from selected lines than in cages with males from control lines in one situation in which territorial behavior was expressed (Hoffmann and Cacoyianni 1989). In contrast, lines had the same mortality when males were held individually in cages so that there were no territorial interactions. Thus one cost of an increased level of territorial aggression may be a decrease in life span.

It should eventually be possible to relate differences in territorial aggression between natural populations to the types of selection pressures they have experienced. There are genetic differences in territorial success between natural populations of *D. melanogaster* from different geographic locations (Hoffmann 1989), but the selective factors responsible for this variation are not known. One problem with geographic comparisons is that genetically based differences in body size between populations can influence territorial success (Hoffmann 1989), in contrast to size variation within populations. Genetic divergence for size at the geographic level will be influenced by selection pressures unrelated to territoriality, making geographic variation in territorial success difficult to interpret. Nevertheless, territorial defense is not observed on some resources (e.g., Taylor and Kekic 1988), and this should enable the identification of situations in which nearby *D. melanogaster* populations with similar mean body sizes have been exposed to different selection pressures for territorial aggression.

Summary

D. melanogaster males from a population show different levels of territorial aggression. Some of this variation is associated with the age of males and their prior experience. When individuals of a similar age are compared, success in territorial encounters is associated with residency and body weight.

Territorial success can be readily altered by artificial selection. Genetic factors influencing territorial success are not associated with body weight. Parent-offspring comparisons indicate that genetic factors can also contribute to variation among field-collected flies. Males from lines selected for increased territorial aggression have greater mating success in some situations but not in others. Increased territorial success seems to be associated with a reduction in longevity.

Acknowledgments

I thank Chris Boake, Susan Riechert, and an anonymous reviewer for constructive criticisms of an earlier version of this chapter.

10

Genetic Analysis of Cannibalism Behavior in *Tribolium* Flour Beetles

Lori Stevens

An adult and eggs of the confused flour beetle, *Tribolium confusum.*

The evolution of behavior, like the evolution of any trait, is determined by selection on the trait and the genetic basis of the trait. Thus, in order to understand the evolution of a behavior, one needs to understand the genetic basis of phenotypic differentiation. This chapter examines the genetic basis of differences in cannibalism behavior between four genetic strains of the confused flour beetle, *Tribolium confusum.*

In this chapter, previously published experiments using four cannibalism strains of *T. confusum* are reviewed with an emphasis on the application of quantitative genetic methodologies to the study of behavior (Stevens and Mertz 1985; Stevens 1986, 1989, 1992). The first section describes the cannibalism strains. A genetic analysis of cannibalism behavior follows, in which three basic questions are considered: (1) Is the between-strain phenotypic variance in cannibalism genetic in origin? (2) How many genes are involved in the genetic differences between strains? (3) What are the genetic correlations among the several forms of cannibalism studied? These are questions that may often be asked by behaviorists concerned with interpopulational phenotypic differentiation and its evolution.

In answer to the first question, the phenotypic differences between the populations are in fact genetic, as compared with environmental, in origin. The answer to the second question, the number of genes affecting a character, tells us about the potential phenotypic distribution of the trait. With two alleles per locus and allele frequencies of 0.5 at each locus, the proportion of a population with each of the extreme phenotypes is $(^1/_2)^n$, where n is the number of loci. For a trait controlled by one locus and two equally frequent alleles, the proportion of individuals with extreme high or low phenotypes is 0.25. As the number of genes controlling a trait increases, fewer individuals exhibit extreme phenotypes. Not only is the proportion of individuals in a phenotypic class affected by the number of loci controlling a trait, the number of phenotypic classes is also a function of the number of loci. If we assume all loci have equal effects and no dominance, the number of phenotypic classes is $(n + 2)$. Note that the number of phenotypic classes increases linearly with the number of loci, whereas the proportion of individuals in a particular phenotypic class increases exponentially. The final question considered is the degree to which the three cannibalism behaviors studied are genetically correlated. This question is of interest because it determines the extent to which the traits are able to evolve independently and the rate of response to selection. If two traits show a high degree of genetic correlation, selection on one trait will produce a correlated response in the other. Selection for similar cannibalism rates (e.g., high adult and high larval cannibalism of eggs) will be facilitated by a positive genetic correlation, and inhibited by a negative correlation. Similarly, selection for different cannibalism rates (e.g., high adult cannibalism of eggs and low larval cannibalism of eggs) will be inhibited by a positive genetic correlation and facilitated by a negative correlation.

In the final section, the evolution of social behavior in general is considered. Although my experiments constitute a genetic analysis of cannibalistic behavior, the work can be viewed as a model system for studying the evolution of social behavior, i.e., the evolution of interactions among individuals. Behaviorists often focus on the benefits (or costs) of traits at the individual level (Krebs and Davies 1991, assuming that interference behaviors with little or no benefit to the perpetrator are absent (or rare) and that there is a single optimum rate of expression of interference behaviors that may be environmentally mediated. Hamilton (1964a,b, 1970) predicted that interference behaviors such as cannibalism should be selected against in the absence of a benefit to the individual, but when such a benefit exists, the behavior should evolve to a single environmentally dependent optimum. A single optimum is also predicted by a polygenic model with additive genetic effects (Aoki 1982; Crow and Aoki 1982) and by models of frequency-dependent selection of quantitatively inherited traits (Harper 1989). However, social behaviors, which by definition are performed within conspecific groups, are likely to have consequences at multiple levels of population structure. Empirical studies of cannibalism in *T. confusum* (Park, Mertz, and Petrusewicz 1961; Park, Leslie, and Mertz 1964; Park et al. 1965; Stevens and Mertz 1985; Stevens 1986, 1989) suggest that group-level effects and population structure are important in the evolution of cannibalism. When group- and individual-level effects are important, studies that emphasize individual selection and ignore higher levels of selection are inadequate. Theories that incorporate population structure and interactions among individuals, such as Wright's shifting balance theory (Wright 1977) or Griffing's theory of natural selection incorporating interactions among individuals (Griffing 1981) are more appropriate for studying the evolution of social behavior. An appropriate approach to studying behavior would be to partition selection into group- and individual-level components using a methodology such as contextual analysis (Heisler and Damuth 1987; Goodnight, Schwartz, and Stevens 1992).

THE "CANNIBALISM" STRAINS OF *T. CONFUSUM*

Flour beetles are holometabolous insects with egg, larval, pupal, and adult life stages. For this species, the egg stage lasts about 5.5 days, the larval stage about 20 days, and the pupal stage approximately one week (Park, Mertz,and Petrusewicz 1961). The median time to adult death is 250 days posteclosion (Mertz, Park, and Youden 1965). Cannibalism in *T. confusum* involves mobile stages (adults and larvae) eating quiescent stages (eggs and pupae). Three cannibalism traits, adults eating eggs, larvae eating eggs, and adults eating pupae, are responsible for the bulk of cannibalistic predation in this species and thus were the focus of my studies.

The four cannibalism strains, bI, bII, bIII, and bIV, were each derived during the mid-1950s through 3–4 episodes of brother-sister mating from a single 25-year-old laboratory strain known as b+ (Park, Mertz, and Petrusewicz 1961). The purpose of the inbreeding was to obtain novel strains with differing competitive characteristics. Initially there were 158 such stocks, and 4 were chosen on the basis of visual examination to exhibit productivity differences. Subsequent assays confirmed that the strains exhibit a tenfold differences in equilibrium population size and have similar birth and death rates and development times (table 10.1). However, cannibalism rates are inversely correlated with population size (tables 10.1 and 10.2); thus bI, bII, bIII and bIV were dubbed the "cannibalism" strains.

The strains have been relatively stable for over 60 generations (Stevens and Mertz 1985). Only three of the twelve cannibalism rates measured changed significantly over this time ($P < .05$): the low-cannibalism bI became even less cannibalistic for adults eating eggs, the intermediate-cannibalism bII also became less cannibalistic for adults eating eggs, and the high-cannibalism bIV strain became even more cannibalistic as a larval egg eater. Thus, although Hamilton and others predict one optimum (Hamilton 1964a,b, 1970; Aoki 1982; Crow and Aoki 1982; Harper 1989), several cannibalism rates were stable and there was no tendency for the strains to converge on a single "optimum" cannibalism rate (see also Roff, chap. 3).

Pattern of Inheritance of Cannibalism Behavior

In my study of cannibalism behavior, I documented that the between-strain cannibalism differences indeed have a genetic basis and examined their pattern of inheritance (Stevens 1986, 1989). This was done by crossing the bI (low) and bIV (high) cannibalism strains (dubbed P_1 and P_4, emphasizing their role as parental strains) and examining the expression of cannibalism in the F_1, F_2, backcrosses, and parental strains. The pattern of inheritance—that is, additive effects, dominance, sex linkage, and maternal

Table 10.1 Life history parameters of the four cannibalism strains of *T. confusum*

Strain	Population Size[a]	Eggs/3 Days[b]	% Hatch[b]	Development Time (days)[b]			Age at Death (days)[c]	
				Eggs	Larvae	Pupae	Males	Females
bI	467	28.8	70.5	5.37	18.96	6.98	306	321
bII	291	27.1	71.5	5.43	20.11	6.48	256	349
bIII	82	24.1	71.0	5.63	18.89	6.78	363	288
bIV	40	31.2	65.0	5.60	19.92	6.72	348	357

[a]Average number of larvae, pupae, and adults in 16-week-old populations initiated with 6 males and 6 females in 4 g of medium. (Data from Park, Mertz, and Petrusewicz, 1961).
[b]Data from Park, Mertz, and Petrusewicz, 1961.
[c]Data from Mertz, Park, and Youden, 1965.

Table 10.2 Cannibalism rates of the four cannibalism strains of *T. confusum*

	Larvae Eating Eggs[a]				Adults Eating Eggs[b]				Adults Eating Pupae[c]			
Strain	1962	1981	Stable[d]	σ^2_G[e]	1962	1981	Stable[d]	σ^2_G[e]	1959	1981	Stable[d]	σ^2_G[e]
bI	50.5	76.0	Yes	No	22.3	9.5	No	Yes	10.2	11.8	Yes	Yes
bII	142.5	141.0	Yes	No	69.3	60.0	No	Yes	10.9	9.8	Yes	No
bIII	214.0	234.0	Yes	Yes	76.0	79.8	Yes	Yes	20.9	17.3	Yes	No
bIV	256.3	288.5	No	Yes	73.2	70.0	Yes	Yes	21.8	22.5	Yes	Yes

Sources: 1959 and 1962 data are from Park et al. 1965; 1981 data are from Stevens and Mertz 1985; σ^2_G (genetic variation) values are from Stevens 1989.

[a]Mean total number of eggs eaten by 50 larvae over the age interval 0–22 days.

[b]Mean number of "marked" eggs eaten in 48 hours by 25 female and 25 male adults approximately 20 days of age.

[c]Mean number of pupae eaten by 10 female and 10 male adults during two 1-week assays.

[d]Values in 1959 or 1962 and those in 1981 are not significantly different: $P < .05$.

[e]Inbreeding experiment detected genetic variation, σ^2_G.

effects—can affect the potential of a population to respond to selection (Fisher 1930; Falconer 1989); therefore, I partitioned the between-cross variation in cannibalism into additive and nonadditive gene effects (see, for example, Mather and Jinks 1977; Falconer 1989). In addition, the crosses were examined for evidence of sex linkage and maternal effects.

Studies of cannibalism in *Tribolium* have compared different species, sexes, life stages, and genetic strains using groups of predators and prey (Rich 1956; Park et al. 1965; Mertz and Cawthon 1973; Cawthon and Mertz 1975; Stevens and Mertz 1985; Craig 1986). My studies were conducted using vials containing flour/yeast medium in darkened incubators regulated at 29° C and a relative humidity of 70%. This environment had constant spatial, climatic, and nutritive components and was easily replicated (Park, Leslie, and Mertz 1964). Quiescent prey were distributed throughout the medium and motile predators were added at the surface. After a specified time, which depended on the life stages being assayed, the contents of the vials were sifted and the numbers of predators and prey were counted and recorded. Although assays of prey vulnerability detected no among-strain differences in susceptibility to cannibalism, the prey were always from the ancestral strain, b+, to ensure that treatment differences were due solely to differences among predators.

Larval cannibalism of eggs was assayed over the entire larval life stage using 50 predators (Stevens 1986, 1989). Every 48 hours, the remaining eggs were counted and replaced with 100 new eggs. This procedure was continued for 22 days, by which time most of the larvae had pupated. There were six replicates per treatment. Assays of adult cannibalism of eggs are confounded by the fact that adults oviposit and cannibalize eggs during an assay period. Therefore, I estimated the rate of egg cannibalism by adults by introducing a known number of dyed red eggs and counting the remainder at the end of the assay period (Rich 1956). Fifty adults (sex ratio unity) 30 days posteclosion and 100 dyed red eggs for 48 hours were used in each of six replicates per treatment. Adult cannibalism of pupae was assayed with 20 predators (sex ratio unity) and 20 prey for 7 days. The assay was performed when the adults were 1 week old and again 2 weeks later using the same adults (then 4 weeks old). The data were the total numbers of pupae eaten during both assays. There were six replicates per treatment. However, five replicates (four F_2s and one F_1) were lost.

For each type of cannibalism, assays of the parental and F_1 beetles

Table 10.3 Components of the mean of a quantitative character

	Mean Phenotype		
Group	*m*	*d*	*h*
P_1	1	−1	0
P_4	1	1	0
F_1	1	0	1
F_2	1	0	0.5
B_1	1	−0.5	0.5
B_4	1	0.5	0.5

Source: Mather and Jinks 1977.

were used to estimate three parameters: the midpoint between the two parents, m (also referred to as the midparent); the additive effect, d; and the dominance deviation, h. The mean value of any group (parental, F_1, F_2, or backcross) can be described in terms of these components, as outlined in table 10.3. The joint scaling test (Cavalli 1952; Mather and Jinks 1977) was used to estimate the genetic parameters. The parameters are estimated from the means of all available groups. Expected group means are calculated from the estimates of the components. The additive-dominance model is then tested by comparing the observed means with their expected values.

For each group, an equation can be formulated equating the components of the model with the group mean. Each equation is then weighted by the inverse variance of the group mean with the result that the contribution of a cross to the estimation is determined by its precision. For example, with F_1 populations the equation is:

$$\bar{x}_{F_1}/\mathrm{var}(x) = (m + d)/\mathrm{var}(x).$$

When more than three populations are available, the equations are combined to give three equations whose solution is a least squares estimate (Mather and Jinks 1977, 37–40). By matrix inversion, the three equations give estimates of the parameters, along with the standard error of the estimate, which can be used to determine whether an estimate is statistically different from zero.

The adequacy of the additive-dominance model was tested

using the minimum chi-square method (Kempthorne 1957): the deviation squared is divided by the variance of the mean and summed over all crosses. The degrees of freedom for this test is (number of group means −3), so the goodness of fit cannot be determined with fewer than four crosses; in other words, the two parental types and the F_1 are sufficient to estimate the parameters but not to test the goodness of fit. For adult cannibalism of eggs and larval cannibalism of eggs, only the parental and F_1 beetles were assayed, and the fit of these parameter estimates to the model was not tested. Assays with pupal prey are less labor-intensive, and three additional groups (the F_2 and two backcrosses) were assayed. With the additional degrees of freedom for adults eating pupae, the joint scaling test was used to test the fit of an additive-dominance model. Lack of fit of the additive-dominance model would be evidence for sex linkage or maternal effects. For larvae eating eggs and for adults eating eggs, additional assays using backcrosses were done to test for sex linkage (Stevens 1986). The results of these two experiments could not be combined to use the joint scaling test because different assay conditions were used and because environmental effects influenced the expression of adult cannibalism of eggs (see below).

The results are shown in figure 10.1. In this figure, the *x*-axis is the average proportion of bIV alleles in a particular group. In the bI strain, none of the alleles are from the bIV strain; in the bIV strain, all of the alleles are from bIV. In the F_1, half of the alleles are from bIV, and in the F_2, on average, half the alleles in an individual are originally from bIV. In a backcross to bIV, an average of 3/4 of the alleles are from bIV, and in a backcross to bI, an average of only 1/4 of the alleles are from bIV.

For larvae eating eggs, the estimate (and 95% confidence interval) of the midpoint between the two parents, *m*, (fig. 10.1a) is 139.9 (139.7–140.2); of the additive effect, *d*, 109.6 eggs (109.3–109.9); and of the dominance deviation *h*, −2.75 eggs (−2.76–−2.74). The dominance estimate is essentially zero, as it is only 2.5% of the additive effect. Additional assays using the reciprocal backcrosses to each parental strain indicate that the character is not sex-linked (Stevens 1986).

For adults eating eggs, (fig. 10.1b), the estimate (and 95% confidence interval) of *m* is 33.8 (31.9–34.8); of *d*, 22.7 eggs (21.9–23.5); and of *h*, 3.4 (2.8–4.0). All three estimates are significantly different from zero. The dominance deviation is small relative to the additive effect, about 15%. Results from additional assays of reciprocal

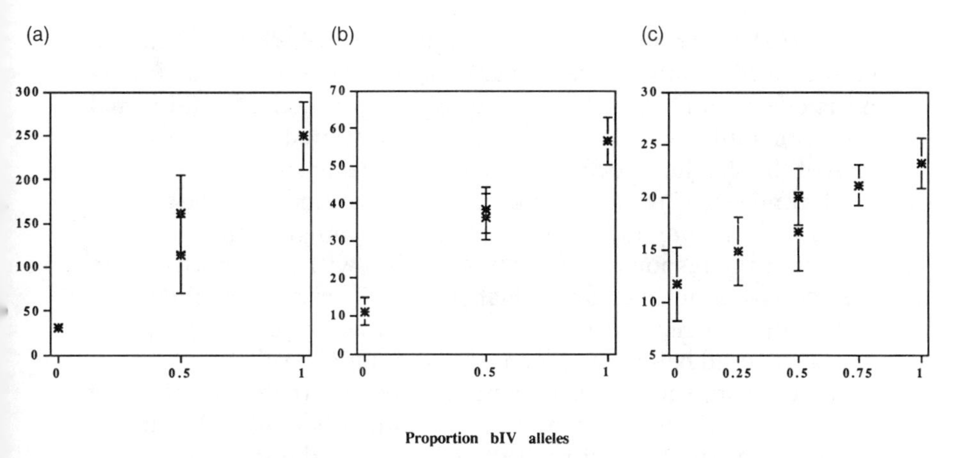

Fig. 10.1 Cannibalism rates from the genetic crosses: (*A*) larvae eating eggs, (*B*) adults eating eggs, and (*C*) adults eating pupae. Means ± SD are shown.

backcrosses to each parental strain indicate that the character is not sex-linked (fig. 10.1b, proportion bIV alleles 0.25 and 0.75; Stevens 1986). The unexpectedly low value for backcrosses to bI (proportion bIV alleles 0.25) is likely to be the result of some uncontrolled environmental difference between experiments. Such environmental effects are likely to affect the precise estimate of a parameter; however, they are not likely to affect the relative ranking of the groups. Thus the estimates of *m*, *d*, and *h* vary between experiments; however, the relative positions of P_1, P_4, and F_1 do not.

For adults eating pupae (fig. 10.1c), the estimate (and 95% confidence interval) of *m* is 16.6 (9.6–23.7); that is, 42% of the prey were cannibalized. The population means range from 11.7 (29%) for bI to 23.2 (58%) for bIV. The estimate of *d* is 6.2 pupae (2.4–9.9), and of *h*, 2.7 (−1.2–6.6), or 44% of the additive effect. The estimates of the midparent and the additive effect are statistically different from zero, but the dominance deviation is not. The minimum χ^2 test (Mather and Jinks 1977, 37–40) reveals that the additive-dominance model is adequate: sex linkage and maternal effects are nonsignificant ($\chi^2 = 3.8$, df = 3, $P = .28$).

In summary, crosses between the high and low cannibalism strains (bI and bIV) confirm that the between-strain differences (see table 10.1) have a genetic basis (Stevens 1986, 1989). The between-strain cannibalism differences are due, on average, to additive genetic effects, and the traits are autosomally inherited. Additional assays of adult cannibalism of eggs demonstrated that the expression of these behaviors is influenced by the environment.

Natural selection at the individual level acts on additive genetic variance (Falconer 1989), and Fisher's fundamental theorem predicts that the response to natural selection will be proportional to the additive genetic variance (Fisher 1930). These results show that cannibalism behavior can be regulated by additive genetic effects and thus could be selected for at the individual level.

In addition, these results demonstrate that there was additive and/or epistatic genetic variance for cannibalism in the b+ ancestral strain at the time of the inbreeding, after the strain had been under laboratory culture for some 25 years or 75 generations. The cannibalism rates of b+ are intermediate to those of bI and bIV, suggesting that individual selection favored intermediate cannibalism in the b+ ancestral strain.

One Gene of Large Effect or Polygenic Control?

Because such large differences in cannibalism were achieved after only 3–4 generations of inbreeding, it seemed possible that the between-strain differences might result from a single gene. In addition, because the high-cannibalism strain exhibited high rates for all three types of cannibalism and the low-cannibalism strain exhibited low rates for all three types of cannibalism, it was possible that the three types of cannibalism were controlled by the same gene(s). Pleiotropy, the condition in which a gene has more than one phenotypic effect, is a common property of major genes, but its role in quantitative genetics has received little consideration (Barton and Turelli 1989; Falconer 1989). The number of genes and genetic correlations are of interest because they affect the number of cannibalism phenotypes and genotypes possible in a population. Thus the number of loci and the genetic correlations provide information about the potential for evolution both within the cannibalism strains and in the ancestral population.

Estimating the Number of Loci: Methods

Statistical methods. Estimates of the number of loci controlling a metric character are complicated by factors such as linkage and unequal gene effects. Consequently, one actually estimates the equivalent number of freely segregating genetic factors. Castle (1921) and Wright (1952, 1968) developed a methodology for estimating the number of "segregating factors" between two homogeneous parental lines that was generalized to a method for heterogeneous parental populations (Lande 1981a). Discussions of similar methodologies can be found in several papers (Pearson 1904; Fain 1978; Mayo, Hancock, and Baghurst 1980; Morton 1982; Mitchell-Olds and Rutledge 1986; Mitchell-Olds and Bergelson 1990). Recently, ways of correcting inherent biases in Wright's methods have been described, along with ways to estimate the sampling variance (Zeng, Houle, and Cockerham 1990; Zeng 1992). These authors demonstrate that the expected number of loci without linkage is estimated to be one-third the actual value, and caution that "even in the best of circumstances, information . . . is very limited and can be misleading." The best of circumstances include populations that differ by, say, 10 phenotypic standard deviations, large (>200) sample sizes, and replicating estimations (Zeng 1992). An alternative strategy is to estimate quantitative trait loci using molecular markers (Edwards, Stuber, and Wendel 1987; Weller, Soller, and Brody 1988). However, this approach is limited to species with genomes suitably mapped at the molecular level.

Wright's technique for estimating the number of segregating factors involves comparing the variance of the cannibalism phenotype in segregating generations (F_2 and backcrosses) with the variance in the parental strains and their F_1 hybrids. The estimate assumes: (1) that the environmental variance is equal in the parental, F_1, F_2, and backcross populations, (2) that all loci have equal effects, (3) no linkage, (4) no dominance, and (5) no genetic interactions (e.g., no epistasis). The number of segregating factors will be underestimated if the environmental variance is smaller in the F_2 and backcross populations. Violation of assumptions 2–5 will also result in an underestimate. Thus this method estimates the minimum number of loci. In the absence of recombination, the haploid number of chromosomes ($n = 9$ in *T. confusum*) sets an upper limit on the estimate of the number of loci.

I used Lande's (1981a) generalization of Castle and Wright's original methodology (Castle 1921; Wright 1952, 1968) to estimate the number of loci contributing to the cannibalism differences between the high- and low-cannibalism strains. The components of the phenotypic variance of a quantitative character for the parental, F_1, F_2, and backcross populations are presented in table 10.4.

Table 10.4 Components of the phenotypic variances of a quantitative character

Group	Variance
P_1	$\sigma^2_{G_1} + \sigma^2_E$
B_1	$0.75\sigma^2_{G_1} + 0.25\sigma^2_{G_4} + 0.5\sigma^2_S + \sigma^2_E$
F_1	$0.5\sigma^2_{G_1} + 0.5\sigma^2_{G_4} + \sigma^2_E$
F_2	$0.5\sigma^2_{G_1} + 0.5\sigma^2_{G_4} + \sigma^2_S + \sigma^2_E$
B_4	$0.25\sigma^2_{G_1} + 0.75\sigma^2_{G_4} + 0.5\sigma^2_S + \sigma^2_E$
P_4	$\sigma^2_{G_4} + \sigma^2_E$

Source: R. Lande, 1981, The minimum number of genes contributing to quantitative variation between and within populations, *Genetics* 99:541–53.

The variance in each parental population can be divided into genetic, $\sigma^2_{G_1}$ and $\sigma^2_{G_4}$, and environmental, σ^2_E, components. The F_1 population is expected to have half the genetic variance of each parental strain and an equivalent amount of environmental variance. The F_2 has half the genetic variance of each of the parental strains as well as a component of variance, the segregation variance, σ^2_S. The segregation variance is the variance resulting from genetic differences between the two parental strains. For example, with one locus, two alleles, and no dominance, 25% of the F_2 would be like P_1, 50% would be like F_1, and 25% like P_4. The variance from this distribution of progeny in the F_2 is the segregation variance. It will be largest when only one locus is responsible for the differences between strains, and it will decrease as the number of loci increases. An example of calculating the expected segregation variance for one locus is outlined below.

To estimate the segregation variance one needs to know the variance among individual F_2 progeny. Although previous assays had been conducted with groups of predators, in determining patterns of inheritance it is the phenotypes of individuals that are of interest. Individuals in isolation exhibit very low cannibalism rates, and the between-strain differences were not expressed when individuals were assayed singly. This is another example of the effect of the environment on the expression of cannibalism. Because cannibalism is a group-dependent behavior, assaying F_2 individuals was not feasible for this behavior, so a method of progeny testing was developed (fig. 10.2). Individuals whose phenotype was of interest (designated testor males) were mated to P_4 females and

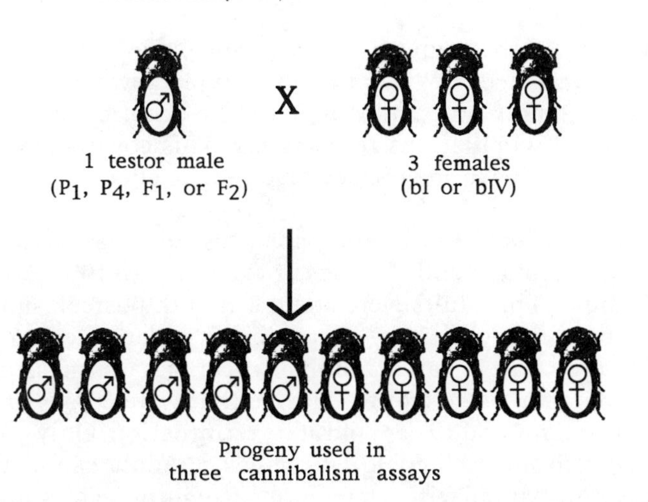

Fig. 10.2 The breeding design used to estimate the number of loci responsible for the different cannibalism phenotypes. Although adult progeny are shown, both adult and larval progeny were assayed.

their progeny were assayed. Testor males (as opposed to females) were assayed because previous assays (see above and Stevens 1986) had detected maternal effects and no sex linkage. This suggested that differences among females might be due to maternal and/or genetic effects, whereas differences among males were likely to result only from genetics. P_1, P_4, and F_1 testor males were also assayed to provide a basis for estimating the segregation variance. The entire experiment was replicated in the sense that the testor males were subsequently mated to P_1 females and a second estimate of the number of loci was made for each trait.

From each testor male different statistically independent groups of progeny were used to estimate the three types of cannibalism. These data were then used to estimate the genetic correlations for the three types of cannibalism. The genetic correlations may arise from pleiotropy (one gene affecting more than one cannibalism phenotype) or linkage disequilibrium. Covariance matrices were calculated using the mean phenotype for each of the three cannibalism traits from a single male's progeny. The covariances have components analogous to the variance components in table 10.4 (Lande 1981a), and the proportion of the covariance due to genetic correlations is:

$$\frac{\mathrm{Cov_S}}{\mathrm{Cov_{F_2}}} = \frac{\mathrm{Cov_S}}{(0.5)\mathrm{Cov_{G_4}} + (0.5)\mathrm{Cov_{G_4}} + \mathrm{Cov_E} + \mathrm{Cov_S}}$$

where Cov_s is the segregating covariance, Cov_E is the environmental covariance, Cov_{F_2} is the phenotypic covariance in the F_2 population, and Cov_{G_1} and Cov_{G_4} are the genetic covariances in the P_1 and P_4 populations respectively. This covariance ratio includes both additive and nonadditive genetic effects.

Assay conditions. Adult cannibalism of eggs was assayed with groups of 10 females and 10 males in 4 g flour with 100 marked eggs for 72 hours. The adults were about 1 month posteclosion. From each cross, five different, statistically independent replicate groups were assayed. Larval cannibalism of eggs was measured over the entire larval stage using 40 larvae that were 3 days old at the start of the assay and 19 days old at its termination. Only one group of progeny from each male was assayed. Adult cannibalism of pupae was measured with 5 female and 5 male predators and 20 unsexed prey for 10 days. In experiments assaying larvae eating eggs and adults eating pupae with the progeny of testor males mated to P_4 females, there were no differences between strains. Therefore, for these experiments I did not estimate the number of segregating factors.

Prior to statistical analysis, the variances were tested for homogeneity and transformations were used when appropriate. Two estimates of the segregation variance were made by considering the components of variance shown in table 10.4:

$$\sigma_S^2 = \sigma_{F_2}^2 - (0.5\sigma_{F_1}^2 + 0.25\sigma_{P_1}^2 + 0.25\sigma_{P_4}^2) \tag{10.1}$$

and

$$\sigma_S^2 = \sigma_{F_2}^2 - \sigma_{F_1}^2. \tag{10.2}$$

If the differences between two strains are due to one gene, then when F_2 males are mated to P_1 females, the possible proportions of bIV genes in the progeny are 0, 0.25, and 0.50; with P_4 females, the proportions are 0.5, 0.75, and 1.0. The probabilities of these proportions are the same as those for the F_2 of any monohybrid cross, 0.25, 0.50, and 0.25 respectively. The predicted number of prey eaten as a function of the proportion of bIV genes is calculated from the linear regression of cannibalism on the proportion of bIV genes in the P_1, P_4, and F_1 populations. The expected mean of the F_2 is calculated, and the expected segregation variance is the sum of the deviates squared, weighted by their probability. The variance is expected to decrease as the number of segregating factors increases because fewer individuals have extreme phenotypes.

Because large sampling variances are associated with these estimates (Zeng, Houle, and Cockerham 1990; Zeng 1992) I did not statistically compare the observed and expected segregation variance.

Estimates of the Number of Loci: Results

Adults eating eggs. The estimate of the number of segregating factors from the assay with bI females is two, and the estimate from the assay with bIV females is three. Although confidence limits are not put on these estimates, the fact that they are similar supports the hypothesis that there are only a few genes responsible for most of the differences between strains. Since a violation of the assumptions of the model leads to an underestimate of the number of genes, and the estimates suggest that at least two genes are involved, it is unlikely that only one gene is responsible for the between-population difference.

Larvae eating eggs. The larvae eating eggs data were natural-log transformed to achieve homogeneity. The estimate for larval cannibalism of eggs with the P_1 females is four segregating factors. A second experiment yielded an estimate of five segregating factors (Stevens 1986). Recall that for this species the haploid number of chromosomes is 9 and that in the absence of recombination, this sets an upper limit on the number of loci. Furthermore, if the assumptions listed above are violated, the number of loci is underestimated. These two factors suggest that at least half the chromosomes of *T. confusum* affect larval cannibalism of eggs.

Adults eating pupae. The estimates of the segregation variance for adult cannibalism of pupae were very different. One estimate (using eq. 10.1) suggests that there are infinitely many genes; the other (using eq. 10.2) suggests that there are 2–3 segregating factors. Two possible causes of the different estimates are: (1) there is a genotype by environment interaction such that the environmental variance is not equal in all groups; (2) the variation within the parental strains is causing an underestimate of the number of factors. Lande (1981a) states that variation within the parental stains will result in an underestimate. Additional experiments (Stevens 1986) detected genetic variation for adult cannibalism of pupae within bI and bIV. It is difficult to interpret these vastly different estimates, and therefore, no conclusions are made about the number of genes for adult cannibalism of pupae.

Table 10.5 Estimates of the fraction of covariance between cannibalism traits that is due to genetic correlations

	Trait		
Trait	LEE	AEE	AEP
LEE	—	.91	.37
AEE		—	.06
AEP			—

Source: L. Stevens, The genetics and evolution of cannibalism in flour beetles, genus *Tribolium,* Ph.D. diss., University of Illinois at Chicago, and L. Stevens, 1989, The genetics and evolution of cannibalism in flour beetles, genus *Tribolium, Evolution* 43:169–79.
Note: Values are Cov_S/Cov_{F_2}; see text. LEE, larvae eating eggs; AEE, adults eating eggs; AEP, adults eating pupae.

Genetic correlations. The range of estimates of the genetic correlations is 0.05–0.91 (table 10.5). The largest estimated correlation is that between larvae eating eggs and adults eating eggs: an estimated 91% of the observed phenotypic covariance between the two types of egg cannibalism may result from genetic correlations. The estimates indicate that 37% of the covariance between larvae eating eggs and adults eating pupae has a genetic basis, whereas only 6% of the covariance between the two types of adult cannibalism is estimated to be due to genetic correlations.

Does Selection Favor an Optimum Cannibalism Rate?

As discussed above, several theoretical models concluded that an interference behavior such as cannibalism should be selected against in the absence of a benefit to the individual, but when such a benefit exists, it should evolve to a single environmentally dependent optimum (Hamilton 1964a,b, 1970; Aoki 1982; Crow and Aoki 1982; Harper 1989). In contrast to this prediction, cannibalism rates in four genetic strains of *T. confusum* were found to be remarkably stable in the absence of artificial selection for over 60 generations of laboratory husbandry (see table 10.2; Stevens and Mertz 1985). Although three of the twelve cannibalism rates measured did change during the 20-year interval, there was no tendency for he strains to converge on a single "optimum" cannibalism rate. This finding suggests there may be multiple stable adaptive peaks for cannibalism and provides empirical evidence supporting Wright's shifting balance theory (Wright 1977).

Evolution of Cannibalism

There are three hypotheses for the stability of the between-strain differences in cannibalism:

Hypothesis 1: There was no genetic variation within the strains.
Hypothesis 2: Stabilizing selection was maintaining at least some of the populations at different adaptive peaks for cannibalism.
Hypothesis 3: Selection on cannibalism was negligible. In the absence of genetic drift, the populations maintained the cannibalism rates measured 20 years earlier.

To test hypothesis 1, the four cannibalism strains and their parental strain were used in an inbreeding experiment that screened each strain for within-strain genetic variation (Stevens 1986, 1989); however, the genetic variation was not partitioned into additive and nonadditive components (see below). Inbreeding rather than a selection experiment or half-sib study was used because multiple strains and traits were characterized. From each strain, ten lines were created and maintained by transferring 30 unsexed adults to 50 g of fresh flour at 8-week intervals for seven generations. This was followed by four generations of full-sib mating. Of the ten lines created from each strain, the two lines with the highest and the two with the lowest population size were assayed for the three cannibalism phenotypes. For the twelve cannibalism traits studied (3 types of cannibalism $\times$ 4 strains = 12 traits) eight exhibited genetic variance (see table 10.2, columns labeled σ^2_G). A response to natural selection can only occur when there is additive genetic variance (σ^2_A), but the inbreeding experiment detected overall genetic variance, a combination of additive and nonadditive components (σ^2_{NA}): $\sigma^2_G = \sigma^2_A + \sigma^2_{NA}$. Thus my results do not refute the possibility that there was no additive genetic variance within the strains. However, in combination with the additive genetic variance between strains described above, the results suggest that there was additive genetic variation within the strains.

For those strains that had genetic variation for a particular cannibalism trait in 1981 (indicated by "yes" in table 10.2, columns labeled σ^2_G), I tried to distinguish between hypothesis 2 and hypothesis 3 (Stevens 1989). Because strains that were not genetically variable in 1981 could have been so in 1960, this biases the interpretation toward hypothesis 1 and against invoking selection or neutrality. To differentiate between selection and drift, I used an F test that compares the observed change in variance between populations to that expected by random genetic drift (Lande 1977; Wade 1984; Stevens 1989):

$$F = \frac{\Delta\sigma^2_{\bar{z}}}{h^2\sigma^2 \frac{t}{N_e}}$$

In this equation, $\Delta\sigma^2_{\bar{z}} = \sigma^2_{\bar{z}\ (1981)} - \sigma^2_{\bar{z}\ (\text{ca. } 1960)}$ (the difference between the variance of the strain means in 1981 [Stevens and Mertz 1985] and that around 1960 [Park et al. 1965]), t is the number of generations, N_e is the effective population size, and $h^2\sigma^2$ estimates the additive genetic variance. Very small values of F (two-tailed $P >$ .95) indicate that the change is less than that expected by drift, that is, each strain is being maintained at an adaptive peak by selection; large values of F (two-tailed $P < .05$) indicate that some or all of the strains are being selected away from where they were around 1960.

I did not explicitly measure all of these variables; however, I present the results as an example of how one could test the idea of one versus multiple optima for a behavior. In contrast to studies estimating the number of loci, tests of selection versus drift for behaviors should yield interpretable data with reasonable sample sizes. Such analyses will also provide exciting information about the evolution of behavior.

N_e was estimated to be 600. Stock populations are changed three times a year and are founded by about 600 premated individuals each generation (Stevens and Mertz 1985). Thus 600 is a minimum estimate of N_e. For adults eating eggs and larvae eating eggs, $t = 57$, and for adults eating pupae, $t = 60$. Using a low estimate of N_e will tend to underestimate F and make it less likely that differences will be interpreted as being due to selection. The among-strain variance from the original assays was used as the estimate of the additive genetic variance. This value is the upper limit on the additive genetic variance. Overestimating the additive genetic variance also will tend to underestimate F and will make it less likely that differences will be interpreted as being due to selection. Thus a high F value is strong evidence that selection occurred.

Strains bI and bII showed no genetic variation for larval cannibalism of eggs (see table 10.2). Strain bIV increased in cannibalism and had genetic variation; bIII also had genetic variation, but no change in cannibalism was seen in this strain. The F test was inconclusive ($F = 0.84$, $P > .05$); indicating that the observed pattern could be the result of either weak selection or random genetic drift in a large population.

All four strains showed genetic variation for adult cannibalism of eggs, and the two low-cannibalism strains bI and bII had significantly lower cannibalism rates in 1981 than in 1962. The other two strains had not changed. If the heritability of this trait is greater than 0.75, the F test is not significant, suggesting that the observed changes result from genetic drift; however, if the heritability is less than 0.75, the F test is significant, suggesting that the observed changes result from directional selection. Heritability estimates are usually below 0.75 (see for example, Falconer 1989, 165, table 10.1) and are highest for morphological characters and lower for characters closely connected with fitness (Mousseau and Roff 1987; Roff and Mousseau 1987). Cannibalism behavior is influenced by environmental factors and as a result may have low heritability. Thus it seems likely that the heritability of adult cannibalism of eggs is less than 0.75 and the F test is significant. This suggests that bIII and bIV were being maintained by selection at adaptive peaks while bI and/or bII were being selected to adaptive peaks. There were no changes in adult cannibalism of pupae, although two of the strains exhibited genetic variation for this trait. The observed lack of change suggests that selection is maintaining the strains at adaptive peaks.

Stability, either due to selective neutrality or due to the diminished effects of genetic drift in large populations, is to be expected when populations are at or near adaptive peaks. With frequency-dependent selection leading to multiple stable adaptive peaks, the selection coefficient is a function of gene frequency. Consequently, selection coefficients are large away from adaptive peaks and small near them (Wright 1977). Because of this a population may randomly drift about a particular peak (Wright 1977). The observation that the stability of the cannibalism rates is a result of drift does not exclude the possibility that the strains are at different peaks.

My data support the assumptions of the shifting balance theory (phases 1 and 2: Crow, Engels, and Denniston 1990). The inbreeding program employed by Park to obtain the cannibalism strains produced genetically differentiated populations (Park, Mertz, and Petrusewicz 1961; Stevens and Mertz 1985; Stevens

1986, 1989). The evidence that the strains are at different adaptive peaks for cannibalism was discussed above (see also Stevens 1989, 1992). Several possible genetic mechanisms for the formation of adaptive peaks have been proposed (Wright 1977, 1978). It is not clear whether the adaptive peaks for cannibalism result from frequency-dependent selection, epistasis, a combination of these processes, or an entirely different mechanism.

Phase 3 of the shifting balance theory requires that populations be dynamic through any combination of migration, extinction, and recolonization, resulting in proliferation of genotypes from populations that are at relatively high adaptive peaks. The observed extinctions of single-species populations of the high-cannibalism strain bIV suggest that interdemic selection through local extinction and/or migration would favor an upper limit to cannibalism (Wright 1960). In a previous study, five of sixteen cultures of the high-cannibalism strain bIV became extinct (Park, Leslie, and Mertz 1964). These observed extinctions may result from high cannibalism rates combined with a long postreproductive life. In a population initiated with four adult pairs, a large F_1 population is produced. Cannibalism by this F_1 population is so intense that there is no subsequent recruitment. The long postreproductive period ensures that no young are around when adults die of senescence. Theory predicts that when cannibalism is so extreme, interdemic selection would oppose it because of the increased chance of extinction, not only as a result of extreme cannibalism, but also as a result of small population size (Wright 1960). Thus, these results support Wright's shifting balance theory and suggest that the role of population structure in the evolution of social behavior needs to be considered.

Griffing (1981) also developed a general theory of natural selection that considers the role of population structure in the evolution of behavior. Griffing's theory is an extension of classic genetic selection theory to include interactions among individuals. His results demonstrate that when individuals within a population interact, individual selection does not maximize population fitness. This theoretical conclusion supports my empirical studies of cannibalism: the populations had genetic variation for cannibalism yet selection did not result in maximum population fitness (i.e., maximum population size), and the populations did not converge on a single optimum cannibalism rate.

Griffing examines several restrictions of the general model to elucidate the conditions that would lead to optimal selection results. One of the modifications restricts interactions to related individuals in a population. This is equivalent to adding population structure to his model. In a structured population, selection is more efficient at maximizing population fitness. In terms of the cannibalism strains, this implies that when populations are structured, selection among groups will be effective at producing populations with high fitness.

Thus, theoretical studies combined with my empirical studies on the evolution of social behavior suggest that when the fitness of an individual depends on the distribution of genotypes within the population, population structure and multilevel selection must be considered in order to understand evolution.

Summary

The genetic studies show that cannibalism is a genetically controlled character in *T. confusum.* This behavior has dramatic effects at the population level, contributing to tenfold differences in population size. The three cannibalism phenotypes studied are autosomally inherited and exhibit minor degrees of dominance. The expression of cannibalism is influenced by both biotic and abiotic environmental factors. The differences between the high-cannibalism and low-cannibalism strains seem to be under polygenic control for larvae eating eggs and adults eating pupae; however, adult cannibalism of eggs may be regulated by as few as two genes. Finally, larval cannibalism of eggs and adult cannibalism of eggs show high genetic correlations.

Acknowledgments

I'd like to thank C.R.B. Boake, D. E. McCauley, J. M. Schwartz, and C. J. Goodnight for their comments on the manuscript. Special thanks to Jenny Evans. Funding for preparation of this chapter was provided by NSF Grants BSR-8906347 and DEB-9209695.

11

The Heritability of Mate-Attractive Traits: A Case Study on Field Crickets

Ann V. Hedrick

Male *Gryllus integer,* uncharacteristically posing above ground. (Photograph by C. A. Hedgcock, R.B.P.).

Assumptions and predictions about the genetic basis of sexually selected behavioral traits are an integral part of models for the evolution of mating behavior (Andersson 1987; Heisler, chap. 5), and testing these assumptions and predictions requires a genetic approach to the study of behavior. Because many behavioral traits in natural populations, including those associated with mating, are continuously distributed (Boake, chap. 1), quantitative genetics is often the most appropriate genetic approach to take. Therefore, quantitative genetic techniques can be especially useful in the study of sexual selection.

In this chapter I discuss my research on field crickets as an example of how quantitative genetics can be used to address questions regarding the evolution of sexually selected behavioral traits. The chapter is divided into three major sections. The first section outlines the theoretical background for my research question. The second describes the combination of behavioral and quantitative genetic techniques I used to address this question and gives my results. The third discusses some issues for further study that are raised by my genetic data.

Theoretical Background

A critical assumption in many models for the evolution of female choice is that male traits that are preferred by or attractive to females are heritable (Andersson 1987; Heisler, chap. 5). This assumption first arose as a hypothesis to explain why females exercise mate choice in mating systems in which males offer only gametes to their mates. Since females in these mating systems cannot be choosing mates for any immediate resources males might offer (e.g., food or male parental care), many authors, including Fisher (1958) and Trivers (1972), hypothesized that in this case, females were choosing mates for heritable traits.

Specifically, Fisher (1958) proposed in his "runaway" model for the evolution of female choice that fathers that were attractive to females would pass on heritable, attractive traits to their sons. These sons in turn would be preferred as mates by "choosy" females in subsequent generations. Thus, the reproductive success of preferred males would be enhanced, and the attractive trait would be favored by sexual selection. Since the choosiest females would mate with the most attractive males, the mating preference and the male trait would become genetically correlated, and female mating preferences would evolve as a correlated response to selection acting on the heritable male trait. Fisher's verbal model and his conclusions have been reproduced more recently by several mathematical models for the evolution of female mating preferences (O'Donald 1980; Lande 1981b; Kirkpatrick 1982).

The proposition that mate-attractive traits are heritable has been questioned, however, by Maynard Smith (1978, 172; 1985), who suggested that mate-attractive traits might have low or zero heritabilities. His logic was derived from Fisher's (1958) fundamental theorem of natural selection (the rate of increase in fitness under directional selection is equal to the genetic variance in fitness), which is often interpreted to mean that those traits closely related to fitness will have little additive genetic variance, and therefore low heritabilities (Mousseau and Roff 1987). Maynard Smith (1978, 1985) noted that female choice should exert strong directional selection for preferred male traits, and reasoned that this would deplete additive genetic variance in the traits. Thus, the traits would approach genetic fixation over time, and little or no heritable difference would remain to serve as an evolutionary basis for female choice. Although Maynard Smith offered other reasons why heritability might instead be maintained, his argument has been cited as an objection to the hypothesis that mate-attractive traits are heritable (e.g., Borgia 1979; Dominey 1983). Consequently, the hypothesis has been somewhat controversial (Ryan 1985; Kirkpatrick 1987b).

At the same time, evidence has accumulated in the life history literature for the maintenance of substantial amounts of genetic variance in individual traits related to fitness (Charlesworth 1987). Over the past ten years, both empirical and theoretical developments have demonstrated that fitness-related traits do not necessarily have low heritabilities even if the heritability of overall fitness is low (Charlesworth 1987; Price and Schluter 1991), and have suggested that a variety of mechanisms can maintain genetic variance in them (Dingle and Hegmann 1982; Lande 1982a; Rose 1982; Istock 1983; Charlesworth 1987; Mousseau and Roff 1987). Although most of this work does not address sexually selected traits explicitly (but see Hamilton and Zuk 1982), it implies that such traits might be heritable and that female choice for heritable traits remains a valid hypothesis.

Probably as a consequence of the new perspective offered by the life history literature, a number of authors working on sexual selection theory have assumed that this hypothesis is valid, and have used the heritability of preferred male traits as a basic assumption in their models for the evolution of female choice (Arnold 1985; Boake 1986; Heisler, chap. 5). These models fall into two groups. The "good genes" models (e.g., Hamilton and Zuk 1982; Kodric-Brown and Brown 1984; Andersson 1986; see also Heisler 1984b) propose that females use the preferred male character as an indicator of other heritable traits that enhance viability (i.e., components of fitness that are not sexually selected). In contrast, the "not-necessarily-good-genes" models (Lande 1981b; Kirkpatrick

1982) indicate that females may evolve preferences for heritable traits that need *not* be indicators of heritable, viability-enhancing traits. Despite the differences between these two sets of models, both of them rely on the assumption that preferred male traits are heritable (Arnold 1985; Boake 1986). Thus, empirical evaluation of the models rests upon testing this key assumption, in addition to measuring correlations between the preferred trait and other components of fitness.

Here, I describe a research program in which I tested the assumption that preferred male traits are heritable by using quantitative genetic methods to estimate the heritability of a behavioral trait in a population of field crickets.

Techniques and Results

To test this assumption unambiguously, it is first necessary to show that the following three conditions are met within a single population: (1) Males must show phenotypic variation in a particular trait. (2) The trait must be *repeatable;* i.e., individual males should be relatively consistent in their expression of the trait over time (age), so that when the trait is measured repeatedly on the same individuals, a measurable proportion of the total phenotypic variance in the trait can be ascribed to consistent differences *among* individuals. (3) The trait must be used by females as a *specific* criterion for mate choice. Ideally, one should precisely identify the trait as an important criterion for female choice by holding other male qualities constant and showing that the trait itself affects choice.

Once one has shown that the above three conditions are met within a single population, then the heritability of the male trait should be measured in that *same* population, since estimates of heritability are population dependent (Falconer 1989), female mating preferences and preferred male traits may evolve jointly (Lande 1981b), and models of sexual selection assume that both female mating preferences and preferred male traits are present and genetically variable within a single population (e.g., Lande 1981b).

Choice of a Study Animal

Field crickets (*Gryllus integer,* Orthoptera, subfamily Gryllinae), proved to be especially well suited to this study. First, they met several criteria that are important in any quantitative genetic study:

crickets are easily obtained in large numbers, easy to rear in the laboratory, and have a reasonably short generation time of about 12 weeks. Second, they have a non-resource-based mating system. Male *G. integer* offer females only a small spermatophore (Alexander 1975; Gwynne 1983), so females are unlikely to choose mates on the basis of immediate resources such as food or parental care.

Third, male calls attract sexually receptive females, such that females might discriminate among males on the basis of calls alone. This feature allowed me to use phonotaxis experiments to gain direct information on the specific criteria that females use for discriminating among potential mates (Hedrick 1986). In these experiments, females are offered male calls played back through speakers, and move toward the source of the call they prefer. Phonotaxis experiments are frequently used in mate choice experiments on acoustic insects (summarized in Ewing 1989). They are ideal for testing female choice because they allow one to precisely identify the criterion for choice by isolating different properties of male signals and testing for their separate effects on choice (Hedrick 1986). They also avoid two problems that are frequently encountered in tests of mating preferences: noncooperation of male subjects and dependence of the female's response on the particular male-female interaction.

Data from other species indicated that female crickets had some capacity for discriminating between male calls. Phonotaxis experiments had demonstrated that female crickets can distinguish between signals of their own species and signals of sympatric species (e.g., Walker 1957; Hoy, Pollock, and Moiseff 1982), and two studies showed acoustic preferences within species (Crankshaw 1979; Forrest 1983), although these did not precisely identify the criteria used by females for discrimination. Finally, there was some evidence for the genetic basis of calling behavior in male crickets: Hoy, Pollack, and Moiseff (1982) had reported polygenic inheritance of species-specific characteristics of calls in *Teleogryllus*, and Cade (1981) had found heritable differences in the tendency to call among males of a Texan *Gryllus* species.

Documenting Phenotypic Variation

The first step in my research program was to document variation among individual male crickets in a behavioral trait (some aspect of calling behavior) that might influence female mating preferences. To lend itself to a quantitative genetic study, this trait

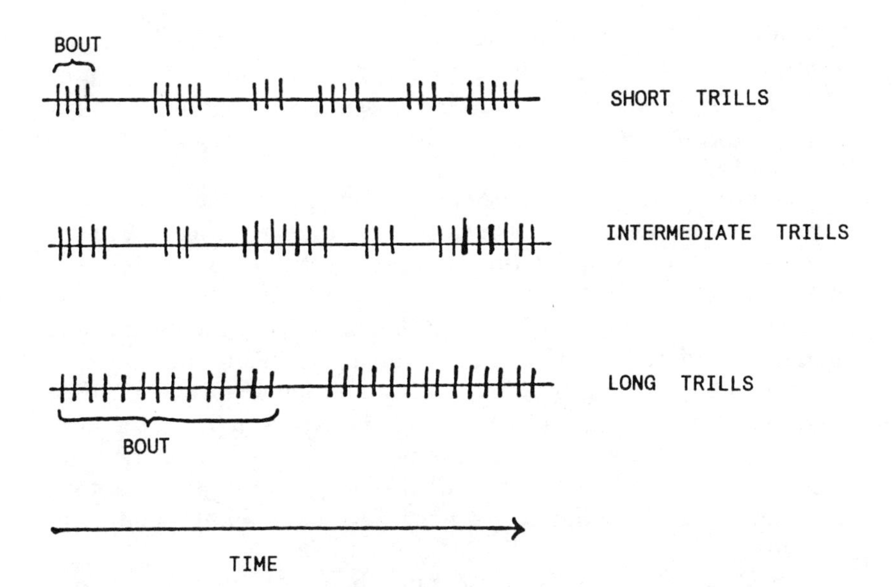

Fig. 11.1 Interindividual variation in the calls of male *Gryllus integer.*

needed to be both easily quantified and continuously distributed in the population.

Initial surveys of male calling behavior in the field revealed promising variation in calls. The call of *G. integer* is a rapid trill. Some males trilled for a long time without stopping. Others broke up their trills frequently with short pauses, and many males were intermediate between these extremes (fig. 11.1). This variation could have been due to a variety of causes, including environmental causes such as temperature.

Methods. To rule out environmental causes of variation and quantify variation among males in their durations of uninterrupted trilling (calling bout lengths), I captured free-living males in Davis, California, and brought them into the laboratory, where 5-minute tape recordings of their calls were made. Since some males could have been satellites, which seldom call but intercept females attracted to calling conspecifics (Cade 1981), I only recorded those males that called readily when placed next to other calling males. Each male was spatially and acoustically isolated from all conspecifics during recording. Recordings were made at 28° C, during peak calling times for laboratory males (approximately 2 to 4 hours after the onset of dark), using a Nagra-E tape recorder and a Sennheiser microphone. Tape speed was 7.5 inches per second. Because my intent was to document phenotypic variation in the population, the 50 recorded males were not a random sample from the population, but were selected on the basis of prior differences in their calls.

Recordings were then played back at one-fourth speed on a Uher tape recorder and durations of calling bout lengths measured with a stopwatch. A bout was defined as a period of calling containing no pause greater than 0.1 second (real time).

To characterize the calling bout distribution of each individual male, I calculated the percentage of his total calling time spent in bouts of different lengths. First, calling bout lengths were classified into 5-second interval categories (i.e., durations of 0–4.9 seconds, 5.0–9.9 seconds, etc.). For each male, I then calculated the total amount of time (= T) that he called during his 5-minute sample. Last, I computed the fraction of T that he spent in bouts of different lengths.

I used percentage of time spent in bouts of different lengths, rather than mean bout length, to characterize calling distributions because means did not accurately reflect interindividual variation in the duration of uninterrupted trilling. Many males inserted several short bouts between very long bouts, making short bouts as numerous as long bouts in their calling distributions.

Results. Analysis of the calls revealed considerable interindividual variation in calling bout lengths (fig. 11.2). Maximum bout lengths for the fifty males ranged from 0.9 to 256 seconds. Variation among males was continuous. However, I placed males in three classes for ease of reference. Fourteen males spent 75–100% of their total calling time in short bouts (under 5 seconds long), and were classified as "interrupters" (fig. 11.2A, $N = 14$ males). Eighteen males spent less than 25% of their calling time in short bouts, and were classified as "long trillers" (fig. 11.2C, $N = 18$ males). All other males were classified as "intermediates" (fig. 11.2B, $N = 18$ males). Thus, the analysis demonstrated that males in this population of *G. integer* vary in their calling bout lengths. Fortuitously, calling bout length was a continuously distributed behavioral trait that could be easily quantified for quantitative genetic study.

Showing That Bout Lengths Are Repeatable

The next step of my research program was to show that calling bout lengths are repeatable; i.e., that individual males are relatively

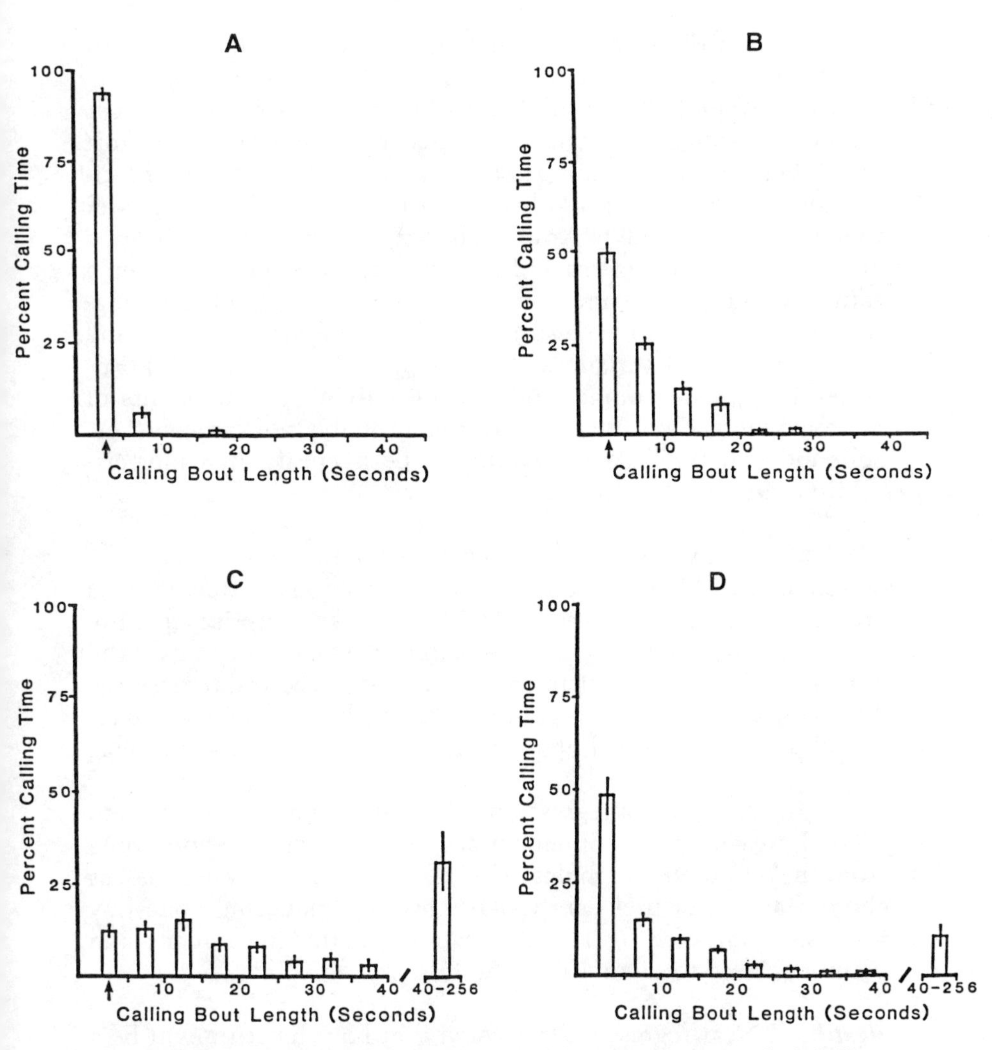

Fig. 11.2 Mean percentage of time spent in each bout length category by (*A*) all interrupters, $N = 14$ males; (*B*) all intermediates, $N = 18$ males; (*C*) all long trillers, $N = 18$ males; (*D*) the entire sample of 50 males (*A, B,* and *C* pooled). Means for the categories were obtained by adding together each male's percentage of time in the category, and then dividing by the number of males in the group. Vertical bars represent $\pm$ 1 standard error. Percentage of calling time spent in bouts <5 seconds long (arrow) was used to describe distributions from individual males (not shown). From A. V. Hedrick, 1986, Female preferences for male calling bout duration in a field cricket, *Behav. Ecol. Sociobiol.* 19:73–77; courtesy Springer-Verlag.

consistent in their calling bout lengths over time (age). Repeatability is a measure that describes the proportion of total phenotypic variance in a trait that is due to consistent differences among individuals (Falconer 1989). Because it sets an upper limit on the heritability of a trait, it can be used to indicate whether estimation of heritability is likely to yield a value significantly greater than zero (Boake 1989b; Arnold, chap. 2). The "repeatability" of a quantitative trait is estimated by making repeated measures on a sample of individuals and then calculating the ratio of the among-individual variance to the sum of both the among-individual and within-individual variances: $R = V_{among}/(V_{among} + V_{within})$ (Falconer 1989). The among- and within-individual components of variance themselves are estimated using analysis of variance (Lessells and Boag 1987). Possible values for both repeatability and heritability range from zero to one.

Methods. To investigate whether calling bout length was a repeatable character, 32 males were recorded again approximately 3 weeks after the first recording was made. Three weeks comprises roughly 50% of a male's reproductive life span in the laboratory. I used the percentage of time spent in short bouts (<5 seconds) to describe and categorize each male's calling distribution, then calculated repeatability as described above (Falconer 1989; Lessels and Boag 1987).

Additionally, linear regression was used to compare the first and second measurements of percentage of time spent in short bouts from the 32 individual males. This provided an easy way to assess the repeated measures data visually before conducting a one-way analysis of variance (fig. 11.3). Note, however, that linear regression cannot be used to calculate repeatability.

Results. Linear regression suggested that bout length might be a repeatable character. First and second measurements from individual males of percentage of calling time spent in short bouts were strongly correlated (fig. 11.3, $r^2 = .76$, $P < .01$, $n = 32$ males; percentages arcsine=transformed), and the regression line did not differ in slope (= 1) or intercept (= 0) from ($Y = X$), the line predicted if calling distributions remained the same over time ($P > .10$ in both cases, t tests).

Calculation of repeatability using components of variance revealed that the percentage of calling time spent in short bouts was highly repeatable ($t_{intraclass} = 0.85$, $P < .001$). Thus, calling bout length is a repeatable trait in this population of *G. integer.*

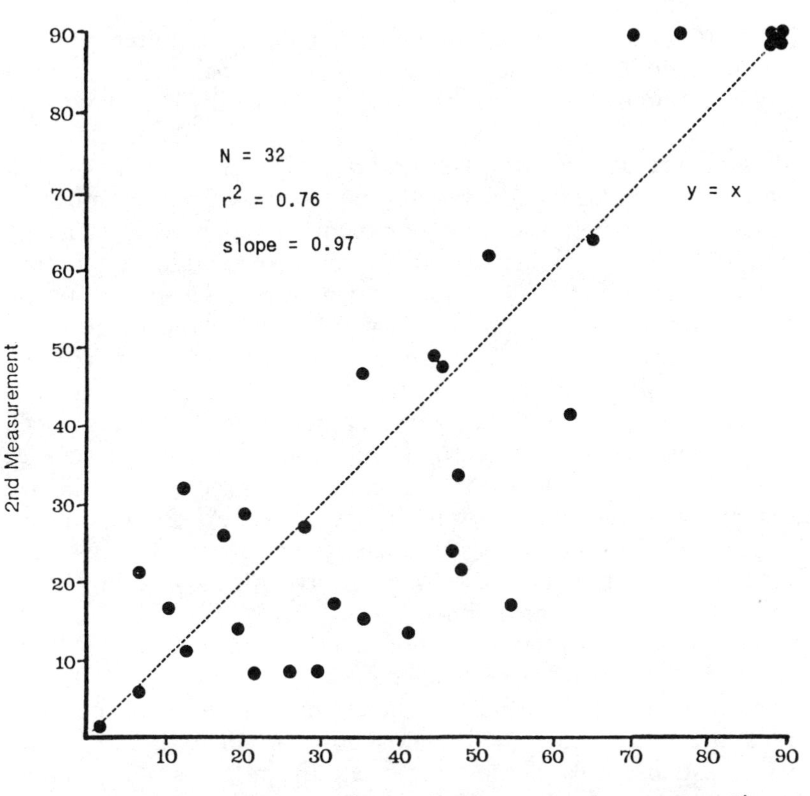

Fig. 11.3 Repeated measurements for each of 32 different males of the percentage of calling time spent in short bouts (percentages arcsine transformed). Second measurements (y-axis) were made 3 weeks after first measurements (x-axis). The dashed line corresponds to ($y = x$), the line predicted if calling distributions remained the same. The actual regression line formed by the data is ($y = 0.97x - 2.66$). The slope of this line (0.97) does not differ significantly from one and is significantly different from zero. From A. V. Hedrick, 1986, Female preferences for male calling bout duration in a field cricket, *Behav. Ecol. Sociobiol.* 19:73–77; courtesy Springer-Verlag.

Demonstrating That Females Use Calling Bout Length as a Criterion for Mate Choice

To examine whether calling bout lengths would affect female mating preferences, I conducted phonotaxis experiments. Additionally, I collected data on mating behavior to establish that the effect of preferential phonotaxis would not be countered by other female mating preferences that might be exhibited at later stages of courtship. I operationally defined a female mating preference as any behavior by a female that increases the probability of her mating with some males rather than others (following Halliday 1983).

Phonotaxis experiments: General methods. At the start of each phonotaxis trial, a female was confined in a small test chamber (10 cm × 4 cm × 4 cm) made of wire screening, with holes at each end. This chamber was then placed in the center of a small, straight tunnel (165 cm × 4 cm × 4 cm), also made of wire screening. For the first 7 minutes of the trial, screen guards blocked the female's exit from the test chamber. During the first 2 of those 7 minutes, the room was quiet; during the next 5 minutes, I conducted simultaneous playbacks of two different male calls from speakers placed at opposite ends of the tunnel. The calls were played back at matched, natural sound intensities (72 dB spl at 1 m) from two Nagra-E tape recorders with built-in speakers. After five minutes of playback, I raised the guards and allowed the female free access to the entire tunnel for the subsequent five minutes, during which the playbacks were continued.

Each female was tested only once, and the apparatus was lined with clean paper between trials so that females could not follow chemical cues. Calls were switched to opposite sides of the apparatus between trials to guard against position effects. I scored a positive response only if the female left the test chamber, walked down one arm of the tunnel, and contacted the endscreen on that side within the playback period.

Test 1: Discrimination by females between interrupter and long-triller calls. The first choice test was designed to determine whether wild-caught females (which were not necessarily virgins) preferred a call with long bouts to a call with short bouts, or vice versa. For this test, two tapes were made that were matched in sound level. One tape was of an interrupter male, and one of a long triller male. Although these tapes represented extreme calling types, preliminary surveys of calling males had indicated that interrupters and long trillers frequently call from positions adjacent to one another in the field.

Ten of eleven wild-caught females walked toward the source of the long-triller call, rather than the interrupter call. The probability of this result occurring by chance alone is .006 (binomial test).

Test 2: Control for other call properties and female experience. The second choice experiment was designed to rule out the possibility that female preferences in the first experiment reflected either the females' prior mating experiences or preferences for some property of the calls other than bout length (e.g., sound frequency). Laboratory-reared virgin females were used to eliminate possible bias due to mating experience, and I controlled for properties of the call other than bout length by making duplicate tapes of a long-triller call. One of these was recorded normally, but when I recorded the other, I erased portions of the call at regular intervals to simulate the call of an interrupter male. No erasures were longer than the gaps in actual interrupter calls. Thus, the tapes differed only in bout length.

When offered a choice between these two tapes, thirteen of fourteen virgin females walked toward the source of the natural, unaltered long-triller call ($P = .001$, binomial test). Thus, females without prior mating experience preferred calls with longer bout lengths even when all properties of the call other than bout length were controlled.

Test 3: Control for tape editing. Finally, I conducted a third choice experiment to eliminate the possibility that in test 2 I had erased some important call component when making the simulated interrupter call. Here, the two tape-recorded calls offered to virgin females were the simulated interrupter call and an actual interrupter call. The calls had the same carrier frequency. If the simulated tape sounded "normal" to females, and also if bout length was the only criterion for preference, females were expected to show no consistent preference for either tape. This expectation was met. Of fourteen virgin females, seven chose the simulated interrupter call, and seven chose the true interrupter call. Therefore, the results of the phonotaxis experiments strongly suggested that females of this population use calling bout length as a specific criterion for mate choice.

Subsequent choice experiments conducted during other studies and using different tapes of male calls have confirmed this result (see Kroodsma 1989). Note that females may assess bout length by "measuring" and responding to the total sound energy in calls over some discrete time interval. Calling bout length generally covaries with sound energy per unit time when both are measured over a 5- to 15-minute interval; consequently, this mechanism could result in a preference for calls with long bouts.

Does Phonotaxis Correspond to Actual Mating?

In mate choice experiments such as this, it is important to demonstrate that the behavioral response being scored (in this case, phonotaxis) corresponds to an actual mating decision (Arnold 1983b). In field crickets, the male call attracts a female from a distance. However, once she reaches the male, he must court her successfully before mating occurs. There are several stages in this latter process (described below) at which females might exercise additional mate choice, potentially countering or weakening the effect of preferential phonotaxis.

To court a female, a male field cricket stands in front of her, switches from calling to a softer "courtship" chirp, and moves slowly away from her. If she is receptive to that male, the female will follow him and eventually mount him (Boake 1983). Therefore, after encountering a male, a female could discriminate against that male by not following or mounting him (Burk 1983). Next, the male must successfully attach a spermatophore. This task can be difficult with a noncompliant female (Burk 1983). Finally, a female may be able to discriminate against a male by removing the spermatophore before the capsule is emptied of sperm (Simmons 1986). For example, in a closely related species, *Gryllus bimaculatus,* fertilization rates are highest during the first 10 minutes after attachment (Simmons 1990). In summary, it is possible that although females preferentially move toward males with long calling bouts, they discriminate against those males at some later stage of courtship, *after* they have reached the calling male.

Methods. To investigate this possibility, I conducted mating trials in the laboratory with 55 long trillers and 30 interrupters. In each trial, a female was placed alone with a male of known calling type. Each female and male was tested only once. Courtship and mating followed a predictable sequence, allowing me to collect data on a series of eleven variables for each trial reflecting male courtship behavior and female response. These data (summarized in table 11.1) were then examined to determine whether males of different calling types differed in their mating success once the female had encountered them.

When the 55 long trillers and the 30 interrupters were compared, no statistically significant differences between them were found in any of the eleven courtship and mating variables (table 11.1). These results suggest that females do *not* discriminate against either long trillers or interrupters once they have reached a male. Thus, phono-

Table 11.1 Data from mating trials

	Long Trillers	Interrupters
Proportion of males that courted	1.00 (55)	0.97 (30)
Proportion of males that females followed	0.91 (55)	0.97 (29)
Proportion of males that females mounted	0.94 (50)	0.96 (28)
Seconds to first mount	124 ± 23 (47)	260 ± 91 (27)
Frequency of successful first mounts	0.51 (47)	0.59 (27)
Frequency of second mounts when first mounts were unsuccessful	0.78 (23)	0.73 (11)
Frequency of successful second mounts	0.56 (18)	0.62 (8)
Seconds to first successful mount	171 ± 43 (47)	272 ± 94 (27)
Proportion of successful matings	0.64 (55)	0.73 (30)
Proportion of matings in which female left spermatophore attached > 10 minutes	1.00 (33)	1.00 (19)
Proportion of matings in which female left spermatophore attached > 20 minutes	0.82 (33)	0.89 (19)

Note: Proportions or means ± standard errors for long trillers versus interrupters are given; total *N*s are given in parentheses. See text for explanation of the variables. In each case, the difference between long trillers and interrupters was *not* statistically significant ($P > .05$, chi-square or t test).

taxis is apparently the stage of reproductive behavior at which females of this population exhibit most of their mate discrimination. This result implies that preferential phonotaxis corresponds to a mating decision.

Measuring the Heritability of Male Calling Bout Length

The experiments above demonstrated that calling bout length varied phenotypically among males, was repeatable, and was used by females as a specific criterion for mate choice. Therefore I proceeded to estimate the heritability of this behavioral trait using methods of quantitative genetics.

Choice of quantitative genetic method for estimating heritability. Of several possible methods (Arnold, chap. 2) for estimating the heritability of calling bout length, I chose a father-son regression. Using this method, one regresses the mean values of sons within families against the values of their fathers, and heritability is estimated as twice the slope of the regression line (Falconer 1989).

This method was preferable to a selection experiment or a half-sib analysis for several reasons. First, the father-son regression provided a very straightforward test of the hypothesis that fathers that are preferred by females pass on attractive traits to their sons. Second, it was more efficient than a selection experiment, which would have required rearing and measuring many crickets over several generations. In selection experiments, it is advisable to select simultaneously in two directions (e.g., for longer and shorter bouts) and to replicate each selection line (Falconer 1989). The number of individuals selected to begin each line should not be too small (<10) to guard against inbreeding depression, but the intensity of selection depends on the percentage of measured individuals retained as parents: selection leads to more rapid change among lines when this percentage is small (Falconer 1989). Thus, if the top and bottom 10% of sampled individuals were used to begin two "high" and two "low" lines of 10 individuals each, 200 crickets would have to be sampled initially. Moreover, the selected lines would have to be sampled and maintained over four to five generations. In contrast, by using the father-son regression method, I was able to calculate heritability with just 136 males (38 fathers plus 98 sons), and obtained heritability estimates in a single generation.

The half-sib method was also poorly suited to my study. This method involves mating each of a number of males to several females, and then measuring the offspring of each female (Falconer 1989). However, breeding designs requiring multiple male matings can contain pitfalls for mate choice studies. If females have strong mating preferences, it may be very difficult for the researcher to mate an "unattractive" male with numerous females. Failing to use those males, or repeatedly offering them to different females until they are finally accepted as mates, could introduce assortative mating into the breeding design, and this can bias heritability estimates derived from sib analysis (Falconer 1989). In contrast, the father-son regression technique requires only one mating per male with a randomly chosen female. This was relatively easy to achieve with *G. integer* because most mate discrimination apparently oc-

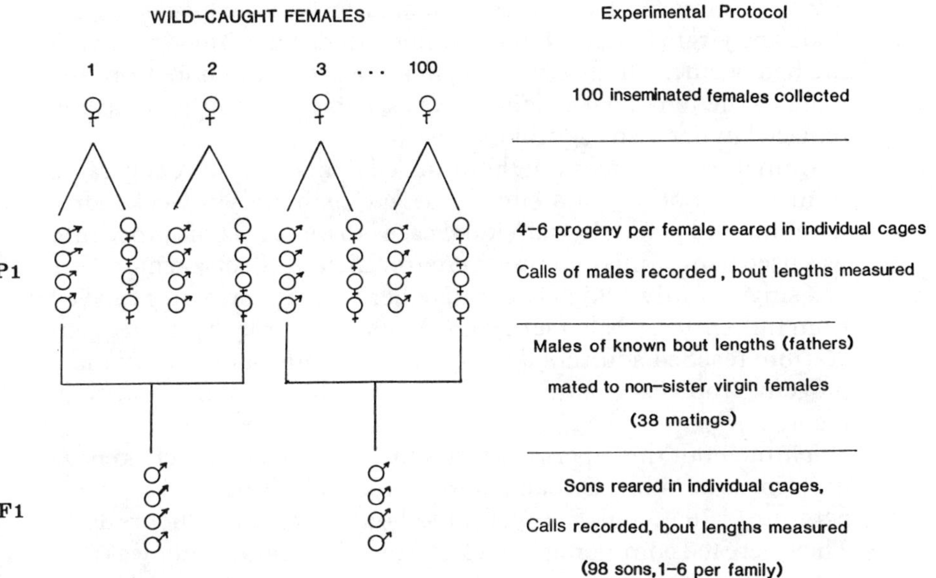

Fig. 11.4 Experimental protocol for obtaining heritability estimates. The parental generation (P_1) provided fathers and the offspring generation (F_1) sons for the father-son regression.

curs at phonotaxis (see above), although it could be difficult in other species.

Collection and rearing of crickets. To obtain the parental generation for the experiment, 100 gravid females were collected from a field site in Davis, California, and placed in individual cages for oviposition. One to six immature offspring from each of these females were separated from their mothers and siblings approximately 4 weeks after hatching (at one-fourth to one-half adult size) and placed in individual cages, where they matured. These offspring became the parental generation (P_1) for the experiment (fig. 11.4).

Calls of adult P_1 males were recorded using prior methods, and then one P_1 male from each of 38 families was mated to a non-sister virgin female chosen at random. All pedigrees of mated pairs were unique; each female was mated only once and was unrelated to every other mated individual. Mating was encouraged by placing the male and virgin female alone together; after I determined that mating had occurred (I observed transfer of a spermatophore from the male to the female), the female was separated from the male and replaced in her own cage for oviposition.

Immature progeny from the arranged P_1 matings were separated from their mothers and siblings at approximately 4 weeks after hatching and placed in individual cages to mature. This procedure was used to minimize common environmental effects on members of a single family, which can inflate heritability estimates derived from full-sib data (Falconer 1989). When these progeny (the F_1 generation) reached adulthood, I recorded the calls of one to six male progeny from each family (N = 38 families; 98 male progeny recorded).

Throughout the experiment, conditions in all cages were kept as uniform as possible. Crickets were reared and maintained at a temperature of 28° C and on a cycle of 12 hours light and 12 hours dark. They were fed commercial chick starter ad libitum; water was provided continuously in shell vials stoppered with cotton.

Recording and analysis of calls. I made a 5-minute tape recording of each male's call, using my previous methods. As before, I defined a bout as a period of calling containing no pause greater than 0.1 second (real time), and I characterized each male's calling distribution by measuring the proportion of his total calling time that he spent in short bouts 0–4.9 seconds in duration. Hence, heritabilities were estimated for the same character (percentage of calling time spent in short bouts) for which female preferences had been investigated (Hedrick 1988).

Estimation of heritability. I estimated the heritability (defined as additive genetic variance divided by total phenotypic variance; Falconer 1989) of the percentage of calling time spent in short bouts for the F_1 generation using two different methods. First, square root-transformed values of the percentage of time spent in short bouts by fathers (P_1) and their sons (F_1) were used in a father-son regression ($N = 38$ fathers, 98 sons). Families varied in size, so I weighted families by family size according to Falconer's (1989) modification of the formulas developed by Kempthorne and Tandon (1953). Data transformation was necessary because linear regression models assume that the mean and variance of distributions are uncorrelated (Weisberg 1980). As in the present case, this assumption is sometimes violated by behavioral measurements collected on natural populations. However, the problem can often be corrected by data transformation. Although the arcsine transformation is generally recommended for percentage data such as mine (Sokal and Rohlf 1981), a plot of studentized residuals versus fitted values revealed nonconstant residual variance in arcsine-transformed percentage of the time spent in short bouts (Weisberg 1980; note that this problem was not detected in the data set used to calculate repeatability). I therefore conducted the father-son regression using square root-transformed values of the percentages of time spent in short bouts; this transformation corrected the problem of nonconstant residual variance.

I was also able to estimate heritability using the intraclass correlation coefficient (t) between full sibs. Employing this method, heritability is estimated as twice the intraclass correlation coefficient, but the estimate may be inflated by nonadditive genetic variance and common environmental variance (Falconer 1989). Thus, it is not as reliable as the regression method, although it provides a convenient way to check an estimate derived from regression. I calculated the intraclass correlation coefficient between full sibs with square root-transformed data collected from the full-sib brothers within F_1 families ($N = 26$ families with two or more male offspring; 86 males total).

Results: Heritability estimates. The estimate of heritability from the father-son regression was significantly greater than zero (table 11.2), as was the estimate from sib correlation. Moreover, the estimate from regression was similar to that from sib correlation

Table 11.2 Heritability estimates

	Heritability Estimate	Standard Error of Estimate	N	Significance Level
Father-son	0.75	0.25	38, 98	$P < .005$
Full-sib	0.69	0.24	26, 86	$P < .005$

Note: Values are for square root-transformed data. Standard errors were computed according to formulas given in Falconer 1989. *N*s given are the number of families, followed by the total number of sons or brothers. Significance levels are from the corresponding regression analysis (in the case of father-son data) or analysis of variance (in the case of full-sib data). Untransformed values of the percentage of calling time spent in short bouts ranged from 1.5%–100% for fathers, and 1.0%–100% for sons.

(father-son regression, 0.75; sib correlation, 0.69). Averaging the estimates from regression and correlation gave a heritability estimate of 0.72. These results demonstrate that male traits on which female mating preferences are based can be heritable.

Issues for Further Study

Evidence for the Heritability of Mate-Attractive Traits

The heritability estimates for calling bout length in *G. integer* support a fundamental assumption of many theoretical models for the evolution of female choice. Although behavioral traits such as calling bout length are often cited as criteria for female choice (e.g., Gibson and Bradbury 1985; Klump and Gerhardt 1987), their heritability in natural populations is virtually unknown. However, Moore (1990b) documented heritable variation in behavioral elements of courtship in a laboratory population of cockroaches (*Nauphoeta cinerea*). Additional evidence for the heritability of mate-attractive traits is provided solely by studies of morphological traits (Endler 1983; Carson 1985; Houde 1987); these traits tend to be easier to measure and less inherently variable than behavioral traits, which are frequently context dependent. A major challenge for further work on the heritability of mate-attractive behaviors will be to find behavioral traits in males that are easy to quantify, repeatable, and serve as important criteria for female choice even when other male qualities are held constant (through either experimental or statistical means).

Heritability Estimation in the Laboratory

One drawback to estimating heritability in the laboratory, where estimation is most feasible, is that laboratory measurements of heritability can overestimate heritability in the field if changes in environmental variables alter phenotypes and/or affect genotypes differentially (Mitchell-Olds and Rutledge 1986; Prout and Barker 1989; Riska, Prout, and Turelli 1989). Thus, although laboratory estimates can reveal additive genetic variance within populations, they may be of limited utility for inferring the outcome of selection in the field. To solve this problem, several authors have suggested that heritability in the field can be approximated by regressing characters of laboratory-reared offspring on those of their wild-caught parents (Coyne and Beecham 1987; Prout and Barker 1989).

Another way to address the problem is to identify important environmental variables and examine their effects on the trait in question. For example, in *G. integer,* one environmental variable that might fluctuate in the field and influence calling bout lengths is the amount of food available to developing crickets. Therefore, I conducted a preliminary study on the effect of food availability during development on calling bout length. Adult males reared on ad libitum diets did not have significantly different calling bout lengths from their full-sib brothers reared on a low-food regimen, although their body sizes differed. Apparently, low-food conditions do not change bout lengths even though they affect adult body size. The effects of other environmental variables that might influence bout lengths of male *G. integer,* such as rearing temperature, are not known.

Testing Additional Assumptions

As described above, my results support a key assumption of models for the evolution of mating preferences. However, they also provide a foundation for tests of additional assumptions that have remained largely untested to date. These include assumptions regarding both the specific "decision rules" used by females to discriminate among potential mates with different genotypes, and the correlations between preferred, heritable male traits and other phenotypic and genetic characteristics of males. In many species, empirical investigation of these topics has been hampered by the difficulty of identifying heritable, preferred traits in males. Therefore, the Davis population of *G. integer* presents a rare opportunity to examine these topics in more detail, as discussed below.

Female mating decisions. Although theoretical models for the evolution of female choice critically depend on the specific "decision rules" females use to discriminate among potential mates, these rules are poorly understood. For example, under field conditions, females might not always mate with the preferred male genotype (as is sometimes assumed in models, e.g., Lande 1981b; Kirkpatrick 1982) because of search costs, such as the time, energy, or risk of predation involved in traveling to prospective mates (Wilson and Hedrick 1982). Search costs and the direct selection they may impose on female mating preferences have received increasing attention in theoretical models (Pomiankowski 1987a; Kirkpatrick 1987a; Real 1990; Kirkpatrick and Ryan 1991), but to date, very little attention from empiricists (Real 1990). In *G. integer*, however, recent evidence suggests that females make trade-offs in their mating decisions between preferences for particular male genotypes and the perceived risk of predation involved in traveling to males with those genotypes (Hedrick and Dill 1993). Mate searching exposes female crickets to predation risk, particularly when they are moving through open areas (Sakaluk and Belwood 1984; Sakaluk 1990). Consequently, female *G. integer* avoid moving through open space, and prefer to move through cover or crevices. We found that when given a choice between moving across open space toward a long-triller call versus moving through cover toward an interrupter call, females often "compromised" and chose the interrupter call in cover. These results indicate that females may adjust their mating decisions in response to the perceived risk of predation along paths to prospective mates. Therefore, they will not necessarily mate with the preferred male genotype when considerations such as predation risk enter into their mating decisions.

Females also may experience difficulties in assessing male genotypes in the field, for example, if many males are calling simultaneously (Gerhardt and Klump 1988) or male density is very low. Careful behavioral study will be necessary to determine the relative importance of genetic versus more immediate factors in the mating decisions of females in *G. integer* and other species.

Correlates of heritable, preferred male traits. Many models propose that heritable, preferred male traits offer cues to females about other male characteristics. Specifically, the "good genes" models (e.g., Kodric-Brown and Brown 1984; Andersson 1986; see also Heisler 1984b) assume positive genetic correlations between heritable, preferred traits such as calling bout length and other components of fitness (e.g., development time, dominance, vigor, longevity). Other models (e.g., Lande 1981b; Kirkpatrick 1982) do not require these positive genetic correlations, and suggest that female preferences can actually evolve for male traits that are selectively neutral or selected against by natural selection. For example, in *G. integer,* males with longer calling bouts might be more conspicuous to predators, and thereby suffer higher mortality rates than males with shorter bouts. Distinguishing among these alternatives will require further use of quantitative genetic techniques, as described by Boake (1986).

The Maintenance of Genetic Variance in Mate-Attractive Traits

Since females in the Davis population of *G. integer* prefer calls with long bouts, and since both length is heritable, how is genetic variance in calling bout length maintained in this population of crickets? Below, I consider several possible explanations for my results.

First, as mentioned above, genotype-environment interactions and other environmental influences on the phenotype could mean that heritability in the field is not as high as my estimates from the laboratory (Riska, Prout, and Turelli 1989). Thus, much of the additive genetic variance in bout length that was detected in this study may not be exposed to selection under many field conditions (Istock 1983; Dingle 1984).

Second, the evolutionary impact of female choice on calling bout length might be diminished by limitations on female choice in the field. These limitations could result from the direct costs of mate searching, such as the time, energy, and risk of predation involved in traveling to preferred mates (see discussion above), or from problems in assessing male genotypes (Wilson and Hedrick 1982; Pomiankowski 1987a; Kirkpatrick 1987a). Also, if noncalling males intercept females on their way toward preferred males, some genes for bout length will remain unselected within the noncalling portion of the population each generation. Although additive genetic variance can be maintained by polygenic mutation and recombination alone, the magnitude of additive variance that can be maintained increases as selection intensity decreases (Lande 1976a; Turelli 1984). Therefore, the factors above could help preserve additive genetic variance in bout length by decreasing the intensity of selection imposed on this trait by female choice.

Finally, balances among selection pressures could moderate the effect of strong, directional female choice on male bout length and allow genetic variance to persist. For example, variance could be maintained by fluctuations in selection pressures that favor different genotypes at different times or in different environments (Felsenstein 1976), or by negative genetic correlations between calling bout length and other components of fitness such as development time or longevity (Lande 1982a; Rose 1982). Determining the ways in which genetic variance in preferred male traits such as bout length is maintained is a challenging but worthwhile objective for future studies on the evolution of mating preferences. In these studies, both behavioral and quantitative genetic perspectives and methods will clearly play an important role.

Summary

Many theoretical models for the evolution of female choice rely upon the assumption that male traits that are preferred by females are heritable. I tested this assumption in a Californian population of field crickets (*Gryllus integer*). Males of this population showed phenotypic variation in their calling bout lengths, even under standard laboratory conditions. Analysis of calling bout length over time (age) for individual males indicated that this behavioral trait was highly repeatable. A series of phonotaxis experiments demonstrated that females were preferentially attracted to calls with longer bouts, and mating trials suggested that preferential phonotaxis corresponded to a mating decision. The heritability of calling bout length was significantly greater than zero, showing that preferred male traits can be heritable. Important issues for future study include the specific "decision rules" used by females to discriminate among potential mates with different genotypes, and whether females necessarily mate with preferred male genotypes when considerations such as predation risk enter into their mating decisions.

12

Quantitative Genetics of Locomotor Behavior and Physiology in a Garter Snake

Theodore Garland, Jr.

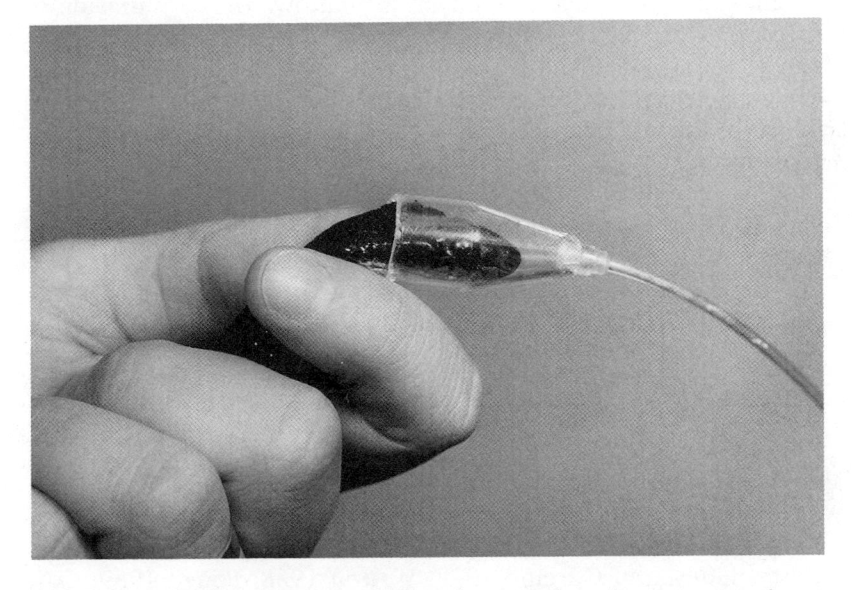

Black racer (*Coluber constrictor*) wearing a plastic mask for determination of maximal oxygen consumption and energetic cost of locomotion (Walton, Jayne, and Bennett 1990; photograph courtesy of B. C. Jayne). A similar mask, but smaller, was used in the present study for measuring newborn garter snakes.

In this chapter I demonstrate the use of quantitative genetic analyses for studying the evolution of certain behavioral traits related to locomotor capacities and underlying elements of morphology, physiology, and biochemistry. I studied the common garter snake (*Thamnophis sirtalis*), one member of a genus that has served as an increasingly common model in studies of individual variation and its genetic bases. The strengths of this study include (1) its use of a natural population, (2) its focus on behavioral and whole-animal performance traits thought to be of considerable ecological and selective importance (e.g., antipredator display, maximal crawling speed, endurance), and (3) its inclusion of measures of other traits at lower levels of biological organization (e.g., heart size, blood hemoglobin level, tissue enzyme activities) that may have mechanistic links with traits at the whole-animal level, but are of less direct selective importance.

Quantitative genetic theory offers some (not always straightforward) predictions concerning how heritable traits of differing selective importance ought to be. In addition, existing knowledge of mechanisms from exercise physiology allows the formulation of some a priori predictions as to which traits are likely to show genetic couplings. These couplings may be estimated as genetic correlations and tested for statistical significance. Genetic correlations are important because they can significantly constrain (or facilitate) the course of multivariate evolution, depending on the form of natural selection (Arnold 1981c, 1987a, 1988, 1992b; Arnold, chap 2; Falconer 1989; Cheverud and Moore, chap. 4; Roff, chap. 3). The importance of studying multiple traits when attempting to identify possible constraints has been emphasized by Pease and Bull (1988; see also Lynch, chap. 13).

Relevance of Exercise Physiology for Behavioral Evolution

Many behaviors involve locomotion (which is itself a behavior). Capacities for locomotion may therefore place constraints on behavioral repertoires or on the frequency or intensity with which elements of repertoires can be employed (e.g., Halliday 1987; Hertz, Huey, and Garland 1988; Walton 1988; Pough 1989; Garland, Hankins, and Huey 1990; Garland and Losos 1994). Whether such constraints actually occur in any particular situation depends on the physical intensity of the behavior (e.g., rate of ATP use by particular muscles) in relation to the capacity of the relevant muscles to produce ATP aerobically and/or anaerobically, and potentially on a variety of other psychological, physical, physiological, and biochemical factors (Brooks and Fahey 1984; Astrand and Rodahl 1986; Garland 1993). The important point here is simply that many behaviors depends on locomotion.

Whenever behavior involves locomotion, we can expect correlated evolution of behavior and relevant locomotor capacities

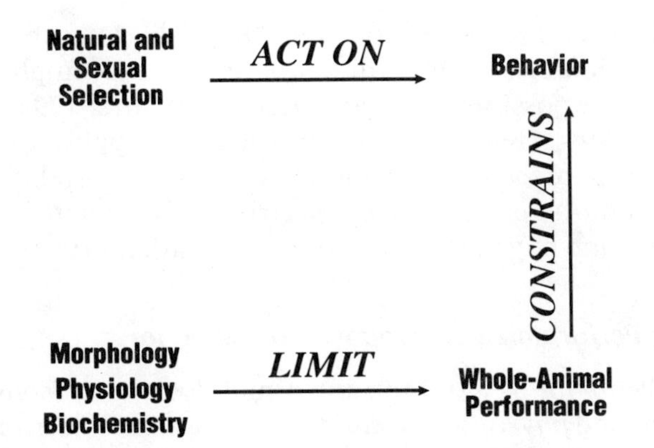

Fig. 12.1 Conceptual relationships of morphological, performance, and behavioral traits in relation to natural selection (original concept by Arnold [1983a], modifications by Garland, Bennett, and Daniels [1990]; Garland and Carter [1994]; Garland and Losos [1994]). Many other arrows might be added to such a diagram, and their significance can be addressed empirically through the use of path analysis (Arnold 1983a; Garland and Losos 1994). Natural selection is seen as acting most directly on behavior, rather than on organismal performance capacities. Thus, behavior may act as a buffer between selection and performance, with morphological, physiological, and biochemical traits being even further removed from the direct effects of selection (Garland, Bennett, and Daniels 1990). Whether lower-level traits, such as morphology, ever have direct effects on fitness is a difficult conceptual and empirical issue (Garland and Losos 1994).

(cf. Lynch, chap. 13, on thermoregulatory behavior in mice). "Coadaptation" (sensu Huey and Bennett 1987) refers to correlated evolution (sensu Martins and Garland 1991) that is caused by correlated selection pressures. Given that we expect behavior and appropriate locomotor capacities to be under correlational selection (e.g., Brodie 1992), then coadaptation per se of behavior and locomotor capacities would be expected (cf. fig. 12.1). We would also expect coadaptation of the various morphological, physiological, and biochemical characteristics that determine locomotor capacities (but see Garland and Huey 1987). Complementary to ideas about correlated evolution are the concepts of constraints and "trade-offs" (Maynard Smith et al. 1985; Pease and Bull 1988; Arnold 1992b; Roff, chap. 3). Given the physical properties of the materials with which organisms are constructed and the finiteness of their environments, we can expect both internal (e.g., physiological, biomechanical, genetic) and external (e.g., behavioral, ecological) constraints on the ways in which multiple phenotypic traits can coadapt. Formulating more specific predictions about correlated evolution can be quite difficult, and many complementary approaches are possible (e.g., see Garland and Carter 1994; Garland and Losos 1994). Here, I focus on the utility of applying quantitative genetic approaches to predictions about the correlated evolution of locomotor performance capacities and associated elements of behavior and physiology (cf. Dohm and Garland 1993).

Defining Performance as Opposed to Behavior

I define *behavior* as anything an animal does (or, in some cases, fails to do), and *physiological capacity* as the ability of an animal to perform a particular act under forced conditions (i.e., with maximum motivation). For example, endurance capacity may be defined operationally as the length of time an animal can run on a treadmill before fatigue, with fatigue being defined as failure to maintain the work rate due to preceding work (Simonson and Weiser 1976). Measuring endurance capacity in such a way may be difficult if one cannot clearly determine when fatigue occurs. Motivation can be a very problematic confounding factor (Bolles 1975). How, for example, can one be sure that an animal is actually performing to its physiological limits? In humans it is well known that differences in motivation can lead to large differences in performance (e.g., the better the competition, the better the maximum performance). But in many wild animals, including fishes (references in Garland and Adolph 1991), amphibians (e.g., Moore and Gatten 1989), lizards (e.g., Garland 1984; Huey et al. 1984; Garland and Losos 1994), and small mammals (Djawdan 1993), fatigue is relatively easy to identify and sometimes occurs shortly prior to complete exhaustion (e.g., loss of righting response in lizards: Huey et al. 1984). Supplementary physiological measurements (such as blood lactate levels) may be used to help gauge whether animals have performed to their physiological limits (Arnold and Bennett 1984; Djawdan 1993; discussion in Garland and Losos 1994).

Physiological capacities so measured fall within the general term *organismal performance.* In two papers that seem to have crystallized the thoughts of many, Huey and Stevenson (1979) and Arnold (1983a) argued that measures of organismal performance are crucial links between morphology, physiology, or biochemistry, on the one hand, and ecological importance—or fitness—on the other (see also Price and Schluter 1991). In particular, Arnold

(1983a) argued that selection acts directly on performance capacities rather than on lower-level traits such as morphology or physiology. Garland and Losos (1994) have argued further that Arnold's morphology → performance → fitness paradigm should be modified to include behavior, as shown in figure 12.1. In this expanded scheme, behavior is seen as a potential "filter" between selection and performance capacities (Garland, Bennett, and Daniels 1990; Garland and Carter 1994).

Defining and eliciting maximal physiological performance from wild animals tested in the laboratory is possible for many species (but see discussion in Garland and Losos 1994). Determining whether animals in nature commonly or rarely behave in ways that tax their maximal physiological capacities is a separate problem. Unfortunately, few quantitative empirical studies have been designed explicitly to address this fundamental question (Halliday 1987; Hertz, Huey, and Garland 1988; Walton 1988; Pough 1989; Garland, Hankins, and Huey 1990; Garland 1993; references therein).

Genetic Correlations, Evolutionary Constraints, and Complementary Predictions

Quantitative genetic analyses can be used to address questions concerning the correlated evolution of behavior and physiology. This approach is complementary to studies of individual (e.g., Garland 1984; Bennett 1987; Bennett, Garland, and Else 1989; Garland, Hankins, and Huey 1990), interpopulation (Arnold 1981b,c, 1988; Garland and Adolph 1991), and interspecific (Brooks and McClennan 1991; Harvey and Pagel 1991; Martins and Garland 1991; Garland et al. 1993) variation and covariation, as well as to experimental manipulation of physiological capacities or hormonal levels to assess their effects on behavior (e.g., references in Wingfield et al. 1990; Garland and Adolph 1991; Crews 1992; Garland and Losos 1994).

Understanding the genetic basis of natural variation allows one to draw inferences about the past history of selection acting on a population and to make specific predictions about the likely future results of hypothetical patterns of natural selection (cf. Lynch, chap. 13). Current genetic "architecture" (sensu Broadhurst and Jinks 1974; Mather and Jinks 1982) is both the result of past selection and the determinant of future responses to selection. For example, a high heritability suggests that a trait has not been subject to strong selection in the past and would respond to future selection. Conversely, a low heritability suggests that a trait may have been subject to strong past selection (although many other possibilities exist: see Discussion; Garland, Bennett, and Daniels 1990; Price and Schluter 1991) and would respond only slowly to future selection. Genetic correlations may result from past selection pressures (Cheverud 1984b, 1988b; Brodie 1992, 1993) and also may indicate whether genetic response to future selection will be impeded or facilitated by the prevailing polygenic architecture. When estimates of multivariate phenotypic differences among related taxa are available, the genetic covariance matrix can be used to reconstruct the net forces of selection necessary to have produced these differences, although this requires the genetic parameters to have remained constant during divergence (e.g., Arnold 1981a,c, 1988, 1990, 1992b; Arnold, chap. 2; Lande 1988; Turelli 1988a,b; Brodie 1993).

We can also use knowledge of behavioral and physiological mechanisms to generate predictions about genetic correlations, and vice versa (cf. Henderson 1989; Garland and Carter 1994). In particular, speed and endurance might be expected to show a "trade-off" because of the way muscles are constructed and the way they function (discussion in Garland 1988). This hypothetical trade-off should be evidenced as a negative genetic correlation. Moreover, two different behavior patterns, one of which requires speed and the other endurance for execution, should also show a negative genetic correlation and hence a trade-off. Thus, individuals good at one behavior would necessarily be less good at the other. Under this scenario, positive selection for one behavior pattern would result in a negative correlated response in the other.

Rationale for Studying Garter Snakes

Garter snakes (genus *Thamnophis*) are New World natricine snakes, closely related to the New World water snakes (*Nerodia*) (Lawson 1985; A. de Queiroz and R. Lawson, pers. comm.). They represent a relatively recent radiation within the family Colubridae, and just over twenty species are currently recognized. Many species of garter snakes can be locally abundant and, in many populations, relatively large numbers of gravid females can be captured. This fact has been exploited by a number of workers who study individual variation in neonates and/or the genetic basis of phenotypic traits (e.g., Herzog and Bailey 1987; Herzog and

Burghardt 1988; Herzog, Bowers, and Burghardt 1989b; Schwartz and Herzog 1993; review in Brodie and Garland 1993).

Thamnophis sirtalis is an excellent study organism because of (1) its broad geographic distribution, (2) the wealth of background information and ongoing ecological studies on this and related species (e.g., Arnold 1981a,b,c; Arnold, pers. comm.; Kephart 1982; Kephart and Arnold 1982; references therein), and (3) the ease with which quantitative locomotor and behavioral data can be collected. The common garter snake is an ecological generalist with the broadest geographic range of any North American reptile. Geographic variation in scale counts and color patterns (on which the taxonomy is based; Ruthven 1908) is extensive, and has been used to designate eleven subspecies (Fitch 1965; Fitch and Maslin 1961; Christman 1980; Stebbins 1985; see also Arnold 1988). Electrophoretic studies reveal extensive protein variation, both within and among populations of *T. sirtalis* and among the species of *Thamnophis* (Dowling et al. 1983; Lawson 1985; Dessauer, Cadle, and Lawson 1987; Schwartz 1989; Schwartz, McCracken, and Burghardt 1989; A. de Queiroz and R. Lawson, pers. comm.). The morphologically based subspecific designations do not necessarily correspond to electrophoretic patterns (R. Lawson, pers. comm.) nor to quantitative analyses of coloration, scalation, or internal anatomy (Benton 1980). Geographic variation in behavioral traits has also been demonstrated in *T. sirtalis* (Burghardt 1970; Herzog and Schwartz 1990; Arnold 1992a; see also Drummond and Burghardt 1983; Arnold 1981a,b,c on *T. elegans*). Recently, Schwartz (1989; Schwartz, pers. comm.) has demonstrated differences in scale counts, chemical prey preferences, aggregation and antipredator behaviors, critical thermal minimum, and protein polymorphisms between two populations from Wisconsin and Michigan.

This chapter focuses on antipredator displays, a behavior that is easily quantified and heritable (Arnold and Bennett 1984; Garland 1988; Brodie 1992, 1993). Antipredator displays can be scored at the end of endurance trials, and afford the opportunity to study variation in a behavioral character as it relates to underlying variation in performance capabilities, physiology, morphology, and biochemistry. Antipredator behaviors may also evolve in concert with appropriate color patterns (Brodie 1989b, 1992, 1993; discussion regarding *T. sirtalis* in Fitch 1965), and color pattern variation seems often to be related to crypsis in snakes (King 1987; Greene 1988).

Methods for This Study

Animal Collection and Husbandry

During the summer of 1984, Steve Arnold and I collected gravid garter snakes at his study sites near Eagle Lake, California. These were returned to Al Bennett's laboratory at the University of California at Irvine and housed individually under standardized conditions (further details are presented in Garland 1988; Garland, Bennett, and Daniels 1990; Garland and Bennett 1990; Dohm and Garland 1993). Newborn snakes were housed individually and tested over a period of several weeks. Because we were measuring several traits on each of many individuals, logistics precluded us from measuring all individuals at the same age (see below).

Measurement of Behavior, Performance, and Physiology

Measurement protocols can be found elsewhere (Garland 1988; Garland and Bennett 1990; Garland, Bennett, and Daniels 1990) and will not be repeated in detail here. Snakes were tested for maximal sprint crawling speed on two consecutive days, then for treadmill endurance at 0.4 km/h on two consecutive days, and finally for maximal oxygen consumption on two consecutive days ($\dot{V}O_2max$) while snakes wore a lightweight transparent mask and crawled on the treadmill (Garland and Bennett 1990). Antipredator displays (Arnold and Bennett 1984) were scored on a scale of 0–9.9 (Garland 1988) after snakes were exhausted at the end of endurance trials; higher scores indicate more offensive or aggressive behavior (e.g., striking and biting). Thus, each of the four whole-animal traits was scored twice to assess repeatability. All measurements were made at 30° C, near the mean body temperature for *T. sirtalis* when active in the field (see Garland 1988; Jayne and Bennett 1990b; Schiefflen and de Queiroz 1991; references therein). Snakes were then sacrificed for measurement of ventricle and liver masses, blood hemoglobin content, aerobic and anerobic enzyme activities (Garland, Bennett, and Daniels 1990), and ultimately scale counts (Dohm and Garland 1993).

Multiple Regressions to Reduce Maternal Effects

Estimates of heritabilities and genetic correlations based on partitioning variance and covariance among versus within families of full sibs can be inflated by maternal effects (Falconer 1989; Arnold, chap. 2; Cheverud and Moore, chap. 4), as well as by nonadditive genetic effects (see Brodie and Garland 1993, table 8.1). Multiple regression techniques can be used to partly remove these effects (Garland 1988; Brodie 1989b, 1992, 1993; Garland and Bennett 1990; Garland, Bennett, and Daniels 1990). More specifically, Garland (1988) suggested that in reptiles, which have indeterminate growth, many maternal effects may be mediated through maternal size and/or condition. For example, larger dams tend to give birth to larger offspring. Variation in dam size may be partly genetically based, but much of it will be due to variation in age and/or past nutritional history. To the extent that variation in dam size is due to nongenetic effects (e.g., age, nutrition) and affects offspring characteristics, estimates of genetic variances and covariances based on full-sib analysis of variance (or offspring-on-dam regression) will be inflated. On the other hand, if the variation in dam size is genetically based, and if this variation is genetically correlated with the offspring trait of interest, then removing phenotypic correlations between the offspring trait and dam size will probably remove some of the genetic variance, thus leading to *under*estimation of heritabilities (Tsuji et al. 1989; Brodie and Garland 1993).

Many of the traits we have studied (e.g., endurance, metabolic rate, heart size) correlate strongly with body size (e.g., Garland 1988; Garland and Bennett 1990; Jayne and Bennett 1990a; review in Garland and Losos 1994; see also Travis, chap. 8). In turn, offspring size is correlated with dam size (r = .34–.46: Garland 1988, table 2). Moreover, estimates of heritabilities based on raw trait values exceeded unity for speed and endurance (Garland 1988), clearly indicating inflation. Therefore, I used multiple regression to control statistically for variation due to offspring mass and snout-vent length, maternal mass and snout-vent length, and litter size. For some variables, I also used the amount of mass lost between birth and testing as an independent variable (Garland, Bennett, and Daniels 1990). Residuals from multiple regression equations were used in all quantitative genetic analyses. Heritabilities calculated for residuals never exceeded unity.

Another factor inflating among-family variance components was age at testing. For logistical reasons, not all offspring could be tested at the same age. Thus, for example, endurance and antipredator display were measured when snakes were 6–56 days of age (Garland 1988). In addition (see Discussion in Garland 1988; Garland and Bennett 1990), offspring were not fed. Thus, both ontogenetic and starvation effects would contribute to among-family variance. We therefore used age and age squared at time of testing as additional independent variables. Fasting for such periods of time is not an unusual situation for snakes in nature. Finally, for the enzyme activities, we also used assay batch (coded as a series of dummy variables) as an independent variable.

Results

Repeatabilities

Day-to-day repeatability was highly significant ($P < .0001$) for each of the four whole-animal traits. Pearson product-moment correlations between trials were .674 for antipredator display, .802 for speed, .696 for treadmill endurance, .882 for whole-animal $\dot{V}O_2max$, and .796 for $\dot{V}O_2max$/g body mass (Garland 1988; Garland and Bennett 1990). Corresponding intraclass correlation coefficients were virtually identical, although $\dot{V}O_2max$ showed a slight (about 3%) decrease from day 1 to day 2. In general, these repeatabilities set an upper limit to heritabilities (Boake 1989b; Brodie and Garland 1993; Falconer 1989).

Statistical Control for Maternal Effects

Table 12.1 shows the significant independent variables for the four whole-animal traits. The greatest amount of variance explained was for the physiological trait $\dot{V}O_2max$, due primarily to its strong correlation with body mass. Interestingly, about 5% of the variance in antipredator display is explained by litter size, with individuals from larger litters tending to exhibit somewhat more aggressive or offensive displays. The biological significance of this correlation is obscure. Residual values for the four organismal traits were approximately normally distributed (e.g., Garland 1988, fig. 1).

Heritabilities

Table 12.2 presents heritabilities based on twice the among-family components of variance (Garland 1988; Garland and Bennett 1990; Garland, Bennett, and Daniels 1990). Somewhat surprisingly, the three organismal performance traits (speed, endurance, $\dot{V}O_2max$) show high heritabilities. The biochemical traits show the lowest heritabilities, and the "morphological" traits (hemoglobin, ven-

Table 12.1 Significant predictors of organismal characters

Character (multiple r^2)	Body Mass	Snout-Vent Length	Dam Mass	Dam Length	Litter Size	Age and Age^2
Antipredator display (4.8%)					+4.8	
Sprint speed (32.5%)		+17.1	−1.0	+1.3		13.1
Endurance (35.7%)	+23.8	+1.9			+4.1	7.8
$\dot{V}O_2max$ (71.1%)	+42.7					23.9

Note: Based on stepwise multiple regression analyses ($P < .05$). Values are partial r^2 (percentage), preceded by sign of partial regression coefficient. Partial r^2 due to age and age^2 have been summed for simplicity.

Table 12.2 Estimated heritabilities based on analysis of variance of presumed full-sib families

Character (# families, # individuals)		h^2 and 95% Confidence Interval
Behavior		
Antipredator display (46, 249)		0.19 < **0.41** < 0.71
Organismal performance		
Maximal crawling speed (46, 249)		0.33 < **0.58** < 0.88
Treadmill endurance (46, 249)		0.44 < **0.70** < 1.00
$\dot{V}O_2max$ (45, 245)		0.62 < **0.89** < 1.19
	Mean = 0.72	
Morphology		
Hemoglobin (45, 244)		0.38 < **0.63** < 0.94
Ventricle mass (45, 244)		0.19 < **0.41** < 0.70
Liver mass (45, 245)		0.36 < **0.61** < 0.91
	Mean = 0.55	
Biochemistry		
Liver citrate synthase (45, 242)		0.02 < **0.21** < 0.50
Liver pyruvate kinase (45, 242)		0.32 < **0.58** < 0.90
Ventricle citrate synthase (45, 241)		−0.14 < **0.01** < 0.26
Ventricle pyruvate kinase (45, 242)		0.05 < **0.26** < 0.56
Muscle citrate synthase (45, 244)		−0.08 < **0.09** < 0.34
Muscle pyruvate kinase (45, 242)		0.01 < **0.19** < 0.46
	Mean = 0.22	

Source: Garland 1988; Garland and Bennett 1990; Garland, Bennett, and Daniels 1990.
Note: Residuals from multiple regression equations were used (see text and Table 12.1) 95% confidence intervals are from Bulmer's (1980, 84) algorithm.

tricle mass, liver mass) show intermediate heritabilities, similar to that for antipredator behavior ($h^2 = 0.41$). Analysis of variance indicates that the mean heritabilities of these three categories of traits differ significantly (Garland, Bennett, and Daniels 1990).

Tests for Major Genes

A crude test for the presence of genes with large effect involves comparing the variability of families that, on average, have high or low versus intermediate scores (references and discussions in Garland 1988; Garland and Bennett 1990). Briefly, if "major genes" are present, then families with extreme phenotypes will be fixed for one or the other allele at the major gene locus, and consequently will show less variability than do intermediate families, which will represent a mixture of genotypes. For the four whole-animal traits, Levene's tests (Conover, Johnson, and Johnson 1981; in this application, an analysis of variance based on absolute deviations of each individual's value from its family mean) indicate that families do differ in variability, but quadratic regressions indicate that intermediate families are actually *less* variable.

Phenotypic and Genetic Correlations

Phenotypic correlations were generally low. Figure 12.2 and table 12.3 show phenotypic correlations between antipredator display and each of the other three whole-animal traits. Antipredator display is weakly positively correlated with both speed and endurance, but not with $\dot{V}O_2max$. Importantly, none of the lower-level traits (e.g., blood hemoglobin content, enzyme activities) are significant predictors of variation in antipredator display, in either bivariate correlations or a multiple regression (Garland, Bennett, and Daniels 1990). Unexpectedly, the residuals for speed and endurance are positively correlated, both phenotypically and genetically (Garland 1988). As expected, endurance and $\dot{V}O_2max$ are positively correlated. Also as expected, $\dot{V}O_2max$ is positively correlated with ventricle (heart) mass ($r = .27$; Garland, Bennett, and Daniels 1990). In multiple regression analyses, only 3–4% of the variation in speed and endurance could be explained by lower-level traits (Garland and Bennett 1990; Garland, Bennett, and Daniels, 1990). This r^2 is much lower than in some species of lizards (Garland 1984; reviews in Garland and Bennett 1990; Garland and Losos 1994). The unexplained variation in (repeatable) measures of speed and endurance may be due to correlations with variation in other morphological and physiological characters, such as scale counts or tail length (Arnold and Bennett 1988; Jayne and Bennett 1989; Dohm and Garland 1993), and/or behavioral factors, such as motivation.

Discussion

Why Are Organismal Traits So Heritable?

A phenotypic trait cannot respond to either direct or correlated natural selection, nor will it undergo genetic drift, unless its narrow-sense heritability is greater than zero. Selection tends to

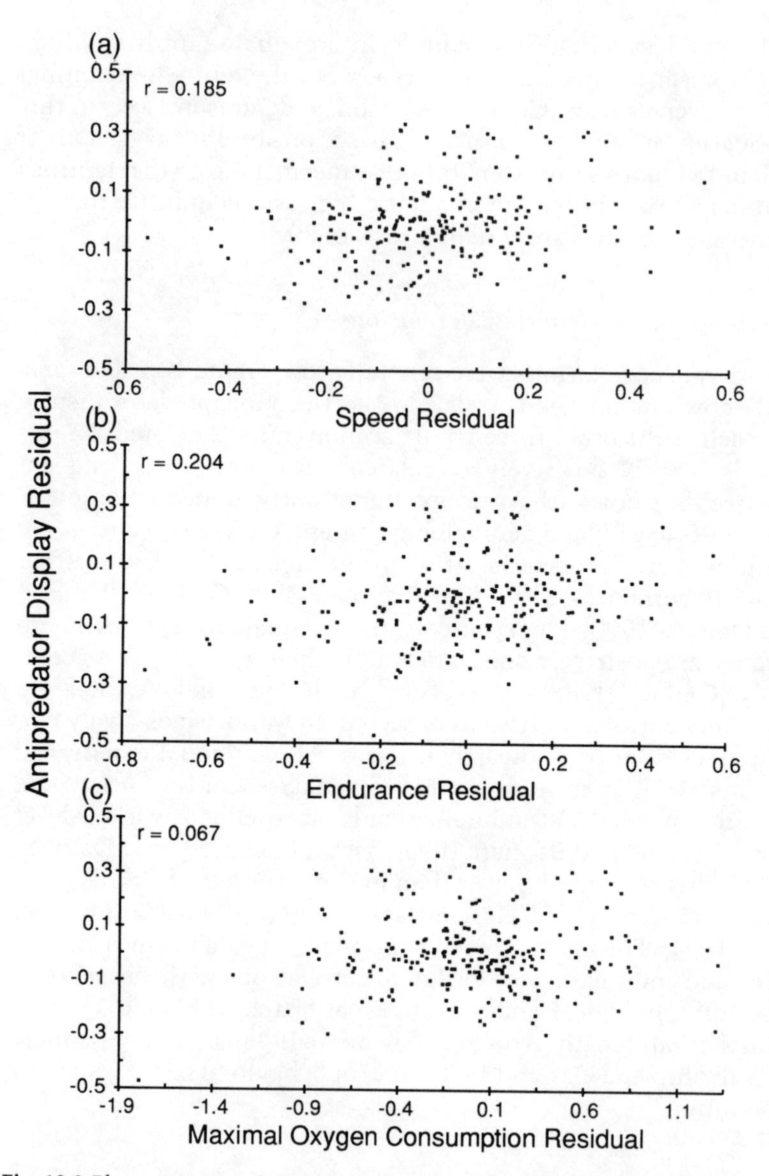

Fig. 12.2 Phenotypic correlations between antipredator display and the three whole-organism performance traits: (*a*) speed, (*b*) endurance, and (*c*) $\dot{V}O_2$max.

Table 12.3 Phenotypic, genetic, and environmental correlations between traits measured at the whole-animal level

	Correlation	Sprint Speed	Treadmill Endurance	$\dot{V}O_2max$
Antipredator display	Phenotypic	0.185*	0.204*	0.067
	Genetic	0.407	0.240	−0.056
	Environmental	−0.029	0.171	0.390
Sprint speed	Phenotypic		0.359*	−0.012
	Genetic		0.585*	−0.159
	Environmental		−0.006	0.470
Treadmill endurance	Phenotypic			0.166*
	Genetic			0.098
	Environmental			0.546

Note: Phenotypic, genetic (maximum likelihood: Shaw 1987; Garland 1988; Garland and Bennett 1990), and environmental (maximum likelihood) correlations (from top to bottom within each trait) are based on residuals from multiple regression equations (see text and table 12.1).
*$P < .05$ by two-tailed test; sample sizes are given in table 12.2.

eliminate additive genetic variation through fixation of favored alleles, while mutation tends to add genetic variation each generation. A mutation-selection balance may therefore be reached such that the heritability of a trait reflects, in part, the relative strength of past selection and mutation. Thus, traits that have been subject to particularly strong selection may be expected to show low heritabilities (recent reviews in Rose, Service, and Hutchinson 1987; Lande 1988; Turelli 1988a; references in Arnold 1990; Garland, Bennett, and Daniels 1990; but see Price and Schluter 1991). Empirical support for this idea comes from the observation that life history traits generally show lower heritabilities than do morphometric, physiological, or behavioral traits (Mousseau and Roff 1987; Roff and Mousseau 1987; Willis, Coyne, and Kirkpatrick 1991; Partridge, chap. 6; references in Garland, Bennett, and Daniels 1990; Price and Schluter 1991).

Measures of whole-animal performance abilities, such as locomotor speed and endurance, are also thought to be subject to relatively strong selection (Huey and Stevenson 1979; Arnold 1983a; Pough 1989; Jayne and Bennett 1990b; Garland and Carter 1994; Garland and Losos 1994). Thus, we might expect them to show relatively low heritabilities, as compared with lower-level traits. In addition, measures of organismal performance are composite traits, determined by multiple underlying traits, and thus are subject to additional sources of environmental variance (as argued by Price and Schluter [1991] for life history traits). Because environmental variance occurs in the denominators of the formulae for heritability, this alone could lead to relatively low values. Contrary to these expectations, however, my three measures of locomotor performance and antipredator display all showed relatively *high* heritabilities (table 12.2). I will briefly consider several possible explanations for this finding (see Garland, Bennett, and Daniels [1990]).

First, as composite characters (cf. Riska 1989; Price and Schluter 1991), measures of organismal performance will incorporate mutational input of genetic variance at loci affecting all lower-level characters that affect performance. This alone could lead to more equilibrium genetic variance in organismal performance than in lower-lever traits. Second, narrow-sense heritabilities will be overestimated if nonadditive genetic effects and/or maternal effects are present (see below). However, such overestimation would not necessarily explain why performance traits should show higher heritabilities than do the enzyme activities. Third, the population we studied might not be at genetic equilibrium, possibly due to relatively recent disturbance by such human activities as cattle ranging and logging (Garland, Bennett, and Daniels 1990; Brodie 1992, 1993). Fourth, locomotor performance might show a trade-off with some other components of fitness, such as growth rate or litter size (cf. Rose, Service, and Hutchinson 1987; Price and Schluter 1991; Partridge, chap. 6; Roff, chap. 3). However, Jayne and Bennett's (1990b) data indicate *positive* correlations between locomotor performance and survival. Unfortunately, we have no data on other fitness components (such as litter size or growth rates) for the newborn snakes we studied. Fifth, perhaps a nonheritable trait, such as nutritional status, affects both locomotor performance and fitness in nature (cf. Price, Kirkpatrick, and Arnold 1988).

Finally, perhaps selection on locomotor performance simply is not as strong as conventional wisdom might dictate. Jayne and Bennett (1990b) report significant positive directional selection acting on both speed and endurance for yearling snakes but *not* for newborns during their first few months of life. Selection intensities were similar to those reported by other workers in a variety of contexts, but it is not clear what level of genetic variance would be expected to obtain. Brodie (1992) also showed no significant selection acting on speed or on distance crawling capacity in *Thamnophis ordinoides* during their first year.

Several factors might account for relatively weak selection on the locomotor performance of young snakes. First, as suggested in figure 12.1, behavior may act as a "filter" between selection and performance (Garland, Bennett, and Daniels 1990; Garland and Carter 1994; Garland and Losos 1994). Selection acts on what an animal actually does, not on what it can or could do. If, for example, all individuals in a population employed stationary antipredator displays, then variation in locomotor performance might be irrelevant. Second, the effectiveness of a behavior involving locomotion may be strongly context dependent, influenced by the substrate, temperature, recent feeding, and type and experience of predator or prey involved (cf. Garland and Arnold 1983; Arnold and Bennett 1984; Ford and Shuttlesworth 1986; Herzog and Burghardt 1988; Brodie 1989a, 1991, 1992; Herzog, Bowers, and Burghardt 1989b). Third, variation in context also might lead to lower repeatability for field locomotor performance and behavior than is apparent in laboratory studies (Boake 1989b). For example, variation in temperature can cause variation in locomotor performance or behavior through direct biochemical (Q_{10}) effects on muscle and nerve function (e.g., Bennett, Garland, and Else 1989; Bennett 1990; references in Huey and Stevenson 1979; Garland and Losos 1994). Also, several species of garter snakes and lizards are known to alter their defensive behavior at low temperatures, after recent feeding, or when gravid (Arnold and Bennett 1984; Herzog and Bailey 1987; Seigel, Huggins, and Ford 1987; Brodie 1989a; Schiefflen and de Queiroz 1991). Lower field repeatability would have the effect of lowering the selection intensity on a trait; it is as if the selective agent cannot precisely define the phenotype (Boake 1989b; Arnold 1990; Brodie 1991, chap. 1; Brodie 1992). Fourth, perhaps individual variation in laboratory measures of locomotor performance (on a smooth racetrack or treadmill) simply does not reflect the variation or covariation that would occur under natural conditions; laboratory and field performance may not be tightly correlated either phenotypically or genetically (cf. Arnold and Bennett's [1984] descriptions of antipredator displays in the field versus the laboratory). For example, since low temperature reduces sprint speed and may cause more aggressive antipredator displays (see Arnold and Bennett 1984; Schiefflen and de Queiroz 1991), the effective correlation between these two traits in the field—where temperatures of active snakes show considerable variation—might be negative in contrast to the positive correlation observed at a constant temperature of 30° C (see fig. 12.2). Fifth, as suggested by Jayne and Bennett's

(1990b) results, selection on locomotor performance may act primarily on older snakes, whereas our heritability estimates were for neonates; the genetic correlation between locomotor capacities at birth and at older ages is entirely unknown. Genetic correlations between elements of locomotor performance measured at different ages are certainly likely to be less than unity, such that selection acting at one age may have little effect on genetic variances of the corresponding behavior at other ages (Arnold 1990). Finally, perhaps selection acts more strongly on absolute than on relative performance (cf. Huey and Bennett 1987, 1105; Bennett 1991, 2); all of our analyses have been on residual scores, correcting for variation related to size and age (this computation of residuals was necessary to remove artifactual sources of variation). Resolving the foregoing possibilities will be a challenge for behavioral and physiological ecologists.

Why Don't Speed and Endurance Show a Trade-off?

Although I have argued that knowledge of behavioral or physiological mechanisms may allow predictions of genetic correlations, the one example in the present study—the trade-off between speed and endurance hypothesized on physiological grounds (Garland 1988)—is refuted by empirical evidence (see table 12.3; see also Jayne and Bennett 1990a). Brodie (1993) also reported significant positive phenotypic and genetic correlations (both between .3 and .5) between sprint speed and distance crawling capacity (treadmill endurance was not measured) in both of two populations of *T. ordinoides.*

One line of evidence I cited (Garland 1988) as suggesting a necessary trade-off was the fact that world-class sprinters and marathoners are different individuals with different physiques and muscle fiber types. But perhaps this example is misleading. World-class athletes are a highly selected subsample of the human population (cf. Wallace 1991). If we consider a random sample of humans, it may be that speed and endurance are actually *positively* correlated—that is certainly what I remember from physical education classes! Perhaps it is only when we consider the subset of the very best athletes that there is a negative correlation between speed and endurance. This argument is illustrated graphically in figure 12.3, and is reminiscent of the antagonistic pleiotropy model for the evolution of negative genetic correlations between major components of fitness (e.g., Rose, Service, and Hutchinson 1987;

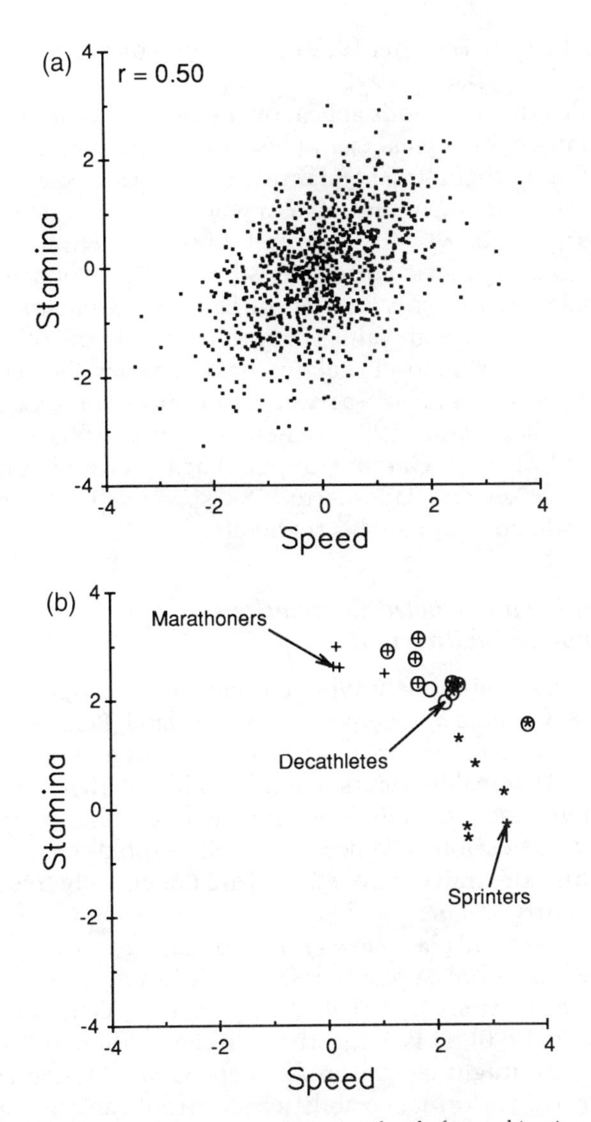

Fig. 12.3 (*a*) One thousand data points drawn randomly from a bivariate normal distribution with a correlation of +.50, means of zero, and standard deviations of unity. (*b*) Most extreme 1% ($N = 10$) of points in terms of speed (asterisks), endurance (plus signs), or their sum (circles). Selection based on these criteria is not always mutually exclusive; thus, the total N is less than 30. Within this highly selected subpopulation, analogous to that represented by world-class athletes (cf. Wallace 1991), speed and endurance actually show a negative correlation, suggesting a trade-off.

Cheverud 1988b; Falconer 1989); a different but related argument is presented by Wallace (1991).

The idea that trade-offs appear only in extreme subsamples may apply to many biological examples. For instance, it has been suggested that mathematical and musical ability are positively correlated in the human population as a whole, but are uncorrelated in geniuses (J. F. Crow, pers. comm.). Similarly, Hiraizumi (1961) found negative genetic correlations between rate of development and female fertility only in lines of *Drosophila melanogaster* exhibiting relatively high values of both traits. On the other hand, a comparison of two closely related lizard species showed a physiologically based trade-off between capacities for speed and endurance (Huey et al. 1984), whereas comparisons of multiple species of lizards (T. Garland, unpublished data) and of mammals (Garland, Geiser, and Baudinette 1988) spanning broad phylogenetic ranges do not show such a trade-off.

Phenotypic and Genetic Correlations: Constraint or Facilitation?

The fact that some phenotypic correlations are significant (Garland 1988; Garland and Bennett 1990; Garland, Bennett, and Daniels 1990) means that some correlated selection (Arnold, chap. 2; Brodie 1992) probably occurs in nature. The fact that some genetic correlations are also significant means that some correlated responses to selection will occur as well (Arnold, chap. 2; Brodie 1993). Thus, the traits I have studied are not entirely free to evolve independently.

Antipredator display shows no significant genetic correlations with the other measured traits, and only two significant phenotypic correlations (see table 12.3, fig. 12.2; Garland, Bennett, and Daniels 1990). This is surprising because different extremes of this behavior might be expected to depend on different aspects of physiological performance abilities (cf. Arnold and Bennett 1984; Garland 1988, 1993; Bennett 1991; Garland and Losos 1994). One might expect "correlational selection" to favor particular combinations of antipredator behavior and locomotor performance capacities (Garland 1988), and these combinations may change with age and size (Brodie 1991, chap. 1), leading to coadaptation (cf. Huey and Bennett 1987) as well as to the evolution of appropriate genetic correlations. In fact, Brodie (1989b, 1993) has detected correlational selection for combinations of antipredator behavior and color pattern.

Apparently, the observed genetic correlations do not place important "constraints" on the joint evolution of the traits I have studied. In fact, the positive genetic correlation between speed and treadmill endurance may actually *facilitate* adaptive evolution, given that positive directional selection may occur on both (Jayne and Bennett 1990b). If both speed and endurance are generally under positive selection, then perhaps the positive genetic correlation between them is partly the result of past selection pressures (cf. Cheverud 1984b, 1988b; Brodie 1992, 1993). Although comparative data for other snakes are scarce (see references in Garland 1988; Garland and Losos 1994), garter snakes appear to be neither exceptionally fast nor enduring; they have not (yet?) evolved to phenotypic extremes at which physiological trade-offs may become ineluctable (see end of previous section).

The significant genetic correlations agree with intuitive reasoning in one case (positive between $\dot{V}O_2$max and heart size: Garland, Bennett, and Daniels 1990) but not in the other (negative between speed and endurance: Garland 1988) (see also Dohm and Garland 1993 for examples involving scale counts). The existence of unexpected genetic correlations may suggest underlying mechanisms different from those that had previously been imagined (cf. Arnold 1981a, 505–6; Arnold 1981c; Garland 1988, 345; Garland and Carter 1994).

The constancy of genetic correlations over evolutionary time should depend in part on their origin. Genetic correlations due to linkage disequilibrium, for example, should dissipate rather quickly (Lande 1980a, 1988; Falconer 1989). Those due to fundamental biochemical, physiological, or biochemical interactions or constraints should last longer (cf. Maynard Smith et al. 1985; Clark 1987, 935; Roff, chap. 3; Arnold, 1992a). Brodie (1993) reported no significant differences in the genetic variance-covariance matrices for two populations of *Thamnophis ordinoides*, based on speed, distance crawling capacity, tendency to reverse crawling direction, and color pattern (see Brodie and Garland 1993 for other examples with snakes).

Limitations of Full-Sib Heritabilities and Genetic Correlations

This study is based on information only from sets of full sibs because we do not know paternity and hence cannot identify half sibs. Although mothers were available, all of the measured traits change ontogenetically such that any measurements of the dams would be affected by age (unknown for our dams), body size, and uncertain ontogenetic repeatability (Garland 1985; Herzog and Burghardt 1988; van Berkum et al. 1989; Arnold 1990; Jayne and Bennett 1990a; Brodie 1991, chap. 1; Shaffer, Austin, and Huey 1991). Thus, newborn offspring-on-dam regressions would be unreliable. Captive breeding programs are possible (Arnold 1981b) but are difficult for several reasons: garter snakes may store sperm, they do not breed as reliably in the laboratory as do *Mus* or *Drosophila,* and they take 2–4 years to reach sexual maturity. The foregoing constraints may apply to many natural populations; therefore, sets of full sibs may be the only relatives from which information will be available. In this section I review the reasons that estimates of genetic parameters based only on full-sib data may differ from estimates based on measures of other relatives (Arnold, chap 2).

Full-sib data alone do not allow estimation of either narrow-sense or what is conventionally defined as "broad-sense" heritability (Arnold, chap. 2; Brodie and Garland 1993). Instead, twice the among-family component of variance estimates additive genetic variance plus half of the dominance variance plus half of the epistatic variance plus twice the common family environmental effects (including maternal effects) (Falconer 1989; Cheverud and Moore, chap. 4). Estimates of genetic correlations from full-sib data are also potentially inflated by influences of dominance, epistasis, and environmental factors (Falconer 1989).

Unfortunately, we do not have a good idea of how large the nonadditive genetic effects might be for traits we studied in garter snakes, nor do we know whether they might differ in magnitude among traits (see also Schwartz and Herzog 1993; Dohm and Garland 1993). Thus, we have no reasonable way to correct for the potential upward biases their presence might introduce into estimates of (narrow-sense) heritabilities and genetic correlations. On the other hand, two recent reviews suggest that full-sib heritability estimates often are not much larger than narrow-sense heritabilities (Roff and Mousseau 1987; Mousseau and Roff 1987; but see Dohm and Garland 1993 on scale counts in the present snakes).

We were able to reduce the influence of potential common family environmental effects via husbandry: specifically, newborns were separated and housed in separate containers on the day of birth, so the problem was eliminated from that point on. For both logistical reasons and to minimize age variation, testing of individuals was done in blocks of six families. Although this may have allowed more common family environmental effects to creep in, we do not consider this a likely possibility, and age was used as an independent variable when computing residuals. In general, any environmental effect experienced uniquely by all members of a single family is of concern, because it will tend to make them deviate from the overall population mean (Boake, chap. 14). As our heritability estimates are based on the among-family component of variance, they may be inflated by common family environmental effects.

In our study, as in several others of garter snakes, these common maternal environments ended on the day of birth. Additionally, all dams experienced a common set of controlled laboratory conditions for about one month prior to giving birth. These two procedures will not, of course, remove possible effects of dam age and size, although the latter might reduce variation in dam condition and hence its effects. Our use of residuals controlled statistically for the effects of dam size, condition, age, and litter size. In all cases, heritabilities estimated for residual traits were lower than those estimated for raw traits, suggesting that we were at least partly successful. On the other hand, this may also have removed some genetic variance from the traits of interest (Tsuji et al. 1989; Brodie and Garland 1993).

Heritability estimates for organisms raised in the laboratory should tend to be higher than for those raised in nature (Riska, Prout, and Turelli 1989; Willis, Coyne, and Kirkpatrick 1991). This potential problem is reduced in the present study because newborn snakes were studied. Moreover, they experienced more than half of their development inside dams raised under field conditions. In any case, no simple correction procedure is possible. The technique described by Riska, Prout, and Turelli (1989; see also Schwartz and Herzog 1993) for estimating "field heritabilities" with offspring-on-parent regressions was not possible in the present study because offspring were not raised to their parent's age prior to measurement.

A final possible confounding factor is multiple paternity. Under multiple paternity, some individuals within a litter are half sibs rather than full sibs, and the among-family component of variance should be multiplied by a factor greater than two (Falconer 1989). Schwartz, McCracken, and Burghardt (1989) demonstrated multiple paternity in over half of the litters from two midwestern populations of *Thamnophis sirtalis*, and calculated that treating all families as if they were composed of full sibs would lead to underestimating heritabilities by a factor of 1.35. Thus, our values will have underestimated heritabilities if multiple paternity was present.

Although we do not know exactly how the various factors listed in this section may have biased our estimates of genetic parameters, such estimates represent an important first step toward understanding the inheritance of behavior in natural populations (cf. Arnold 1981a,b,c, 1988; Arnold, chap. 2).

Suggestions for Future Research

Although garter snakes have long generation times, they are often easy to capture in large numbers and can be maintained relatively easily in captivity. Thus, future studies may be able to use other breeding designs to estimate true narrow-sense heritabilities and genetic correlations. Estimation of these parameters would go a long way toward answering many of the questions raised above. One could also use breeding designs or population crosses (e.g., Arnold 1981b,c) to test for the presence of directional dominance (cf. Lynch, chap. 13). Traits that have been subject to directional selection are predicted to show directional dominance in the favored direction (Broadhurst and Jinks 1974; Mather and Jinks 1982); thus, we might predict that both speed and endurance will show dominance for higher performance. Directional dominance for antipredator display could suggest whether past selection had favored either offensive (aggressive) or defensive behavior. On the other hand, ambidirectional dominance would suggest that past selection had been stabilizing (cf. Rose, Service, and Hutchinson 1987; Travis 1989b).

For predicting the course of microevolution in nature, estimates of genetic parameters for animals raised in the field are necessary. Field heritabilities are expected to be lower than for laboratory-reared animals because environmental sources of variance should be much greater in nature (Riska, Prout, and Turelli 1989; Schwartz and Herzog 1993).

Compared with garter snakes, few organisms offer such rich opportunities for integrative studies of multivariate phenotypic evolution (but see Lynch, chap. 13). Their antipredator displays, open-field behavior (e.g., Herzog and Burghardt 1986), color patterns, and scale counts are particularly noteworthy because all may interact through the common pathway of locomotor abilities (Arnold and Bennett 1984, 1988; Garland 1988). Thus, recent work by Brodie (1989a,b, 1992, 1993) focuses on the functional and genetic integration of antipredator behavior and color pattern in *Thamnophis ordinoides*. Arnold (pers. comm.) is investigating the constancy of genetic variance-covariance matrices for scale counts in natricine snakes. Finally, Dohm and Garland (1993) have measured scale counts in the same individuals studied herein, thus allowing extension of our present comparisons to morphological traits of a very different type. Further comparisons of correlated selection and genetic correlations for different types of traits—such as foraging or antipredator behavior in relation to speed and stamina—would be of particular interest (Arnold 1988; Brodie 1992, 1993).

The relatively long life span of garter snakes means that information on the ontogenetic consistency of individual differences at the phenotypic level—and of parameters in the genetic variance-covariance matrix—is crucial for a complete understanding of multivariate evolution (e.g., Arnold 1981a, 1990). The available data for reptiles indicate that repeatabilities of behavioral and locomotor traits may be statistically significant for relatively long portions of the life span (e.g., 1 year), but that they are generally much lower than day-to-day repeatabilities (Arnold and Bennett 1984, 1988; van Berkum et al. 1989; Jayne and Bennett 1990a; Brodie 1991, chap. 1). Limited results of Jayne and Bennett (1990a) suggest that heritabilities of locomotor performance may decrease with age. Herzog and Burghardt (1988) show that family differences in antipredator responses can be consistent for at least 1 year in captivity, but do not provide heritability estimates. Again, garter snakes would be good models for such studies, because they exhibit many traits that change ontogenetically and some that do not (e.g., scale counts: Arnold 1988; Dohm and Garland 1993).

Finally, we need information concerning how often and under what conditions animals actually use their maximal abilities. Similarly, we need to determine the response of free-living animals to natural predators. How often is speed an important determinant of escape success? Is maximal speed ever used in foraging? Do the answers to these questions depend on the type of predator or prey involved? Answering such questions is not easy, because many animals—including garter snakes—are difficult to observe in nature and because the events of interest are relatively rare, thus requiring long and continuous periods of direct observation. Nonetheless, direct observations of individual animals (yielding "quantitative ethograms") could allow us to get much closer to understanding how, exactly, selection in nature occurs (cf. Pough 1989; Garland and Losos 1994). Integrating information on selection and its agents with knowledge of inheritance under natural conditions is an obtainable goal for some organisms.

Summary

Morphological, physiological, and biochemical traits determine organismal capacities for locomotion, which in turn set ultimate limits within which normal behavior must be accomplished. This hierarchy suggests that behavioral traits or measures of whole-animal "performance" capacities (e.g., maximal speed, endurance) will be of more direct ecological and selective importance than are lower-level morphological or physiological traits (e.g., heart size, enzyme activities). In turn, quantitative genetic theory predicts that, in populations at genetic equilibrium, traits of greater selective importance should exhibit lower heritabilities. Heritabilities are important because they determine how rapidly (if at all) traits can respond to natural (or artificial) selection. A. F. Bennett, C. B. Daniels, and I have therefore compared heritabilities of thirteen traits representing different levels of biological organization by studying offspring born to wild-caught, gravid *Thamnophis sirtalis.*

Quantitative measures of the four whole-animal traits (antipredator display, sprint speed, treadmill endurance, maximal oxygen consumption) were highly repeatable on a day-to-day basis. Prior to genetic analyses, we computed residuals from multiple regression equations in an attempt to statistically remove variation related to body size, dam size, litter size, and age (as well as assay batch for enzyme activities). This treatment is presumed to have reduced the magnitude of maternal effects mediated through dam size and/or condition, but may also have removed some genetic variation. Contrary to theoretical expectations, the organismal performance traits (speed, endurance, $\dot{V}O_2max$) generally showed higher broad-sense heritabilities (h^2 = 0.58–0.89) than did morphological, physiological, or biochemical characters (h^2 = 0.01–0.63), with antipredator display showing an intermediate heritability of 0.41. Phenotypic correlations among traits were generally low, and antipredator display was not significantly correlated with any lower-level trait. However, a few morphological and biochemical traits did correlate significantly with the three measures of organismal performance. Significant genetic correlations existed between speed and endurance (+.59–a surprising result), between $\dot{V}O_2max$ and relative ventricle mass (+.64), and between liver and ventricle pyruvate kinase activities (−.55). Because antipredator display does not appear to be genetically correlated with any other measured trait, it should be relatively free to evolve independently.

Acknowledgments

I thank C.R.B. Boake and A. V. Hedrick for the invitation to participate in the symposium that spawned this book. Much of this research was done in collaboration with A. F. Bennett and C. B. Daniels, with assistance from S. J. Arnold, who provided initial motivation and encouragement for this work, as well as study animals from his field sites. C.R.B. Boake, E. D. Brodie III, M. R. Dohm, R. E. Jung, J. M. Schwartz, and an anonymous reviewer provided helpful comments on the chapter, and E. D. Brodie III kindly provided a copy of his unpublished Ph.D. dissertation. Bruce C. Jayne graciously supplied the snake photograph. Financial support was provided first by NSF Grant DEB-8214656 to T. G. and A. F. Bennett, then by other NSF grants to A.F.B., and most recently by NSF Grants BSR-9006083 and BSR-9157268 to T. G., as well as by the University of Wisconsin Graduate School.

13

Evolutionary Inferences from Genetic Analyses of Cold Adaptation in Laboratory and Wild Populations of the House Mouse

Carol Becker Lynch

Characteristic nests built by males from lines selectively bred for differences in nest building; from left to right, mice are from the high, control (randomly bred), and low lines.

The study of cold adaptation in the house mouse, *Mus domesticus*, has proved to be an ideal system for combining direct ecological information about selective forces with deductions about the selective history of traits obtained from genetic analysis. The primary selective force influencing cold adaptation is temperature, which can be easily manipulated under laboratory conditions. Rodents have been standard models for physiologists for some time, and traits involved in their temperature regulation have been well described (e.g., Hart 1971; Cabanac 1975). Both laboratory and natural populations of the house mouse are capable of breeding under quite cold conditions (Laurie 1946; Barnett et al. 1975; Berry, Bonner, and Peters 1979; Marsteller and Lynch 1987a,b), and the species is not photoperiodic (Bronson 1979); thus, temperature is the major cue for seasonal adjustments, given sufficient food availability (Bronson 1979; Marsteller and Lynch 1987a,b). Not only is *M. domesticus* the best known mammalian genetic system, but wild mice are also easily bred in the laboratory, interbreed freely with laboratory mice, and the generation time (3 months) is relatively short. Only with *Drosophila melanogaster* and *M. domesticus* can we take advantage of the existence of both genetically defined laboratory populations and widely distributed natural populations so that hypotheses generated in one sphere can be tested in the other. This availability of complementary laboratory and natural populations, exploited so effectively by *Drosophila* geneticists, is unique to *M. domesticus* among vertebrate organisms.

I have used quantitative genetic analyses of genetically well defined laboratory populations to develop specific predictions about evolutionary adaptation, and have tested these predictions on natural populations. Adaptation is not usually a matter of changing one or two genes, or even one or two traits, but can involve the whole organism. For this reason I measured a variety of traits, involving morphology, physiology, and behavior, in an attempt to "sample" the organism.

The most important genetic parameters for predicting adaptation (response to natural selection) are the heritabilities of and genetic correlations among the traits of interest. Heritability (h^2) describes the proportion of total observed variation in a single trait due to average effects of alleles that are passed from parents to offspring. Thus, heritability predicts the univariate response to selection acting on that trait. By definition, natural selection acts to improve fitness, and because only *additive* genetic variance can be depleted by selection, heritability estimates can also be used retrospectively in speculation on the relative intensity of prior natural selection on the measured trait. Similarly, genetic correlations (r_A), based on additive genetic covariance between traits, describe the extent of pleiotropy, or the extent to which traits are influenced by the same genes. Such a correlation accounts for the biological interdependence of two traits, and is a valuable tool for understanding the genetic basis of adaptive trait complexes such as cold adaptation. For mechanistic studies, genetic correlations define the path from genes to behavior, and can be used to help identify physiological mechanisms underlying behavioral variation (e.g.,

Lynch, Lynch, and Kliman 1989; Lynch and Lynch 1992; Bult et al. 1992).

Genetic correlations may place constraints on adaptation if selection for one of the traits would produce a maladaptive response in another; this is referred to as "antagonistic pleiotropy." For example, large size in mice is directly related to fitness through larger litters and better lactational ability of females (Falconer 1977) and improved competitive ability in territorial encounters between males (DeFries and McClearn 1970). However, selection for increased body size in mice also results in decreased activity and reaction time, and probably adversely affects the ability to escape predators (Falconer 1977). Thus, most wild populations of house mice, with some exceptions on islands with few or no predators, are smaller than their laboratory counterparts (Berry, Bonner, and Peters 1979).

Predictions from Laboratory Populations

A major question relating to the genetic basis of the evolution of behavior was whether estimates of additive genetic variance and covariance among traits associated with cold adaptation, obtained from extensive analyses of laboratory populations, could be used to predict the differences observed among natural populations of house mice sampled from a temperature cline along the East Coast of the United States. Of these traits, those that possess substantial heritabilities in the absence of antagonistic pleiotropy should be easily altered by natural selection. Those with negligible heritabilities should not. The house mouse is a useful model for this test because it is not a native species, and its relatively recent introduction to North America means that we are observing the effects of relatively short-term natural selection.

Because actual selection gradients were not measured, and the natural populations were not the ones from which estimates of genetic parameters were obtained, it was possible to make only qualitative predictions. However, because the effect of temperature on the observed traits has been studied in this and other laboratories (e.g., Lynch et al. 1976), it was also possible to predict the direction of phenotypic response. The measured traits, and the bases for including them, were body weight, because it is often argued that relatively larger animals are better cold-adapted because they lose less heat (Mayr 1963); thermoregulatory nesting, because small mammals build larger nests in response to cold (references in

Bouchard and Lynch 1989); body temperature, because lower temperature could result in lower metabolic demands in cooler climates (Scholander et al. 1950); and proportional weight of brown adipose tissue, a thermogenic organ (the major source of nonshivering thermogenesis) that increases in size in small mammals exposed to cold (Chaffee and Roberts 1971).

Heritability Analysis

Heritability estimates were obtained from three separate experiments with two different designs. Methods of genetic analysis, especially as applied to behavioral data, are discussed by Roberts (1967a,b), and several methods of heritability estimation of open-field behavior are evaluated by DeFries and Hegmann (1970). In a randomly mating population, heritability may be estimated from the regression of offspring scores on parental means, or from twice the regression of offspring on either parent, because the covariance of parents and offspring contains half the additive genetic variance. This method of parent-offspring regression was used in experiment I, in which the laboratory population came from a highly heterogeneous outbred stock (HS/Ibg) originally derived from a cross among eight inbred strains (see McClearn, Wilson, and Meredith 1970 for the origin of the stock). At the time of this experiment the population had undergone over 50 generations of random mating. There were 225 families represented, with a total of 1,163 mice measured.

Alternative methods are available for line or strain crosses, in which the populations to be crossed are inbred or otherwise genetically differentiated. In a diallel cross, any number of inbred strains are crossed in all possible combinations, including reciprocal crosses. Analysis of variance of such a cross provides information about heritabilities, as well as the extent of dominance for each measured trait. The theory and analysis of the diallel cross is explained by Griffing (1956a,b) and Hayman (1954a,b). Broadhurst (1967) describes diallel analysis with special reference to behavior, and Crusio, Kerbusch, and van Abeleen (1984) provide a worked example. Heritability estimation is based on the concept of "general combining ability," which is the average performance of a strain. General combining ability is estimated as the difference between the mean of the F_1s having that strain as one parent and the overall mean of all F_1s. Variance in general combining ability among the strains gives an estimate of variance in the average effects of alleles carried by each strain, and is equivalent to half the additive genetic variance that would be present in a random-bred population containing the same genes. (See table 13.1 for the equation for heritability.) Analyses of diallel crosses allow a test for linkage so that its effect on inflation of heritability estimates can be noted (Hayman 1954a,b; Lynch and Sulzbach 1984). Experiment II was a diallel cross (strains crossed in all possible combinations, including reciprocals) among the inbred strains BALB/cIbg, C57BL/6J, C3H/2Ibg, and DBA/1BG. Total sample size was 576.

The population in experiment III was an outbred stock derived from the strains in experiment II. Parent-offspring regression was used to estimate heritabilities. Data were available from over 400 families, with a total of 3,253 mice measured.

Details of experiments I–III can be found in Lacy and Lynch (1979), Lynch and Sulzbach (1984), and Lynch, Sulzbach, and Connolly (1988), respectively. In all three experiments, heritabilities were estimated for body weight measured at the beginning and end of the experiment, nesting behavior measured at room temperature (21° C), and lipid-free dry weight of brown adipose tissue measured after 4 to 6 weeks of cold acclimation. Heritability of nesting behavior at cold temperatures (4°–5° C) was estimated in two of the experiments.

Heritability estimates from the three experiments are listed in table 13.1. The two body weight measures and nest building at room temperature all exhibit high heritabilities. In contrast, body temperature and brown adipose tissue weight have very low heritabilities, with the exception of the estimate for brown adipose tissue from experiment III. The uniformly higher estimates from experiment III may be due to the effects of linkage disequilibrium, especially the rather high values for the nesting measures. This population was only a few generations removed from an inbred-strain cross and had not experienced a sufficient number of randomly breeding generations to reach linkage equilibrium. Analysis of the diallel cross indicated nonrandom (linked) distribution of alleles influencing nest building in the inbred populations. Linkage of multiple loci influencing a trait will usually inflate the estimate of additive genetic variance for that trait because, in the absence of independent segregation, additive genetic covariance between linked loci will be included in the estimate of additive genetic variance. Nevertheless, in the absence of antagonistic pleiotropy, body weight and nest building would be expected to respond to selection, while body temperature and brown adipose tissue would not.

Table 13.1 Heritability estimates (±SE) from three experiments for traits related to cold adaptation in laboratory populations of *Mus domesticus*

Trait	Exp. I[a]	Exp. II[b]	Exp. III[c]
Body weight (g) at beginning of experiment	0.30 ± 0.09	0.22 ± 0.08	0.43 ± 0.05
Body weight (g) at end of experiment	0.41 ± 0.05	0.48 ± 0.09[e]	0.55 ± 0.05
Nesting score (g) at room temperature	0.31 ± 0.08	0.28 ± 0.06[d]	0.53 ± 0.03
Nesting score (g) at cold temperature	—	0.18 ± 0.18	0.31 ± 0.04
Body temperature (°C)	−0.20 ± 0.14[d]	0.03 ± 0.02	0.11 ± 0.04
Brown adipose tissue (mg/g body weight)	0.08 ± 0.06[d]	0.09 ± 0.05[e]	0.24 ± 0.09[d]

Source: Reprinted from C. B. Lynch, 1992, Clinal variation in cold adaptation in *Mus domesticus*: Verification of predictions from laboratory populations, *Am. Nat.* 139:1219–36, © 1992 by The University of Chicago.

[a]Estimates from twice the regression of offspring on parents, pooled across sex unless indicated otherwise; population originally derived from a cross among eight inbred strains. Data from Lacy and Lynch 1979.

[b]Estimates from diallel analysis of four inbred strains: h^2 = 2(general combining ability variance)/[2(general combining ability variance) + specific combining ability variance + reciprocal variance + error variance]; average of estimates from each sex unless indicated otherwise. Data from Lynch and Sulzbach 1984.

[c]Estimates as in exp. I; population derived from intercrossing the four strains in exp. II. Data from Lynch, Sulzbach, and Connolly 1988.

[d]Estimate from males only.

[e]Estimate from females only.

These experiments had large sample sizes, so the estimates of low (nonsignificant) heritabilities were quite precise, adding confidence to these predictions.

Artificial Selection for Nesting Behavior

As behavior is an animal's first line of defense against a changing environment, it is not surprising that nesting behavior has proved to be the most interesting trait of those measured; thus, my work has emphasized nest building. In all of these experiments, mice are singly housed, and nest building is measured by placing a pre-weighed roll of cotton batting on the cage lid and weighing the cotton remaining each day for 4 consecutive days. The nest is removed each day, so the "total nesting score" represents the amount of cotton used to build four individual nests. Another experiment showed a very high correlation between total nesting score and nest quality (Lynch and Roberts 1984). This relationship is also obvious from looking at the nests built by the selected lines (see chapter opening figure).

The base population for the selection experiment was the same outbred stock (HS/Ibg) used in experiment I. There were two replicates each of the high-nesting, control, and low-nesting lines. Within-family selection (eight families per line) was used in order to minimize inbreeding (every family contributes to each generation) and to control for the effects of common environment, including maternal effects. After 15 generations of selection (Lynch 1980), the realized heritability was estimated from the regression of the response to selection (measured as the difference between the selected and control lines at each generation) on the cumulative selection differential (difference between the selected parents and their family means, added across generations), and was corrected for the within-family selection design. The standard error was calculated according to Hill (1972). As predicted from the heritability of nest building in the base population (Lacy and Lynch 1979), nest building responded to selection, and the realized heritability (0.28 ± 0.05) agreed well with the previous estimate (experiment I; see table 13.1). The results of 48 generations of selection for nest building are shown in figure 13.1.

Selection on one trait will result in correlated responses in other traits that share a common genetic basis with it (pleiotropy) or are closely linked. After a number of generations of selection, the possible contributions of linkage should be considerably reduced. Replicated, bidirectional selection experiments on a trait of interest are powerful tools for revealing genetic correlations with other traits. Such correlations can provide information about possible constraints on evolution. Genetic correlations can also be used to examine the mechanisms underlying the trait being selected, as all steps in the pathway from genotype to selected phenotype will be affected (e.g., see Bult et al. 1992 for neuroanatomical correlates of differences in nesting).

Correlated responses to selection occurred in body weight and litter size (Lynch 1980), body temperature (Lacy, Lynch, and Lynch

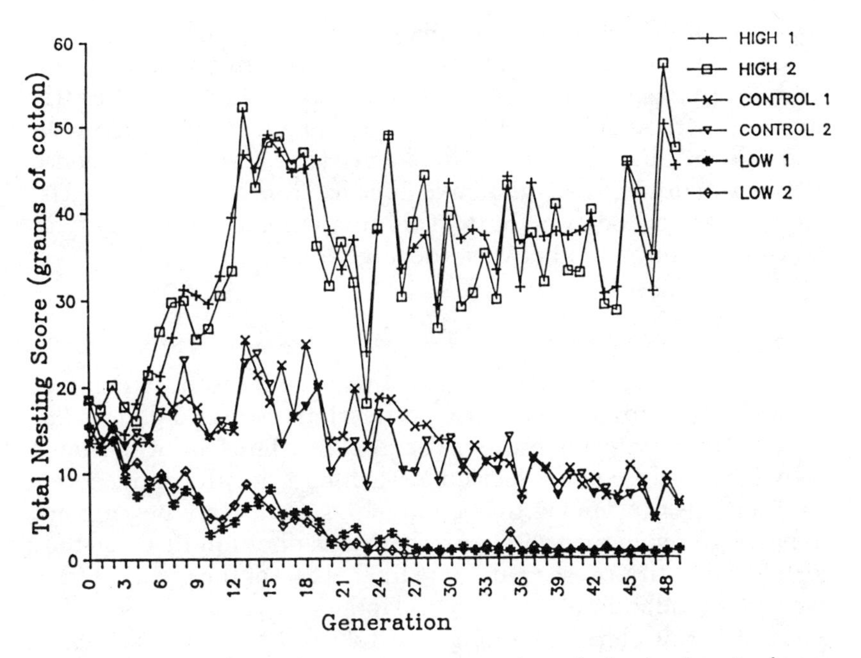

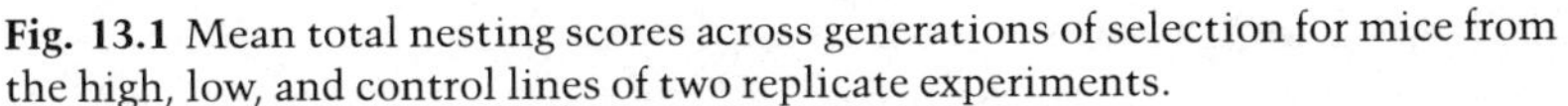

Fig. 13.1 Mean total nesting scores across generations of selection for mice from the high, low, and control lines of two replicate experiments.

1978), and nesting in the cold (Lynch and Possidente 1978; Marsteller and Lynch 1983; Laffan 1989); the values of all four traits increased with increased nest building (positive genetic correlation). Spurious correlations can occur due to the action of drift in the small populations that are often used in selection experiments. Because replicate selection lines in this experiment showed similar correlated responses, the associations appear to represent real genetic phenomena rather than sampling variance.

Correlated Responses to Selection for Body Weight

In selecting for large and small mice, Falconer (1973) produced six replicates each of his large, control, and small lines. After more than 25 generations of selection, the large mice were, on average, twice as heavy as the small mice. The level of replication present in these lines provided us with an exceptional opportunity to examine genetic correlations of other thermoregulatory traits with body weight. Because selection for nest building had resulted in a correlated response in body weight, we anticipated that the reciprocal response would occur if this genetic relationship was at all general. In fact, selection for body weight resulted in a large correlated response in nest building: large mice build significantly larger nests than did small mice (Lynch and Roberts 1984). In addition, brown adipose tissue showed a negative correlated response to selection for body weight (large mice had less brown adipose tissue per gram of body weight than did small mice), while body temperature did not differ between the large and small mice.

Genetic Correlations from Parent-Offspring Cross-Covariance

One disadvantage of relying on correlated responses to artificial selection for estimates of genetic correlations is that only correlations with the selected trait will be revealed. Thus, information is available from the two selection experiments only for traits correlated with nesting and with body weight. Fortuitously, nesting and body weight were the traits with the highest heritabilities, so that correlations with those traits were likely to be of the most interest from the standpoint of predicting adaptation.

Genetic correlations among multiple traits can be estimated from any design that allows heritability estimation, as long as all traits of interest are measured on all individuals. Genetic correlations among traits contributing to cold adaptation were estimated in experiment III (described above and in Lynch, Sulzbach, and Connolly 1988) from cross-covariance of parents and offspring, with standard errors estimated according to Falconer (1989). Not surprisingly, the measures of body weight taken at the beginning and end of the experiment exhibited extensive pleiotropy ($r_A = .88 \pm 0.02$), as did nest building measured at warm and cold temperatures ($r_A = .83 \pm 0.02$). In addition, as shown by the selection experiments described above, morphology and behavior are coupled in evolution, as both nesting measures were significantly genetically correlated with both body weight measures (all correlations about .3). Because females had to be kept for breeding, it was not possible to measure weight of brown adipose tissue on both sexes. The usual procedure is to dissect brown adipose tissue from cold-acclimated mice of similar ages. However, it was not practical to keep males alive until the females had raised a litter. Also, we had evidence that reproduction itself alters brown adipose tissue, so females dissected after they had raised a litter would not provide comparable data. This reduced the precision of the estimates of genetic correlations with brown adipose tissue. However, data from males indicated that the correlation with body weight was negative, as had been shown by the correlated response to selection for body weight. Body temperature did not exhibit the correlation with nest building detected as a correlated response to selection. This discrepancy was most likely due to the very low heritability of body temperature, as additive genetic covariance will not be detected in the absence of additive genetic variance. Although, as mentioned above with regard to experiment III, linkage cannot be discounted as contributing to trait associations, the direction and magnitude of the correlations are consistent with those found in the two selection experiments, indicating that pleiotropy is the major factor.

If natural selection acts through ambient temperature differences, body weight and nest building should exhibit adaptive clinal variation, with more northern mice being larger and building larger nests. Body weight and nesting both had heritabilities sufficient for a relatively rapid response to selection; indeed, both traits have shown large responses in laboratory experiments, in which the realized heritabilities were very close to those estimated in the base populations. The positive pleiotropy between body weight and nest building should enhance response to selection acting through cold ambient temperatures, as increases in both body weight and nest building improve cold adaptation. Furthermore, large size and large nests are both associated with increased fertility (Falconer 1973; Lynch 1980), contributing to increased fitness at cold temperatures (Lynch and Possidente 1978; Marsteller and Lynch 1983). The only correlation that might potentially constrain further adaptive evolution is the negative association between weight of brown adipose tissue and body weight. This association was particularly obvious in the correlated response to selection for body weight described above. This antagonistic pleiotropy, combined with a low heritability, indicates that brown adipose tissue, and therefore the production of nonshivering thermogenesis, are not likely to respond to selection.

Testing the Predictions

These predictions were tested on natural populations of mice from five widely separated geographic regions along the East Coast of the United States: Maine (ME), Connecticut (CT), Virginia (VA), Georgia (GA), and Florida (FL). Within each state except Georgia we obtained mice from two distinct subpopulations, usually from dif-

Table 13.2 Least squares means representing geographic populations of *Mus domesticus* for traits related to cold adaptation

	Population[a]				
Traits	ME	CT	VA	GA	FL
Body weight (g) at 50 days of age	18.5^{A}	16.0^{C}	17.2^{B}	15.4^{D}	15.1^{D}
Body weight (g) at 85 days of age	22.4^{A}	19.7^{B}	19.8^{B}	17.9^{D}	18.5^{C}
Nesting score (g) at 22° C	11.2^{A}	9.4^{B}	8.9^{B}	6.8^{C}	5.0^{D}
Nesting score (g) at 5° C	46.6^{A}	28.9^{B}	27.9^{B}	22.7^{C}	24.3^{C}
Body temperature (° C) at 0300 h	36.78^{B}	37.05^{A}	36.66^{B}	36.61^{B}	37.05^{A}
Body temperature (° C) at 1500 h	35.49^{B}	35.65^{B}	35.58^{B}	34.54^{C}	36.26^{A}
Brown adipose tissue (mg/g body weight)	0.98^{C}	1.16^{A}	1.16^{A}	1.06^{B}	1.05^{B}

Source: Reprinted from C. B. Lynch, 1992, Clinal variation in cold adaptation in *Mus domesticus*: Verification of predictions from laboratory populations, *Am. Nat.* 139:1219–36, © 1992 by The University of Chicago.
[a]ME, Maine; CT, Connecticut; VA, Virginia; GA, Georgia; FL, Florida.
Note: Different superscript letters indicate that means are significantly different ($P < .05$) from one another (Duncan's multiple range test).

ferent towns. The founding populations were live-trapped from unheated areas (food storage areas and hallways) of poultry barns and feed stores. Each geographic population was represented by at least twelve successfully breeding pairs. (Details of this experiment can be found in Lynch 1992).

Table 13.2 contains the least squares means for the five geographic populations, based on three generations of laboratory-reared mice ($N = 760$). Wild-trapped animals were not measured to avoid effects of being reared in different environments. Body weight was generally clinal, with a reversal of CT and VA in young mice and a reversal of GA and FL in older mice. Nesting scores at room temperature showed a perfect rank ordering from north to south, while scores in the cold had one nonsignificant reversal. No other trait showed a consistent clinal ranking, although there were a few significant differences among populations. For resting (daytime) body temperature (at 1500 hours), the two southern populations were different from the three northern ones, but GA mice had the lowest and FL mice had the highest temperatures. For active (nighttime) body temperature (at 0300 hours), CT and FL mice had higher temperatures than the other three populations. ME mice had the least brown adipose tissue. These results support the prediction from the preliminary experiments that body weight and nest building would be adaptively modified, while body temperature and brown adipose tissue would not.

Although estimates of genetic parameters such as heritability are population- and environment-specific, it is possible that such estimates may indicate sufficient biological generality to permit extrapolation between populations. Because natural populations are exposed to complex environmental conditions, qualitative predictions of the response to natural selection may be sufficient for making evolutionary deductions. The predictions tested here were based on a variety of different inbred and outbred populations, and heritabilities and genetic correlations were estimated in a number of different ways. The remarkable agreement among the estimates lent confidence to the predictions. Furthermore, from the many studies that have emphasized a few species of laboratory and domestic animals, and an increasing number of studies on a variety of wild species, it appears that morphological traits are usually highly heritable, and that behavioral traits tend to exhibit significant (if somewhat lower) heritabilities (Mousseau and Roff 1987). In the absence of antagonistic pleiotropy, such traits would be expected to respond to selection, both natural and artificial. Indeed, numerous artificial selection experiments on morphological and behavioral traits have demonstrated this to be true, and examples of adaptive intraspecific morphological and behavioral variation abound.

I have shown that the morphological and behavioral traits of body weight and nesting, which were shown to be heritable in laboratory populations, exhibited a north-south cline in *Mus* populations along the East Coast of the United States, with behavior showing the greatest difference between the northern and southern populations. This finding is consistent with a response to natural selection acting through ambient temperature differences. The

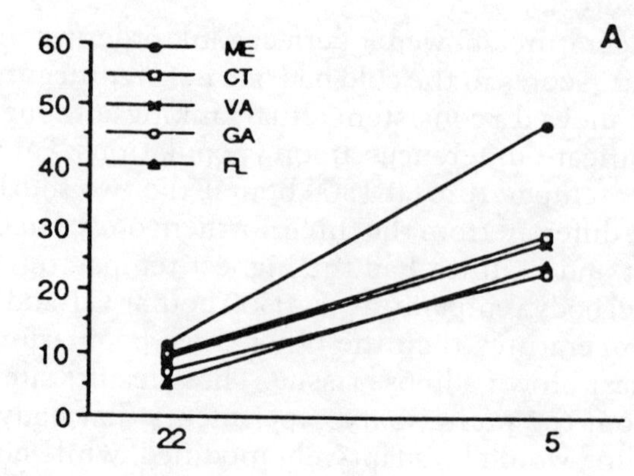

A
60
50
40
30
20
10
0
ME
CT
VA
GA
FL
22
5

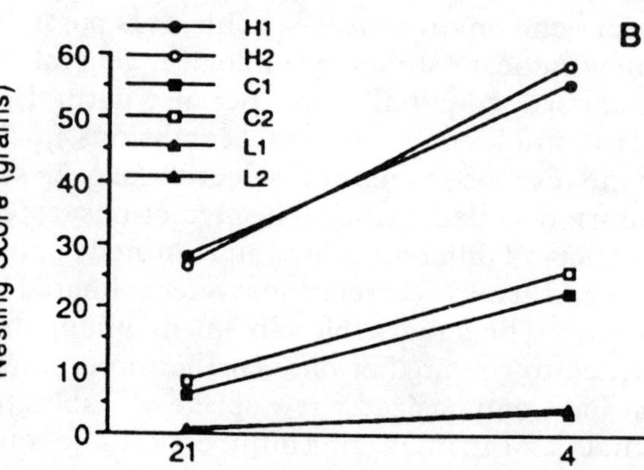

B
Nesting Score (grams)
60
50
40
30
20
10
0
H1
H2
C1
C2
L1
L2
21
4

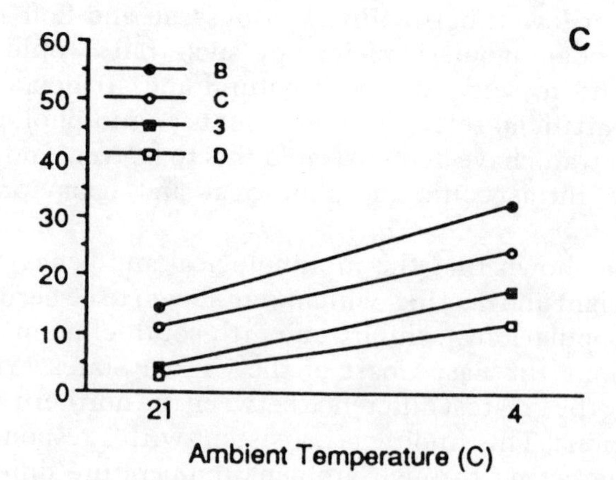

C
60
50
40
30
20
10
0
B
C
3
D
21
4
Ambient Temperature (C)

physiological traits of body temperature and weight of brown adipose tissue, which had heritabilities near zero in laboratory populations, did not exhibit adaptive (clinal) variation. Previously we had suggested that this low level of additive genetic variance might reflect the effects of strong prior selection on metabolic traits early in the evolution of *all* small mammals such that they have been long since pushed to their genetic limits (Lacy and Lynch 1979; Lynch and Sulzbach 1984).

Available information about the genetic correlations among these traits indicated that most were in the "adaptive" direction, with the exception of the relationship between body weight and brown adipose tissue. This negative relationship, combined with the fact that body weight is heritable and, therefore, adaptively modifiable, may explain the counterintuitive observation that Maine mice, which were the largest, had the least brown adipose tissue.

Norms of Reaction for Nesting

The term *reaction norm* refers to trait expression of the same genotype in different environments (Schmalhausen 1949; Gupta and Lewontin 1982). Reaction norms measure "environmental sensitivity" (Falconer 1990) or "phenotypic plasticity," and can be expressed graphically as genotype by environment interactions. Phenotypic plasticity can be treated as a trait under selection, and would be expected to increase (i.e., be associated with increased fitness) when organisms are subject to environments that fluctuate in time and space. As temperature is such an environmental variable in temperate latitudes, and mice respond to differences in temperature by building different-sized nests, nest building is a trait that might be expected to show plastic adaptation, unless, as discussed for other traits, constraints prevent response to selection (Gomulkiewicz and Kirkpatrick 1992).

Norms of reaction for the nesting scores of the geographic populations at two temperatures are shown in figure 13.2a. The reaction

Fig. 13.2 Reaction norms for nesting scores of *Mus domesticus* measured at warm and cold temperatures. (*A*) Five geographic populations (ME, Maine; CT, Connecticut; VA, Virginia; GA, Georgia; FL, Florida). (*B*) Replicated lines selected to their limits for nesting at 21° C (H, high; C, control; L, low). (*C*) Four inbred strains (B, BALB/cIbg; C, C57BL/6J; 3, C3H/2Ibg; D, DBA/1BG). From C. B. Lynch, 1992, Clinal variation in cold adaptation in *Mus domesticus:* Verification of predictions from laboratory populations, *Am. Nat.* 139:1219–36, © 1992 by The University of Chicago.

norm of the Maine population illustrated an effect of temperature (the difference between the nesting scores at the two temperatures) that was significantly greater than the effect on the other populations. Figures 13.2b and 13.2c represent comparable data from laboratory populations. (Data on the selected lines come from Laffan 1989; those on the inbred strains from Lynch and Sulzbach 1984.)

Several points are illustrated by the reaction norms of nest building at two temperatures in the geographic populations. Most important, the similar rank ordering of the populations at warm and at cold temperatures indicate that evolution of nesting at the two temperatures is strongly coupled. This interpretation is consistent with the high genetic correlation found in laboratory populations (Lynch, Sulzbach, and Connolly 1988) and is corroborated by the correlated response to selection for nest building at 21° C (fig. 13.2b; Laffan 1989). Both the laboratory selected lines (fig. 13.2b) and the inbred strains (fig. 13.2c) show the same kind of rank ordering (no crossing), but the lines selected for nesting also show the high variance in the cold exhibited by the geographic populations. Caspari (1958) suggested that increased variability in behavioral phenotypes is an adaptive response; thus one might expect to see the greatest difference between genotypes selected in different environments when they are exposed to the environmental condition that imposes the strongest selection pressure. Falconer (1990) has discussed this expectation and has shown that when both the proximate effects of the environment and the ultimate effects of selection act on a character in the same direction (e.g., both result in higher or lower measured values), then increased environmental "sensitivity" (= plasticity) is expected to evolve. Cold both increases nest building environmentally and provides the selection pressure favoring genotypes that build larger nests, so one would expect that increased thermal sensitivity would evolve in cold environments. This result is shown by the particularly large increase in nest building between warm and cold temperatures exhibited by the Maine mice as well as by the lines selected for high nesting.

Extent and Direction of Dominance

As illustrated above, additive genetic parameters are predictive of response to natural selection; however, for considerations of the relationship of various traits to overall fitness, it is useful to know the amount and direction of genetic dominance (Fisher 1930). There have been a number of different approaches to explaining the evolution of dominance (e.g., Fisher 1928; Wright 1929; Haldane 1932; Dobzhansky 1952; Lerner 1954; Kacser and Burns 1981), but they generally agree that traits related to fitness should exhibit dominance, and that the direction of dominance should be toward increasing fitness. This is above and beyond the relative increase in dominance that might occur as additive genetic variance is exhausted by strong directional selection. It makes intuitive sense that dominant alleles will influence trait expression in the direction of increased fitness, masking the less fit recessive phenotype. This relationship of dominance to fitness, with examples, has been discussed most extensively for behavior by Bruell (e.g., Bruell 1967), and also by Roberts (1967b).

The most straightforward way to look for dominance is to cross inbred strains. This illustrates another advantage of *Mus* as a model system, as a large number of inbred strains are readily available. The means of the parents and F_1s from a cross among four inbred strains are shown in fig. 13.3 for nest building at two temperatures, body temperature, and brown adipose tissue. A detailed analysis of the extent and direction of dominance exhibited by these crosses was obtained from diallel analysis by Lynch and Sulzbach (1984). Nest building exhibits substantial dominance, especially in the cold, where the scores of the F_1s are all higher than those of the highest-scoring parent (overdominance). The fact that there is both more dominance and less additive genetic variance in the cold is consistent with the interpretation that the insulation provided by a nest is more important for fitness in the cold than at room temperature. In addition, the difference in both heritability (see table 13.1) and degree of dominance at the two temperatures indicates the importance of measuring traits in ecologically relevant environments. The physiological traits of body temperature and brown adipose tissue exhibited different genetic architectures with respect to dominance. Body temperature was consistently overdominant in all crosses, while the direction and extent of dominance in brown adipose tissue varied among the crosses. This latter pattern has been referred to as "ambidirectional dominance" and may indicate a past history of normalizing selection (Mather and Jinks 1971).

Data are also available from crosses among the body weight–selected lines developed by Falconer (1973). The objective of crossing these lines was to examine the extent to which dominance contributes to the association between body weight and other traits related to thermoregulation (Lynch, Roberts, and Hill 1986).

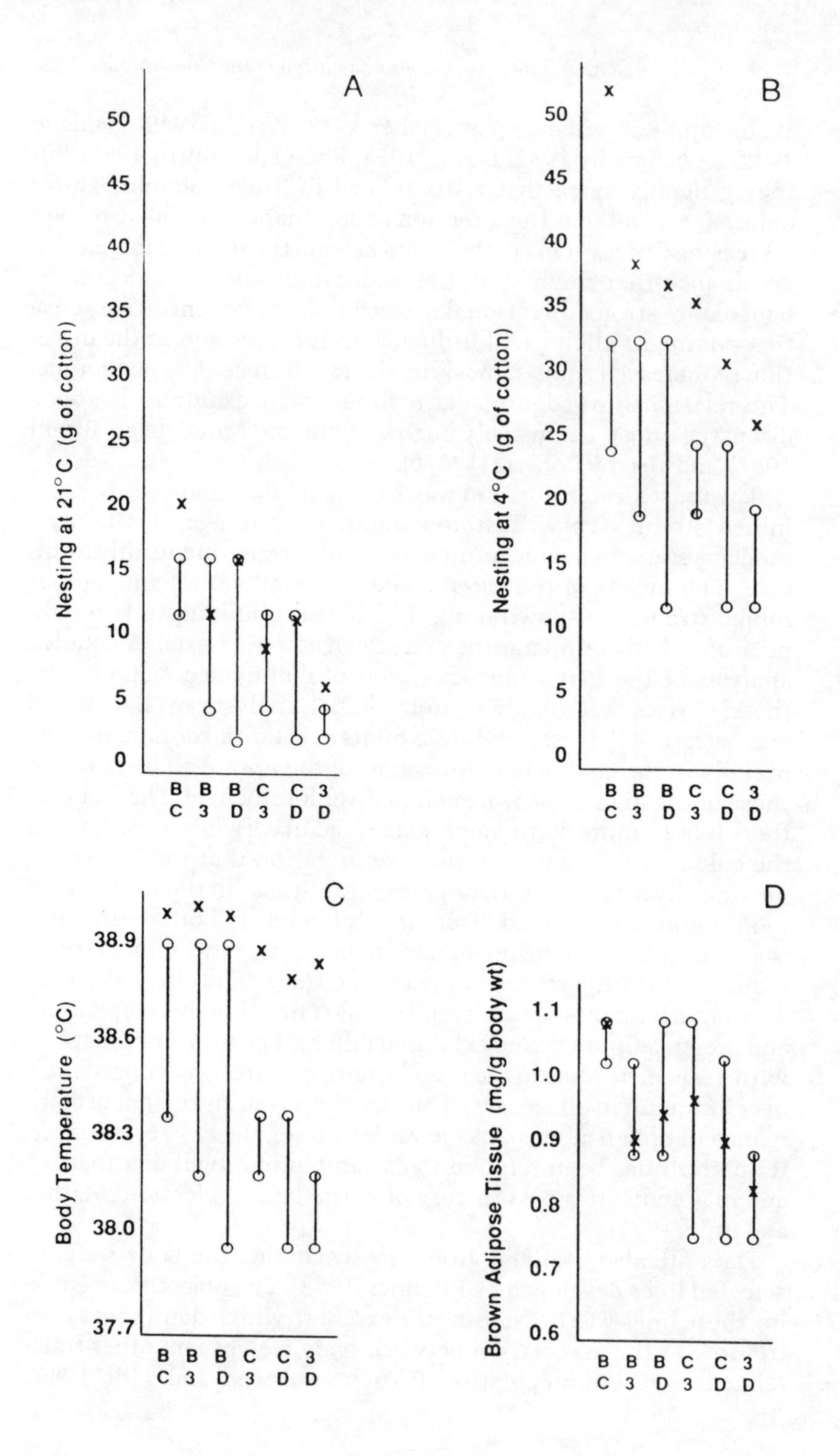
A
B
C
D
Nesting at 21° C (g of cotton)
Nesting at 4°C (g of cotton)
Body Temperature (°C)
Brown Adipose Tissue (mg/g body wt)
50
45
40
35
30
25
20
15
10
5
0
38.9
38.6
38.3
38.0
37.7
1.1
1.0
0.9
0.8
0.7
0.6
B C
B 3
B D
C 3
C D
3 D

Crosses within and between body sizes were made in three blocks, with two of the six replicates crossed per block; e.g., "A" lines were mated within line and with the large, control, and small "B" lines, including reciprocal crosses. Similarly, "B" lines were crossed with "C," and "E" with "F." The details of the crossing design are given in Bhuvanakumar et al. (1985). The results are shown in figure 13.4, where the crosses are plotted against the average crossbred means for body weight, so that the crosses represent increasing numbers of genes for large size going from left to right on the *x*-axis. The pattern of dominance was strikingly similar to that described above for the inbred-strain crosses. Nesting and body temperature both exhibited substantial dominance effects, with the crossbreds generally scoring above the range of the parents, while weight of brown adipose tissue tended to be intermediate between the two parents.

Laffan (1989), in analyzing the causes of limits to selection response, showed that the replicate lines of mice selected for high and low nesting, although phenotypically very similar (see fig. 13.1), were genetically different. (Interestingly, in only one of the four selected lines was the limit due to complete fixation.) These empirical results support the theoretical prediction that in replicated selection the small populations are likely to fix different alleles due to drift. In the case of a cross *between* the two replicate high lines, another between the replicate control lines, and a third between the replicate low lines, the nesting scores of the F_1s were significantly higher than those of the contemporaneous parental lines (Bult and Lynch 1990). Thus, even in cases in which the parents are phenotypically similar with respect to nesting, the crosses show overdominance.

We can assume that the populations of wild mice have become to some extent genetically differentiated because of the significant differences between the populations for the heritable traits described above. We performed a diallel cross among the five geographic populations to get some indication of how much of the genetic variation among the populations is additive as opposed to nonadditive (which we presume is mostly dominance) (Elder 1989). The results are shown in figure 13.5, with the crosses plotted from left to right along the *x*-axis as having increasing numbers of

---

Fig. 13.3 Means of inbred strains (circles) and of their F_1 hybrids (Xs) for nesting at two temperatures (*A*) and (*B*), body temperature (*C*), and lipid-free weight of brown adipose tissue (*D*). Means are averaged over reciprocals, sex, and blocks. B, BALB/cIbg; C, C57BL/6J; 3, C3H/2Ibg; D, DBA/1BG.

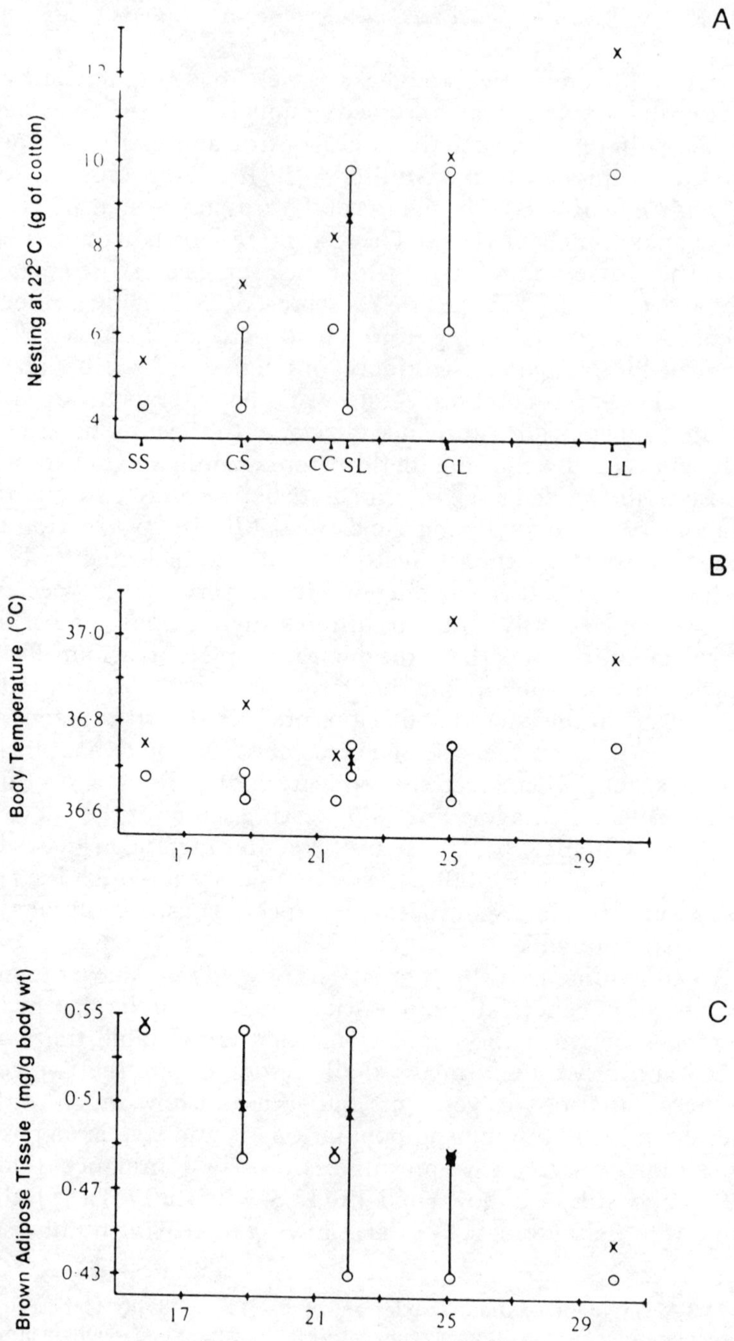
A
Nesting at 22°C (g of cotton)
10
8
6
4
SS
CS
CC
SL
CL
LL
B
Body Temperature (°C)
37·0
36·8
36·6
17
21
25
29
C
Brown Adipose Tissue (mg/g body wt)
0·55
0·51
0·47
0·43
17
21
25
29
6-week body weight (g)

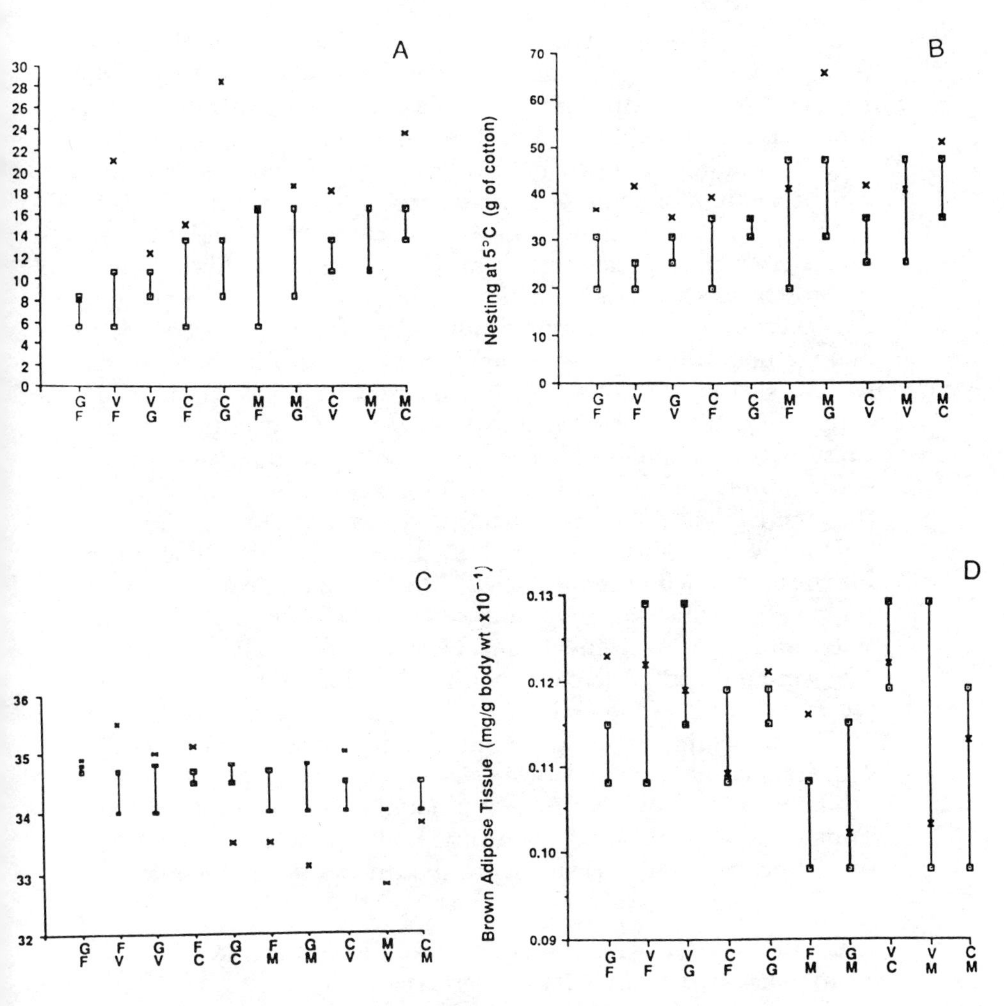

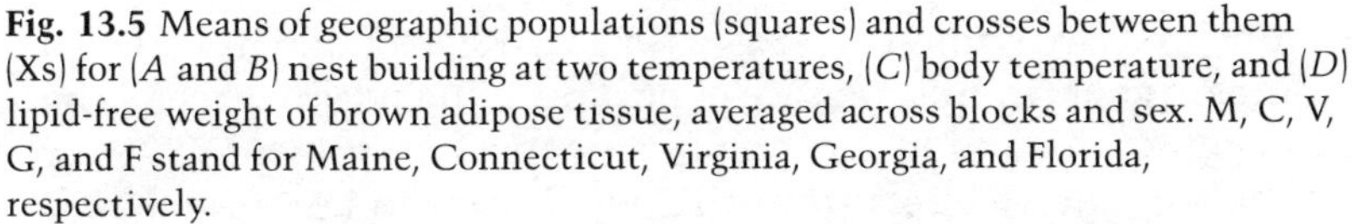

Fig. 13.5 Means of geographic populations (squares) and crosses between them (Xs) for (*A* and *B*) nest building at two temperatures, (*C*) body temperature, and (*D*) lipid-free weight of brown adipose tissue, averaged across blocks and sex. M, C, V, G, and F stand for Maine, Connecticut, Virginia, Georgia, and Florida, respectively.

Fig. 13.4 Means of lines selected for differences in body weight (circles) and crosses between them (Xs) for (*A*) nesting, (*B*) body temperature, and (*C*) lipid-free weight of brown adipose tissue. Means are plotted against the crossbred means for body weight, and are averaged across reciprocals, replicates, blocks, and sex. L, C, and S represent the large, control, and small lines, respectively.

genes from more northern populations. A total of 725 mice were measured. For nest building and brown adipose tissue weight, the results were again similar to those described above for the inbred-strain crosses. Nest building exhibited extensive dominance, and for nest building measured in the cold, seven of eleven crosses showed a greater extent of dominance than at room temperature. The presence of nonadditive genetic variance between the populations indicates that they do carry different alleles for nest building. Weight of brown adipose tissue exhibited ambidirectional dominance. However, unlike in the inbreds and selected lines, dominance in body temperature in the geographic populations was more similar to that of the physiological trait of brown adipose tissue. This could indicate a different genetic influence on body temperature in these natural populations than in the laboratory populations, or could simply be the result of lack of genetic differences between the populations at loci influencing body temperature.

In an earlier study (Lynch 1977), a different approach to examining the extent of dominance for nesting was taken. Wild mice were trapped on farms in the vicinity of Iowa City, Iowa, and brother-sister inbred for five generations. Nesting, body weight, and litter size were monitored across generations to look for inbreeding depression, which is the inverse of heterosis in crosses. While adult body weight did not show significant depression across the five generations, nesting and litter size (a known fitness trait) both exhibited approximately the same amount of inbreeding depression. Thus, another population of wild mice exhibited a genetic architecture of nesting that was similar to that in laboratory populations.

Comparative Genetic Architecture of Nesting in a Northern and a Southern Population

We employed two triple-test crosses (TTC) (Kearsey and Jinks 1968) to compare genetic architecture of nest building *within* each of the two extreme (ME and FL) geographic populations (Elder 1989). For this analysis, individuals from the population to be "tested" (almost always males, because the same individuals must be bred repeatedly) are mated to three genotypes of "tester" females. The females should come from two lines selected divergently for extreme expression of the trait of interest (in this case, nest building) and the F_1 between the selected lines. The progeny from the three types of females, referred to as L_1 for the high expression, L_2 for the low expression, and L_3 for the F_1, are then tested for (in this case) nest building. Analysis of variance among tested males of the sums and differences of their progeny means provides tests for the significance of the additive genetic variance ($L_1 + L_2 + L_3$) and the dominance variance ($L_1 - L_2$). The full genetic model and its derivation can be found in Mather and Jinks (1971) and Kearsey and Jinks (1968). This design is an extension of one presented by Comstock and Robinson (1952), and provides a great deal of genetic information with a relatively small sample (as opposed to the alternative, half-sib analysis). In contrast to the Comstock and Robinson design, the TTC also includes a rigorous test for epistasis (nonallelic interactions). If the epistatic variance component is nonsignificant, then heritabilities can be estimated within the tested populations. In only one other study has the TTC been used as originally described, that is, using extreme genetic lines developed by selective breeding to examine the genetic architecture of the selected trait in natural populations. This was a study of escape-avoidance conditioning and associated behaviors in wild rats by Hewitt and Fulker (1983, 1984), who also provided a clear description of the analysis of the TTC design (Hewitt and Fulker 1981).

We were able to use this design because of the existence of the extreme lines selected for nest building. Because these lines had plateaued, we could assume that they were as close to fixation as possible at the loci influencing nesting. It was later established that the low line used was, in fact, fixed (Laffan 1989). Nest building also was the most divergent trait between the northern and southern natural populations (see table 13.2), indicating that it had probably been under fairly intense directional selection in the ME population. If there had been stronger selection due to cold in the ME population, it should show less additivity and more dominance for nest building than the FL population. The "testers" came from the high and low lines of the selection experiment for nest building and the F_1 between them. The "tested" mice were males from the ME and FL populations, each harem-mated to tester females representing the three genotypes (high, low, and F_1). This was done in four blocks (four different groups of females), resulting in approximately 1,500 F_1 mice tested. Because the analysis of variance of the TTC requires a balanced design, only males that successfully produced offspring with the females in all four blocks (12 different females) could be included in the analysis. This resulted in data for 13 ME and 18 FL males. Nest building was measured at both 21° C and 5° C.

For nest building at both temperatures in both sexes the signifi-

Table 13.3 Significance levels for estimates of genetic variance components and heritability estimates for nesting at two temperatures from triple-test crosses in two populations of *Mus domesticus*

Population	Sex	Additive	Dominance	Epistatic	Heritability
		Nesting at 21° C			
Maine	Females	$P < .02$	NS	NS	0.13
	Males	NS	$P < .05$	NS	
Florida	Females	$P < .01$	NS	NS	0.24
	Males	$P < .001$	NS	NS	
		Nesting at 5° C			
Maine	Females	$P < .02$	NS	NS	0.16
	Males	$P < .05$	$P < .02$	NS	
Florida	Females	$P < .001$	NS	NS	0.23
	Males	$P < .01$	NS	NS	

cance of the additive genetic variance component was greater in the FL than in the ME population (table 13.3). Dominance variance was significant for both nesting measures only in the Maine males. As a result, heritabilities of nesting, averaged across sex, were lower in the Maine than in the Florida mice. Epistatic variance was not significant. The lack of epistasis and the fit of the results to the prediction (the northern mice appeared to have been the more strongly selected for nest building) indicate that the laboratory mice were adequate testers for the wild populations; that is, essentially the same genes that result in high and low nesting in laboratory mice also influence nesting in wild populations.

SUMMARY

Quantitative genetic analysis has provided fruitful insights in its application to understanding the genetic basis of adaptation to cold in the house mouse, *Mus domesticus*. Heritability estimates can be used prospectively, to predict the response to selection, as well as retrospectively, to deduce the history of prior selection. In this study, nest building and body weight had significant heritabilities in a variety of laboratory populations, and both traits responded to artificial selection. The significant clinal variation for both these traits in wild populations is consistent with the interpretation that they have also responded to natural selection acting through differences in ambient temperature. In contrast, the physiological traits of body temperature and weight of brown adipose tissue generally had very low heritabilities in laboratory populations and did not exhibit clinal variation. According to the somewhat simplistic assumptions of Fisher's fundamental theorem of natural selection (Fisher 1930), such low heritabilities are thought to indicate a history of past directional selection on the traits that has effectively exhausted the additive genetic variance. Under the tenets of this theorem, the heritabilities of the traits responding to selection in the wild populations should have been reduced, which was detected by the TTC for nesting in the Maine population.

Knowing the extent of dominance is also useful in describing the relative contribution of different traits to fitness, and the direction of dominance indicates the direction of increasing fitness. Thus, high levels of nest building tend to contribute more to fitness than low levels of nest building.

Genetic correlations describe the extent of common additive genetic effects on two traits, and were used here to predict possible constraints on adaptation. It appears that the negative genetic correlation between body weight and proportion of brown adipose tissue is probably one factor preventing a relative increase in brown adipose tissue in more northern populations.

Acknowledgments

Portions of this research were supported by NIH Grant GM21993, NSF Grant BSR-8414739, NIH RCDA ES00042, NATO Grant 1609, and project grants from Wesleyan University. I thank Mark Courtney for helpful comments on drafts of this chapter, and for many stimulating discussions about population biology.

Conclusions

14

Evaluation of Applications of the Theory and Methods of Quantitative Genetics to Behavioral Evolution

Christine R. B. Boake

Courtship in the Hawaiian fly *Drosophila silvestris.* The male is standing in the head-under-wings posture with his wings extended laterally, preparatory to fanning them. (Photograph by J. Schwartz.)

The chapters in this book have illustrated that quantitative genetic approaches can assist our understanding of the evolution of behavior from both theoretical and empirical perspectives. In this chapter I shall discuss conceptual and methodological issues related to applying quantitative genetics to studies of behavioral evolution, and describe the variety of solutions proposed by the chapter authors, as well as other approaches to resolving the problems.

A major value of quantitative genetic models to the development of evolutionary theory is their ability to describe and predict changes in the phenotype. Because the phenotype is generally all that is available for behavioral biologists to measure, these models may be more appropriate than single-locus models in explaining behavioral evolution. Furthermore, the quantitative genetic assumption that any particular phenotypic trait is controlled by numerous genes in interaction with environmental factors appears to be a more reasonable assumption than that a single gene controls each behavioral trait. Quantitative genetic analyses are based on observations of variation both within and between individuals, observations that behavioral biologists make every day. A major difficulty in applying quantitative genetics to behavioral traits is that behavior so often depends on the context in which it is produced, and consequently the same behavior in the same individual can be influenced by ecology, by heterospecifics, and by conspecifics.

To date, quantitative genetic models for behavioral evolution have focused on sexually selected traits (Heisler, chap. 5), although other subjects with behavioral ramifications, such as maternal effects and life histories (Roff, chap. 3; Cheverud and Moore, chap. 4; Dingle, chap. 7), are receiving attention from theoreticians now. The first value of these models is that they have allowed formal treatments of verbal statements about evolution, such as Lande's formalization of Fisher's process, and have demonstrated outcomes of such models that are sometimes surprising (Heisler, chap. 5). A second value of quantitative genetic models is their clarification of the relative importance of various variables that could be measured. For example, the sexual selection models use a "preference function" which takes into account variability in female mating preferences both within and between females. The emphasis on preference functions has stimulated measurements of individual variation in female behavior (Gerhardt 1991). Models that are not based on quantitative genetics also guide decisions about which variables to measure, but because of their emphasis on phenotypic variation, quantitative genetic models have suggested approaches to measurement that have been uncommon until recently.

The emphasis of quantitative genetics on the genes carried by each individual, and on the interaction between genotype and environment, has led to the prediction that individuals will differ in their behavior. These individual differences may or may not be adaptive; they are a result of the unique genetic constitution, unique developmental background, and unique environment of each individual. The measurement of individual differences, the determination of the causes of the differences, and the assessment of the fitness consequences of such differences are the essence of quantitative genetic analyses of behavioral evolution. As examples, the reproductive behavior of a male sailfin molly depends both on his size (partially genetically and partially nutritionally determined) and on the relative sizes of other males in his group (Travis, chap. 8). The degree of cannibalism shown by flour beetles depends on their genetic background, on the availability of prey, and on the presence of other adults (Stevens, chap. 10). The degree of interruption of the trills of *Gryllus integer* is strongly influenced by genetic factors, and may have appreciable fitness consequences, although the effect on a female's mating decision is also affected by ecological factors (Hedrick, chap. 11). The genetic influences on an individual's behavior may act through neuroanatomy, neurophysiology, hormonal physiology, the ability to learn, and the nature of the decision rules that are used. All of these areas are completely open for studies that take an approach based on individual variation and the influences on such variation.

Empirical uses of quantitative genetics have been valuable in a number of areas, first at the simple level of providing estimates of heritability for certain traits. This has been particularly important in the case of sexually selected traits (Hoffmann, chap. 9; Hedrick, chap. 11), as recent data have demonstrated that in some cases, traits that may be closely related to fitness do have appreciable heritabilities, despite theoretical predictions to the contrary (Maynard Smith 1978, 1985). Measures of heritability help to set bounds on the feasibility of various models: for example, Lande's models are only applicable if the variation in the attractive male trait is heritable, and optimality models also rest on the assumption that genetic variation exists for the traits being modeled (Roff, chap. 3; Hoffmann, chap. 9; Hedrick, chap. 11). The application of quantitative genetics to behavioral evolution has stimulated the development of ways to measure selection because the models rely on a specific formulation of selection (Lande and Arnold 1983; Arnold and Wade 1984; Partridge, chap. 6).

Second, genetic factors can constrain the rate or direction of evolution, and thus quantitative genetics is important for elucidating genetic constraints (see Arnold 1992b for a thorough review). Three kinds of constraints considered in this book are the magnitude of heritability, the sign and magnitude of genetic correlations, and maternal effects. For a given selection intensity, a low heritability will result in slower evolution than a high heritability. Genetic correlations can influence the direction of evolution and the rates at which traits will reach their optima (Via and Lande 1985). However, the degree to which genetic correlations constrain evolution may depend on the time scale involved: if genetic variance-covariance matrices are unstable over evolutionary time (discussed below), then genetic correlations may not be a major constraint (Partridge, chap. 6). Roff (chap. 3) describes the complexities involved in interpreting genetic correlations in terms of evolutionary trade-offs. Cheverud and Moore (chap. 4) and Heisler (chap. 5) provide examples from theory in which maternal effects of genetic correlations can strongly affect the rate and direction of evolution; in some cases maternal effects may prevent a population from reaching an equilibrium. As discussed by Cheverud and Moore, maternal effects have been demonstrated to result in responses to selection that are the opposite of what would be expected on the basis of heritability alone. The need to measure genetic correlations rather than assuming their sign from a knowledge of biology is vividly illustrated by Garland's results, which were very different from what he had predicted (chap. 12).

Third, quantitative genetics can be used for studying adaptation, although definitions of "adaptation" exist that do not involve quantitative genetics (reviewed by Stearns 1992). Such studies can take the form of making and testing predictions about geographic variation in relation to ecological factors (Dingle, chap. 7; Travis, chap. 8; Lynch, chap. 13). Studies of adaptation can be extended to studies of the inheritance of physiological traits and analyses of the relationship of various performance traits to fitness (Garland, chap. 12; Arnold 1981a,b); such studies could be very valuable in assessing the physiological costs and benefits of certain behavioral patterns. Another kind of study of adaptation attempts to explain the maintenance of possibly maladaptive traits, such as cannibalism, in a population (Stevens, chap. 10). Stevens suggests that the maintenance of cannibalism is best explained in terms of Wright's shifting balance theory of interdemic selection. Her research sheds light on the evolution of sociality because it shows that different values of a trait may persist in different populations that are indistinguishable ecologically.

A fourth application of quantitative genetic techniques, and possibly the most exciting, is the use of artificial selection for a particular trait to produce selected lines that can be used in further studies of the trait. This application is totally independent of the debates about the value of quantitative genetic models (reviewed below). Hoffmann's lines of *Drosophila* that were selected for territorial behavior are now being used to examine the costs and benefits of territoriality (chap. 9), and Dingle's lines of milkweed bugs selected for wing length were used to examine the costs and benefits of large and small size in the face of varying food levels (Dingle 1992). Similarly, Partridge is using selected lines of *Drosophila* to elucidate the causes of a correlation between thorax length and male mating success (chap. 6). Mice that were selected for increases or decreases in nest building show neuroanatomical differences (Bult et al. 1992; Lynch, chap. 13). These studies rely on the ability of artificial selection to produce populations outside the phenotypic range of natural populations, and populations with more uniform phenotypes than are normally found. Artificial selection is an ancient form of genetic engineering, and it may be exploited in very modern fashions. For example, one may ask whether replicate lines that have responded to selection have done so through the same mechanisms. In the case of *Drosophila* that have been selected for their resistance to the incapacitating effects of ethanol fumes, lines taken from different natural populations showed significantly different degrees of response (Cohan and Hoffmann 1986). Furthermore, the responses in different lines were due to changes at different loci: in some lines that responded to selection, allele frequencies at the *Adh* (alcohol dehydrogenase) locus changed, and in others they did not (Cohan and Hoffmann 1986). Artificial selection could be an extremely useful tool for neuroethologists, who could examine whether altered behavior always resulted from the same changes in the nervous system. In turn, such data would give evolutionary biologists a deeper understanding of the nature of the evolutionary plasticity of behavior.

The production of quantitative genetic models is far outpacing the tests of their value, but we will not even know how broadly relevant the models are, let alone how accurate their predictions are, unless their assumptions are assessed. This is particularly critical for the burgeoning sets of models of Fisher's process (Heisler, chap. 5), where modeling has the appearance of a cottage industry. The basic model (Lande 1981b) assumed that female mating preferences have no costs or benefits, but Kirkpatrick (1985, 1987a) has shown that if the process of exerting a preference incurs a cost, the line of equilibrium predicted in Lande's model does not appear. Despite the theoretical results having been known for over six years, by the end of 1992 virtually nothing was known about the costs to females of exerting preferences in non-resource-based systems. At the other extreme, some evolutionary outcomes have not been predicted by specific models. This is the case for the existence of population differences in stable levels of cannibalism, a highly antisocial behavior (Stevens, chap. 10), which is predicted in a general way through Wright's shifting balance theory, but is not covered by models. Quantitative genetics has provided a framework for investigating the nature of the genetic differences between populations in levels of cannibalism, but the theory lags behind the data.

Difficulties of Applying a Quantitative Genetic Approach: Conceptual Issues

Quantitative genetic models are definitely useful for modeling and examining evolution, but the limits of their value are the topic of several debates. Two of these issues are, first, whether models of past evolution have any relevance to predictions about the future, and second, how quantitative genetics is relevant to studies of equilibrium populations. The answer to the first question depends on whether or not one takes a uniformitarian view of evolution (Heisler, chap. 5). Of course, a uniformitarian view is necessary regardless of the nature of the models guiding one's evolutionary research; the concept of an evolutionarily stable strategy depends on current phenotypic alternatives and their costs and benefits having been the same in the past as they will be in the future. Finding the limits for a uniformitarian assumption may be tricky: for example, it is reasonable to believe that cricket calling song has attracted females for millenia, but it may not be reasonable to assume that the same features of the song that are important now were important when closely related species were diverging.

The second issue, the relevance of quantitative genetic models and methods to equilibrium populations, is a criticism of attempts to measure additive genetic variation and additive genetic covariation in order to determine whether modern populations are capable of evolving. One extreme assertion is that all modern populations are at evolutionary equilibrium and therefore that no additive genetic variation is expected to be found, obviating quantitative genetic studies. This assertion simplifies the implications of Fisher's fundamental theorem to a great degree, and appears to apply the theorem to any trait with even a slight bearing on fitness. The issue is likely to be much more complex, as I discussed in the first chapter in this book, and expand upon below.

Quantitative genetic models and data address four issues related to evolutionary equilibrium: whether additive genetic variance can be detected at equilibrium, how one determines the equilibrium state of a population, whether the population is at equilibrium, and finally, whether a population has a single equilibrium state. First, assertions that populations are at equilibrium are not unique to criticisms of quantitative genetics, and the debate surrounding the application of quantitative genetic models seems to be simply the most recent version of the selection-neutrality controversy (Lewontin 1974). On one side are arguments that because most populations are close to equilibrium, no genetic variation will be detectable, as a result of the action of selection. The other, "neutral," side asserts that numerous forces could maintain genetic variation within a population, even in the face of selection. Wright (1930) pointed out that in a population at apparent equilibrium, processes such as gene flow, mutation, dominance, and epistasis will allow genetic variation to be maintained. Another reason that genetic variation could persist is that assumptions about the strength of selection are erroneous. For example, male mating signals have often been assumed to be under such strong selection that they will show no measurable heritability. Hedrick (chap. 11) detected significant heritability for calling song structure in male crickets, and pointed out that factors such as the existence of satellite males that can intercept females, and the risk of predation on females, can reduce the strength of selection on male song, allowing the maintenance of appreciable additive genetic variance. Whether or not heritability can be detected often seems to depend on the scale of the experiment; if sufficient resources can be devoted to an experiment to allow large sample sizes, even small heritabilities can be detected. The next step will be to determine the particular reason that detectable heritability was found in any given case.

The definition of the equilibrium state of a population arises from an understanding of its ecology, and is frequently assisted by measurements of selection on various traits (in terms of costs and benefits or in terms of selection differentials or gradients) and by models such as ESS models (Roff, chap. 3). The expected direction and strength of selection on a trait can sometimes be determined by means of long-term studies of single populations or by careful studies of geographic variation (e.g., Grant 1986; Riechert and Maynard Smith 1989). Even short-term analyses may show high variation in selection: Kalisz (1986) found that selection on flowering time in a forest floor herb changed in intensity and sign over a spatial scale of a few meters, and from one year to the next; in this case it would be difficult to describe a population-wide optimum. Lynch's analysis of geographic variation in nest building by mice made use of hypotheses about nest size derived from knowledge of the species' ecology and of climatic variation (chap. 13). Her work emphasizes that a knowledge of behavior alone is insufficient. The fish that Travis studies (chap. 8) have strong differences between populations in the absolute size at which behavioral tactics are switched, suggesting that selective forces could differ between populations. Hoffmann has been using selected lines to measure the costs and benefits of varying degrees of territorial aggression, which could then be applied to develop predictions about the level of aggression at equilibrium (chap. 9).

The question of whether a population is or is not at equilibrium for a particular trait is an empirical one that must be addressed on a case-by-case basis. We rarely know how long a particular natural population has existed, even when we can define what the equilibrium values should be. Populations of colonizing species are unlikely to be at equilibrium. For example, the population density of fungus beetles has significant effects on sexual selection on male horn size, and individual males can experience a range of densities over their life spans as a result of travel between trees and the influences of the weather (Conner 1989). For sedentary species, occasional major environmental perturbations, such as El Niño years, can cause large deviations from the usual pattern of selection. Such perturbations could have a net effect ranging from negligible to dominating, but the net effect is only detectable in the context of long-term studies (Grant and Grant 1989). Geographic variation can be used as a tool in equilibrium analyses: a study of one population can provide a model to be tested with another population that differs in variables used in the model. In the case of spider territorial behavior, riparian populations are much more aggressive than expected in their ecological circumstances, and are not at the predicted equilibrium (Riechert and Maynard Smith 1989).

Not all populations of a species need have the same equilibrium state. The geographic variation reported by Lynch (chap. 13) and Travis (chap. 8) indicates that even if local populations are at or near their optima, these optima differ between populations.

Lynch's results could be explained as adaptive changes; she has clearly demonstrated adaptations for nest building in mice. However, the geographic variation in size-related behavior reported by Travis does not have an unambiguous adaptive explanation. Geographic or environmental differences between populations are not a necessary precondition for the existence of different optima. Populations that are kept in the same environmental conditions can have different stable levels of expression of a phenotypic character, even hundreds of generations after the populations were founded, as shown by Stevens (chap. 10). Stevens' results suggest that when a population can be said to be at or near an optimum, there is no guarantee that it is the only possible optimum.

Quantitative genetic models have often been viewed as representing a different philosophy from optimality models, and as being in opposition to them. Roff discusses complementarities between the two approaches, and emphasizes that studies of evolution would be strengthened by using both perspectives (chap. 3). However, Cheverud and Moore (chap. 4) make a forceful case for preferring quantitative genetic models of evolutionary processes over optimality models. They argue that optimality studies aim to understand the equilibrium state of populations rather than to understand evolutionary processes. This argument in turn reflects the problem of whether the scientist considers populations to be at evolutionary equilibrium or in the process of evolving. Furthermore, Cheverud and Moore illustrate ways in which genetic factors could prevent a population from reaching the predicted equilibrium. Differences in the acceptance of different classes of *models* may reflect differences in the classes of *problems* that are considered interesting; the quantitative genetic perspective of examining the process of evolution as well as the outcome is based on the premise that frequently populations may be displaced from their optima. Cheverud and Moore point out that if a population is found to be away from the optimum defined through an optimality model, the reason may not be that the optimality model is incorrect; they emphasize that evolution may not always optimize. Rather, genetic factors may be precluding the population from attaining the optimum. This reasoning will leave a scientist to make a decision as to how to proceed when data do not conform to the predictions of an optimality model: should the model be adjusted or should genetic measurements be begun?

Finally, two aspects of the genetic assumptions used in quantitative genetic models are the topic of ongoing debates. One of these issues is whether the assumptions of weak selection and ongoing mutation allow the maintenance of sufficient genetic variation to explain the phenotypic evolution that has taken place. This question has been a topic in the genetics literature (e.g., Barton and Turelli 1989). The issue is important to behavioral ecologists because quantitative genetic models of behavioral evolution rely on the mutation-selection balance (e.g., Lande 1981b). However, the empirical studies that are necessary to resolve the question involve measuring mutation rates, a process that is likely to be far more tractable for morphological than for behavioral traits. A related question is whether the assumptions of the constancy of genetic variances and covariances are realistic. The authors in this book who have addressed the question do not agree: Arnold (chap. 2) feels that they are reasonably constant, and Roff (chap. 3) interprets the same papers as showing that the matrices are not constant. Partridge (chap. 6) implies that the matrices are not constant when she argues that comparative studies are inadequate for examining hypotheses about the role of genetic correlations in the evolution of sexual dimorphisms in behavior. Dingle (chap. 7) does not address the question directly, but his data on "migratory syndromes" are remarkable in that they show substantial genetic correlations between certain life history and behavioral traits in migratory populations, and no correlations between the same traits in nonmigratory populations of the same species. It is becoming difficult to accept the assumption of constancy of genetic variances and covariances over evolutionary time. The next step for theoreticians may be to reexamine their models and ask how robust the models are to changes in genetic variances and covariances.

Difficulties of Applying a Quantitative Genetic Approach: Empirical Issues

Several potential major obstacles await scientists who wish to use quantitative genetic methods in their research. These obstacles are sample size, the relation between laboratory measures and evolution in the field, genotype by environment interaction, and the interpretation of the results. For each, I shall describe the problem and consider ways to address it, focusing on the procedures illustrated in this book.

Huge sample sizes are necessary to make accurate measurements of heritability and of genetic correlations, on the order of hundreds of individuals, whether they are allocated as a few individuals per family in many families (Lynch, chap. 13; Klein, DeFries, and Finkbeiner 1973; Klein 1974), or many individuals per family in a few families (Arnold, chap. 2; Garland, chap. 12). Environmental variables must be randomized across families to avoid confounding the genetic and environmental influences on similarities within families. The need to keep track of all the families while rearing them to a uniform age, plus the time involved in measuring even simple traits, makes such studies time-consuming, labor-intensive, and expensive. It is exhausting to measure anything but the most simple behavioral traits on such a scale (personal observation), and rearing has to be carefully choreographed in order to make sure that one can measure all the animals at the appropriate stage in their lives. This kind of attention to sample size is illustrated by Lynch's studies of nest building behavior in mice. Garland found that he could not measure behavior on all the snakes at the same age, so he added age as a covariate in his statistical analyses; his approach made the behavioral measurements feasible, but it did not reduce the sample size.

Arnold pointed out that the common formulation of genetic questions in behavioral evolution is to ask whether additive genetic variance exists, not its precise value. If the scientist simply wishes to bound heritability away from zero rather than needing a precise measure, smaller sample sizes are sufficient. The computing of a sample size for quantitative genetic studies often appears to be rather circular: one needs an estimate of the genetic variation to insert into the equation for the standard error (Robertson 1959a,b). This problem can be resolved somewhat by using repeatability, a phenotypic measure that sets an upper bound on heritability (Falconer 1989; Lessels and Boag 1987). Because heritability is not necessarily as large as repeatability, the sample size may still be underestimated, but the estimate is better than nothing. If repeatability has been measured in the same population and in circumstances similar to those to be used in estimating heritability, its magnitude may be quite useful (Arnold, chap. 2).

Many quantitative genetic experiments have had truncated sample sizes as a result of environmental conditions beyond the scientist's control. The truncation may affect the experiment by simply reducing the sample size, as in the case of my research. My quantitative genetic study of courtship in *Drosophila silvestris* was truncated to 70% of its planned size because introduced yellow jacket wasps underwent a population explosion and devoured the flies in the forest (Aubrey Moore, pers. comm.; the fly population has since recovered). On the other hand, the disaster may tear holes in the design, resulting in variation in the numbers of individuals in each family. For example, Mitchell-Olds (1986) lost plants as a result of frost heaves. When experiments are as large and as demanding of time and space as quantitative genetic studies require, the probability of an unforeseen interruption is very high. If the planned sample size is large enough, the interruption may not make the difference between significant and nonsignificant results. If some families lose more members than other families, the design is referred to as unbalanced, and statistics derived from analysis of variance must be adjusted in ways that are not universally agreed upon (Via 1984). Shaw's (1987) restricted maximum likelihood statistical program is not altered by imbalance; several similar programs have been described (Lange, Westlake, and Spence 1976; Hopper and Mathews 1982; Lange, Weeks, and Boehnke 1988) and these programs are readily available from their authors.

Estimates of quantitative genetic variables usually must be made in the laboratory, but for evolutionary interpretations, we must extrapolate their values to natural populations in the field. The extrapolation can be a problem because estimates of heritability incorporate environmental variance in the denominator (Arnold, chap. 2), and environmental variance is unlikely to be the same in the laboratory and in the field. It is also possible that some traits, particularly physiological and behavioral ones, may not be influenced by the same genes in both environments. An example is a process that is influenced by hormones: hormone production itself is likely to be influenced by the environment. These considerations have led some scientists to assert that heritability estimates can never be applied to any population except the one in which they were made. However, the problems of interpretation are not an unfathomable mystery; the application of some common sense in defining the differences between laboratory and field environments will allow a scientist to draw reasonable conclusions. Clearly this requires doing at least a modicum of field research to understand the environment that the species experiences normally. It may be possible to manipulate certain environmental variables in the laboratory to cover the possible range in the field; for example, Hedrick (chap. 11) reared her crickets under conditions of high and low food availability, and found no effect on their calling behavior. Hoffmann (chap. 9) also extended the environmental range of larval rearing conditions and showed consistencies within lines across rearing environments. It would also be worthwhile to examine the action patterns being studied to determine whether they are affected by the laboratory environment. Lynch (chap. 13) has been extremely successful in applying her studies of the inheritance of nest building behavior, conducted with highly inbred strains in the laboratory, to explain adaptations to latitudinal variation in temperature in natural populations of mice. Her results illustrate that if the appropriate traits are examined, taking into account the biology of a species, the differences between laboratory and field need not be perceived as a black box.

Several other options are available for resolving the laboratory-field problem. One option is to measure "natural heritabilities" in the laboratory (Coyne and Beecham 1987; Riska, Prout, and Turelli 1989; Prout and Barker 1989); another is to conduct pedigree analyses with existing data; and the third is to make the genetic measurements in the field (e.g., Boag 1983; Findlay and Cooke 1983; van Noordwijk 1984). The term "natural heritability" was introduced to describe studies in which wild-caught parents are compared to their laboratory-reared progeny. The breeding design is a parent-offspring regression, and one can often conclude how accurately the laboratory measure reflects the value in the field by comparing phenotypes in the parents and offspring. When wild-caught males are used, they are brought into the laboratory and bred to females from a laboratory stock (Coyne and Beecham 1987; Riska, Prout, and Turelli 1989; Prout and Barker 1989); this ensures that males are transferring only genes to their progeny. In some species, it is possible to bring gravid females into the laboratory and measure their offspring; this technique has the advantage that all the genes come from natural populations. It has the disadvantage that one must assume that random mating took place in the natural population. Furthermore, sperm precedence patterns can complicate the results. For example, garter snakes use copulatory plugs, leading to the assumption that females mate just once (Devine 1975). But electrophoretic data showed that 50% of the females from two populations of garter snakes had been multiply inseminated (Schwartz, McCracken, and Burghardt 1989). Thus, some progeny that had been assumed to be full sibs were actually half sibs. Undetected multiple insemination will lead to an underestimate of heritability.

In long-term studies of natural populations, and in some captive populations, pedigree information may be available (Dingle, chap. 7). Computer programs exist to extract estimates of quantitative genetic parameters from pedigrees (Lange, Westlake, and Spence 1976; Hopper and Mathews 1982; Shaw 1987; Lange, Weeks, and Boehnke 1988). If a few dozen lineages are known, a pedigree analysis may be worthwhile.

Heritability estimates may be made in the field for certain carefully chosen questions, species, and populations, as illustrated by analyses of bird morphology (Dhondt 1982; Boag 1983; van Noordwijk 1984). Although birds have the advantage of keeping their young in one place and of often showing site fidelity as adults, logistical problems can be large. First, recent analyses of nestmates have illustrated that all nestmates may not share the same father or the same mother, as a result of extrapair copulations and of intraspecific nest parasitism (Westneat, Sherman, and Morton 1990). In cases in which the parentage of nestmates is known, a potentially important difficulty for computing quantitative genetic measures is the common environment shared by nestmates, which results in their sharing environmental as well as genetic influences on their phenotype. Shared environmental influences can be removed in cross-fostering experiments (Smith and Dhondt 1980; Dhondt 1982), although such experiments certainly increase the time and labor involved. The possibility that a shared environment will interfere with genetic measurements may not be appreciable for birds: several studies of cross-fostering have shown nonsignificant differences between estimates of heritabilities for morphological traits obtained from cross-fostered birds and from birds reared in their natal nests (Dhondt 1982). However, maternal environmental effects are well known in mammals (Cheverud and Moore, chap. 4); their effects on morphology may be estimated by means of interuterine embryo transfers (Atchley et al. 1991). I am unaware of any cross-fostering studies that have examined the inheritance of behavioral traits in natural populations; for species with learned behavior, the effects of a shared environment could be substantial.

The number of species suitable for field estimates of quantitative genetic parameters is limited to those in which the offspring can be found when they reach the age at which traits must be measured. Evolutionary studies generally include traits that are measured on sexually mature animals rather than only on newborns. Unsuitable species include those with extensive dispersal between birth and first reproduction, and species that cannot be marked as to family origin in a fashion that will persist to adulthood, whereby most arthropods are ruled out. Even for animals that can be marked

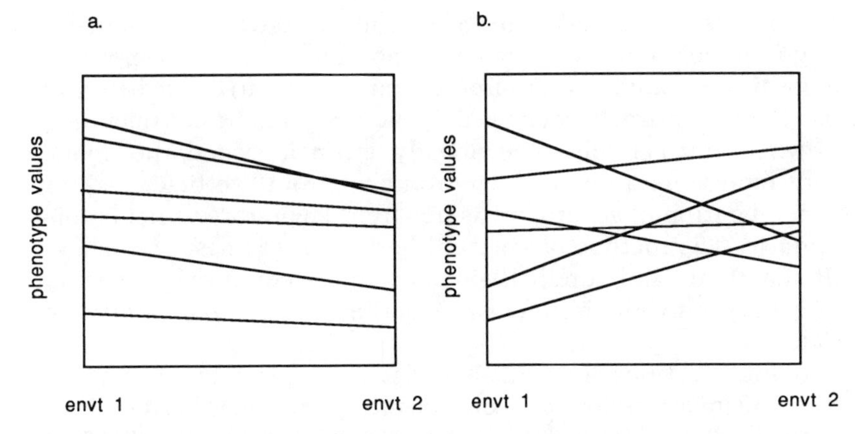

Fig. 14.1 Genotype by Environment interaction. Different environments are represented on the horizontal axis; the vertical axes provide measurements of a phenotypic value. Lines connect values for relatives (clones or siblings) reared in each environment. (*a*) In cases in which genotype by environment interaction is not detectable, the lines do not differ from parallel. (*b*) Genotype by environment interaction is detectable because of the slopes of the lines connecting relatives between environments differ for different families; the lines may or may not cross.

and have small dispersal distances, extensive mortality between birth and maturity can mean that a much larger cohort must be marked initially than will be available for the final measurements. The limitations imposed by these factors explain the prevalence of laboratory studies in this field, and in this book. The species chosen for the laboratory are those that show interesting natural variation in the behavior of interest, that can be persuaded to exhibit the behavior in captivity, and that can be reared in large numbers with reasonable generation times.

The final problem in applying laboratory data to natural populations is genotype by environment interaction. This is a result of different genotypes performing differently in different environments (fig. 14.1). For example, large and small male sailfin mollies use different types of precopulatory behavior, and body size has a genetic component (Travis, chap. 8). But the actual behavior of a molly of a given size depends on whether it is relatively large or relatively small compared with other fish in its population; its behavioral phenotype depends on its social environment. Statistical analyses of genotype by environment interaction are reviewed by Baker (1988). The phenomenon of genotype by environment interaction has two ramifications for the study of the evolution of behavior. First, it is possible that a behavior pattern that is best in one environment is not so valuable in another environment in which a species is found; the interpretation that a particular behavior pattern is "optimal" thus needs to be tempered by consideration of whether it is likely to be globally optimal. Second, the laboratory represents a different environment from the field, making generalization of genetic measures from laboratory to field problematic. The method of "natural heritabilities" described above (Riska, Prout, and Turelli 1989) takes into account the possibility of genotype by environment interaction between laboratory and field.

Some questions about the application of quantitative genetics to evolutionary biology concern the interpretation of heritabilities and genetic correlations (lumped under the term "heritability" for convenience). The results of a quantitative genetic analysis may provide evidence for or against the existence of a nonzero heritability for a particular trait. How will these results be interpreted? When a heritability is small and cannot be bounded away from zero, the interpretation may rely on Fisher's fundamental theorem: a trait with no detectable additive genetic variance could have undergone strong selection in the past, and be at its adaptive maximum, that is, closely related to fitness. However, a statement that there is "no detectable heritability" is subject to caveats about experimental design and sample size (Arnold, chap. 2). If the number of families is small, the data should be subjected to a power analysis to determine the probability of a Type II error (ideally this would be conducted in the planning stage rather than at the end). A statement that a trait has been subject to strong selection in the past would be strengthened if in addition to an absence of detectable additive genetic variance, the presence of dominance variance could be inferred or demonstrated, because dominance variance is an indicator of past selection (Falconer 1989). Dominance variance can be estimated with nested paternal full- and half-sib breeding designs and with diallel crosses, although the potential user should be warned that the assumptions underlying the interpretation of diallel crosses are no less stringent than the assumptions for any genetic analysis (Lynch, chap. 13; Broadhurst and Jinks 1974; Jinks and Broadhurst 1974; Falconer 1989).

Price and Schluter (1991) showed that in the case of life history traits, heritability estimates will necessarily be low, regardless of Fisher's fundamental theorem, because of the number of causal steps from genes to phenotype. At each step additional environmental variance will be introduced. Consequently even if appreciable additive genetic variance for life history traits exists, heritabilities will be low, and they concluded that "heritabilities of single life-history traits cannot be used to indicate whether or not populations are close to equilibrium" (Price and Schluter 1991, 859). They suggested that their argument could also apply to behavioral traits.

When heritability is low, it is nevertheless legitimate to maintain that evolutionary change is not precluded, especially if natural effective population sizes are larger than the experimental sample size. Even heritabilities that are too small to be detected experimentally would allow evolution to proceed, albeit slowly (but evolution does proceed slowly). If an experiment has included measurements of the inheritance of a number of traits, some of which show significant heritabilities, then a comparison of different traits can illuminate the results. If some traits give significant results, then the overall design was capable of detecting significance. Similarly, if similar measures are made on members of two populations, differences in genetic estimates between the populations give valuable information about evolutionary differences (Dingle, chap. 7). The existence of low heritabilities does not simply mean that the trait is uninteresting evolutionarily; such a simplistic conclusion would miss a great deal of the complexities of an issue.

If a heritability is found to be significantly greater than zero, one cannot conclude that the trait is in fact a minor contributor to fitness. First, Wright (1930) pointed out that Fisher's fundamental theorem was appropriate in the absence of mutation, gene flow, and genetic correlations. All these factors can prevent a trait from reaching evolutionary equilibrium, or can slow the rate of approach to equilibrium appreciably (Via and Lande 1985). Second, an allele that improves the fitness effect of one trait could be prevented from becoming fixed if the same allele diminished the fitness effect of another trait. This antagonistic pleiotropy could be responsible for the maintenance of heritability of a fitness-related trait (Rose 1982). When a heritability is significantly greater than zero, one can conclude that the trait is still capable of responding to selection, an interpretation used by Lynch (chap. 13).

Another approach to the problem of interpreting heritabilities is to reduce the emphasis on examining heritabilities. Direct inspection of the additive genetic variances and covariances can be valuable, and both Price and Schluter (1991) and Houle (1992) have proposed that additive genetic coefficients of variation should be examined. These are additive genetic variances standardized by the mean, in analogy to phenotypic coefficients of variation. Houle (1992) has made the case that the additive genetic coefficient of variation is a more appropriate measure than heritability for evaluating the evolvability of a trait. However, in cases in which the goal is to fit an estimate into equations for evolutionary change, heritability may need to be computed because of the formulation of the equations. The same data can be used to compute both heritabilities and additive genetic coefficients of variation, so providing both measures may provide more insight than providing only one.

Considerations for Designing an Experiment to Measure Heritabilities and Genetic Correlations

Cost-Benefit Analysis of the Project

The results of quantitative genetic studies will benefit from extensive planning. The first question must be whether the results will justify the time and effort involved in the project. For evolutionary analyses, the behavioral traits to be studied need to be related to biological questions that are generally considered interesting (Medawar 1979, chap. 3), rather than chosen because they can be measured. Although this appears to be an obvious point, it is often extremely difficult to develop a way to measure behavior repeatably that is not so structured as to take it far from its evolutionary context. The behavior patterns chosen will be compromises between those that are most compelling from a conceptual perspective and those that can be measured, taking into account observability and repeatability. This book provides many examples of methods used to make behavior as repeatable as possible without divorcing it from an evolutionary context. Similarly, the choice of organism may be a compromise between convenience and interest.

A major step in the design stage is to decide on the genetic variables to be estimated, which will affect both the difficulty of conducting the research and the value of the results. Measuring heritability by itself is about as useful as measuring electrophoretic variability, but a lot more costly: the first few times, it is valuable for exploring the nature of the topic, but very soon it becomes more of the same old thing (Arnold, chap. 2). The value of quantitative genetic studies lies in their evaluation of more complex issues, such as adaptation, genetic correlations, maternal effects, and genotype-environment interactions, and in their use of selected lines to open new areas of investigation. All of these studies will provide estimates of heritability, but they reach considerably beyond the simple results. Frequently, the effort needed to evaluate additional factors involves a relatively small increment of time or space above that which will be invested in measuring heritability.

Sample Size and Scaling

The necessary sample size can be computed by referring to formulas in Falconer (1989, 179–84) and Robertson (1959a,b). The actual sample size is likely to reflect a trade-off between the time needed to measure the trait and the ideal sample size that was computed in the planning stages. Clearly it is advantageous to choose traits that are easy to measure and that have high repeatability (Lessels and Boag 1987; Boake 1989b). It is valuable to begin a study with a careful survey of possible traits to measure, including measurements of their repeatabilities and phenotypic correlations. If enough animals are measured, on the order of several hundred, one might be able to argue that one's phenotypic correlations provide an accurate estimate of genetic correlations, and preclude the genetic study (Cheverud 1988a), although this suggestion has been vehemently disputed (Willis, Coyne, and Kirkpatrick 1991).

The scale on which genetic variables will be measured may have a strong influence on the interpretation of the results. Certain behavioral characters such as sensory sensitivity appear to obey logarithmic laws, and various power relationships describe the transformations between different measures of size (Sokal and Rohlf 1981). One may wish to consider whether selection acts on the untransformed measure or on some function of it. Some of the complexities of scaling are discussed by Travis (chap. 8), and the reader is also referred to discussions by Wright (1968), Sokal and Rohlf (1981), and Falconer (1989).

Breeding Design

When one considers the time necessary to score traits, the choice of one- or two-generation breeding designs (sib analyses, parent-offspring regression, diallels), in contrast to artificial selection, rarely makes a large difference. In artificial selection, although only a small proportion of the population is bred each generation, many animals must be scored and the majority rejected. Clearly, a self-scoring test such as geotaxis or phototaxis becomes very appealing. Studies of aspects of social behavior, foraging, or other complex traits will require a major investment, regardless of the breeding design. Artificial selection may require slightly less space than the other designs, although the other designs can be subdivided into several blocks of tractable size, which are conducted in succession.

If genetic correlations are to be estimated, artificial selection may not be the best breeding design, because artificial selection is usually conducted with such small population sizes that change in the allelic frequency of unselected genes is quite likely. This can result in spurious correlated responses to selection. The problem can be mitigated with numerous replicate selected lines: if they all show the same correlated responses, the underlying genetic correlations are likely to be genuine. Alternatively, one can cross the selected lines before testing the animals (Partridge, chap. 6). For the many organisms that have complex behavior but that are expensive or time-consuming to rear, more restricted artificial regimes have been used. Bakker (1986) addressed the limitations of using artificial selection to study genetic correlations by selecting on a number of traits in several lines simultaneously, one trait per line, and using a common control line. In his discussions of Berthold's research on avian migratory behavior, Dingle (chap. 7) described the simplest acceptable selection experiment, using just two selected lines, one selected to increase the value of a trait and the other selected to decrease it. He pointed out that even though the results of such simple experiments are limited in the extent of their interpretation, valuable information can be gained. Hoffmann (chap. 9) was in the enviable position of having large numbers of animals relative to the space available (a major reason why *Drosophila melanogaster* is such a frequent subject of quantitative genetic studies). The interpretation of results from artificial selection experiments depends upon the assumption that the base population was outbred: an inbred population is unlikely to have appreciable additive genetic variation on which selection could act. However, inbred lines can be crossed to produce an outbred population as long as enough generations are allowed before selection so that spurious linkage disequilibrium can be eliminated. The outbred population of mice that Lynch used was developed by crossing inbred strains several decades ago (chap. 13).

Decisions about a breeding design and a sample size have to take into account the recognition that members of the same family must not be reared together. If they are reared in the same environment, then the effects of a shared environment are statistically and causally confounded with the effects of shared genes. Thus, at least two containers (vials, cages, or whatever) are needed to rear members of a family, if more than one member will be scored. This increases the space and effort necessary for rearing. Furthermore, family members must not be housed next to each other because of the possibility that they could experience similar microenvironmental fluctuations. A formal randomization scheme is the best way to assign locations to every individual in the study, which will of course increase the work necessary.

Summary

Quantitative genetic theory has clearly provided novel answers to evolutionary questions. Results of the theory have stimulated empirical research in a variety of areas. The application of quantitative genetic methods to studies of natural populations will always be constrained by logistical difficulties, yet it has enormous value in empirical analyses of evolution in natural populations. The examples in this book demonstrate that the issues to which quantitative genetic models and methods can be applied range from hypotheses about optimality to analyses of the physiological control of behavior. Future quantitative genetic research is as likely to guide theory as to be led by it.

Literature Cited

Note: Numbers in parentheses at the end of each reference refer to the chapters that cite that reference.

Ahlschwede, W. T., and O. W. Robison. 1971. Maternal effects on weights and backfat of swine. *J. Anim. Sci.* 32:10–16. (4)

Alatalo, R. V., L. Gustaffson, and A. Lundberg. 1990. Phenotypic selection on heritable size traits: Environmental variance and genetic response. *Am. Nat.* 135:464–71. (6)

Alexander, R. D. 1975. Natural selection and specialized chorusing behavior in acoustical insects. In *Insects, Science and Society,* ed. D. Pimentel, 35–77. New York: Academic Press. (11)

Alexander, R. D., and G. Borgia. 1979. On the origin of the male-female phenomenon. In *Sexual Selection and Reproductive Competition in Insects,* ed. M. S. Blum and N. A. Blum, 417–40. New York: Academic Press. (4)

Alpatov, W. W. 1930. Phenotypical variation in body and cell size of *Drosophila melanogaster. Biol. Bull.* 58:85–103. (6)

Andersson, M. 1982a. Female choice selects for extreme tail length in a widowbird. *Nature* 299:818–20. (5, 6)

Andersson, M. 1982b. Sexual selection, natural selection and quality advertisement. *Biol. J. Linn. Soc.* 17:375–93. (8)

Andersson, M. 1986. Evolution of condition-dependent sex ornaments and mating preferences: Sexual selection based on viability differences. *Evolution* 40:804–16. (5, 11)

Andersson, M. 1987. Genetic models of sexual selection: Some aims, assumptions, and tests. In *Sexual Selection: Testing the Alternatives,* ed. J. W. Bradbury and M. B. Andersson, 41–53. Chichester: Wiley. (5, 11)

Anholt, B. R. 1991. Measuring selection on a population of damselflies with a manipulated phenotype. *Evolution* 45:1091–1106. (4)

Antonovics, J. 1982. Comment. In *Evolution and Genetics of Life Histories,* ed. H. Dingle and J. P. Hegmann, 235. New York: Springer. (2)

Aoki, K. 1982. Additive polygenic formulation of Hamilton's model of kin selection. *Heredity* 49:163–69. (4, 10)

Arak, A. 1983. Sexual selection by male competition in natterjack toad choruses. *Nature* 306:261–62. (6)

Arak, A. 1988. Callers and satellites in the natterjack toad: Evolutionarily stable decision rule. *Anim. Behav.* 36:416–32. (3)

Arnold, S. J. 1981a. Behavioral variation in natural populations. I. Phenotypic, genetic and environmental correlations between chemoreceptive responses to prey in the garter snake, *Thamnophis elegans. Evolution* 35:489–509. (2, 12, 14)

Arnold, S. J. 1981b. Behavioral variation in natural populations. II. The inheritance of a feeding response in crosses between geographic races of the garter snake, *Thamnophis elegans. Evolution* 35:510–15. (12, 14)
Arnold, S. J. 1981c. The microevolution of feeding behavior. In *Foraging Behavior: Ecological, Ethological and Psychological Approaches,* ed. A. Kamil and T. Sargent, 409–53. New York: Garland Press. (12)
Arnold, S. J. 1983a. Morphology, performance and fitness. *Am Zool.* 23:347–61. (12)
Arnold, S. J. 1983b. Sexual selection: The interface of theory and empiricism. In *Mate Choice,* ed. P. Bateson, 67–107. Cambridge: Cambridge University Press. (1, 5, 11)
Arnold, S. J. 1985. Quantitative genetic models of sexual selection. *Experientia* 41:1296–1310. (5, 11)
Arnold, S. J. 1987a. Genetic correlation and the evolution of physiology. In *New Directions in Ecological Physiology,* ed. M. E. Feder, A. F. Bennett, W. W. Burggren, and R. B. Huey, 189–215. Cambridge: Cambridge University Press. (12)
Arnold, S. J. 1987b. Quantitative genetic models of sexual selection: A review. In *The Evolution of Sex and Its Consequences,* ed. S. Stearns, 283–315. Basel: Birkhauser. (2)
Arnold, S. J. 1988. Quantitative genetics and selection in natural populations: Microevolution of vertebral numbers in the garter snake *Thamnophis elegans.* In *Proceedings of the Second International Conference on Quantitative Genetics,* ed. B. S. Weir, E. J. Eisen, M. J. Goodman, and G. Namkoong, 619–36. Sunderland, Mass.: Sinauer Associates. (12)
Arnold, S. J. 1990. Inheritance and the evolution of behavioral ontogenies. In *Developmental Behavior Genetics: Neural, Biometrical, and Evolutionary Approaches,* ed. M. Hahn, J. Hewitt, N. Henderson, and R. Benno, 167–89. New York: Oxford University Press. (2, 12)
Arnold, S. J. 1992a. Behavioural variation in natural populations. VI. Prey responses by two species of garter snakes in three regions of sympatry. *Anim. Behav.* 44:705–19. (2, 12)
Arnold, S. J. 1992b. Constraints on phenotypic evolution. *Am. Nat.* 140:S85–S107. (2, 12, 14)
Arnold, S. J., and A. F. Bennett. 1984. Behavioural variation in natural populations. III. Antipredator displays in the garter snake *Thamnophis radix. Anim. Behav.* 32:1108–18. (2, 12)
Arnold, S. J., and A. F. Bennett. 1988. Behavioural variation in natural populations. V. Morphological correlates of locomotion in the garter snake *Thamnophis radix. Biol. J. Linn. Soc.* 34:175–90. (12)
Arnold, S. J., and T. Halliday. 1988: Multiple mating: Natural selection is not evolution. *Anim. Behav.* 36:1547–48. (6)
Arnold S. J., and M. J. Wade. 1984. On the measurement of natural and sexual selection: Theory. *Evolution* 38:720–34. (14)
Aspi, J., and A. Hoikkala. 1993. Laboratory and natural heritabilities of male courtship song characters in *Drosophila montana* and *D. littoralis. Heredity* 70:400–406. (9)
Astrand, P. O., and K. Rodahl. 1986. *Textbook of Work Physiology.* 3d ed. New York: McGraw-Hill. (12)
Atchley, W. R., T. Lodgson, D. E. Cowley, and E. J. Eisen. 1991. Uterine effects, epigenetics, and postnatal skeletal development in the mouse. *Evolution* 45:891–909. (4, 14)
Atchley, W. R., and J. J. Rutledge. 1980. Genetic components of size and shape. I.

Dynamics of components of phenotypic variability and covariability during ontogeny in the laboratory rat. *Evolution* 34:1161–73. (4)

Atchley, W. R., J. J. Rutledge, and D. E. Cowley. 1981. Genetic components of size and shape. II. Multivariate covariance patterns in the rat and mouse skull. *Evolution* 35:1037–55. (2)

Atkinson, W. D. 1979. A field investigation of larval competition in domestic *Drosophila. J. Anim. Ecol.* 48:91–102. (9)

Atwood, J. L. 1980. Breeding biology of the Santa Cruz Island scrub jay. In *The California Islands: Proceedings of a Multidisciplinary Symposium,* ed. D. M. Power, 675–88. Santa Barbara, Calif.: Santa Barbara Museum of Natural History. (7)

Austad, S. N. 1984. A classification of alternative reproductive behaviors and methods for field-testing ESS models. *Am. Zool.* 24:309–19. (8)

Baird, R. C. 1968. Aggressive behavior and social organization in *Mollienesia latipinna* (Le Sueur). *Tex. J. Sci.* 20:157–76. (8)

Baker, R. J. 1988. Differential response to environmental stress. In *Proceedings of the Second International Conference on Quantitative Genetics,* ed. B. S. Weir, E. J. Eisen, M. M. Goodman, and G. Namkoong, 492–504. Sunderland, Mass.: Sinauer Associates. (14)

Bakker, T.C.M. 1986. Aggressiveness in sticklebacks (*Gasterosteus aculeatus* L.): A behaviour-genetic study. *Behaviour* 98:1–144. (9, 14)

Barnett, S. A., K.M.H. Munro, J. L. Smart, and R. C. Stoddart. 1975. House mice bred for many generations in two environments. *J. Zool.* (Lond.) 177:153–69. (13)

Barton, N. H. 1986. The maintenance of polygenic variation by a balance between mutation and stabilizing selection. *Genet. Res.* 47:209–16. (5)

Barton, N. H., and M. Turelli. 1987. Adaptive landscapes, genetic distance and the evolution of quantitative characters. *Genet. Res.* 49:157–73. (3)

Barton, N. H., and M. Turelli. 1989. Evolutionary quantitative genetics: How little do we know? *Annu. Rev. Genet.* 23:337–70. (1, 2, 7, 10, 14)

Barton, N. H., and M. Turelli. 1991. Natural and sexual selection on many loci. *Genetics* 127:229–55. (5)

Bauer, S. J., and M. B. Sokolowski. 1988. Autosomal and maternal effects on pupation behavior in *Drosophila melanogaster. Behav. Genet.* 18:81–97. (4)

Beani, L., and S. Turillazi. 1988. Alternative mating tactics in males of *Polistes dominulus* (Hymenoptera: Vespidae). *Behav. Ecol. Sociobiol.* 22:257–64. (8)

Becker, W. A. 1984. *Manual of Quantitative Genetics.* 4th ed. Pullman, Wash.: Academic Enterprises. (1, 2)

Bell, A. E., and M. J. Burris. 1973.Simultaneous selection for two correlated traits in *Tribolium. Genet. Res.* 21:24–46. (3)

Bell, G., and V. Koufopanou. 1985. The cost of reproduction. In *Oxford Surveys of Evolutionary Biology,* ed. R. Dawkins, 83–131. Oxford: Oxford University Press. (3)

Bennett, A. F. 1987. Inter-individual variability: An underutilized resource. In *New Directions in Ecological Physiology,* ed. M. E. Feder, A. F. Bennett, W. W. Burggren, and R. B. Huey, 147–69. Cambridge: Cambridge University Press. (12)

Bennett, A. F. 1990. The thermal dependence of locomotor capacity. *Am. J. Physiol.* (*Regul. Integrat. & Comp. Physiol.* 28) 259:R253–58. (12)

Bennett, A. F. 1991. The evolution of activity capacity. *J. Exp. Biol.* 160:1–23. (12)

Bennett, A. F., T. J. Garland, and P. L. Else. 1989. Individual correlation of morphology, muscle mechanics and locomotion in a salamander. *Am. J. Physiol.* (*Regul. Integrat. & Comp. Physiol.* 25) 256:R1200–1208. (12)

Benton, M. J. 1980. Geographic variation in the garter snakes (*Thamnophis sirtalis*) of the north-central United States, a multivariate study. *Zool. J. Linn. Soc.* 68:307–23. (12)

Bernardo, J. 1991. Manipulating egg size to study maternal effects on offspring traits. *Trends Ecol. Evol.* 6:1–2. (2)

Bernon, D. E., and P. B. Siegel. 1983. Mating frequency in male Japanese quail: Crosses among selected and unselected lines. *Can. J. Genet. Cytol.* 25:450–56. (9)

Berry, R. J., W. N. Bonner, and J. Peters. 1979. Natural selection in house mice from South Georgia (South Atlantic Ocean). *J. Zool.* (Lond.) 189:385–98. (13)

Berthold, P. 1988a. The biology of the genus *Sylvia*—a model and a challenge for Afro-European cooperation. *Tauraco* 1:3–28. (7)

Berthold, P. 1988b. Evolutionary aspects of migratory behavior in European warblers. *J. Evol. Biol.* 1:195–209. (7)

Berthold, P., G. Mohr, and U. Querner. 1990. Steuering und potentielle Evolutionsgeschwindigkeit des obligaten Teilzieherverhaltens: Ergebnisse eines Zweigweg-Selktionsexperiments mit der Monchsgrasmucke (*Sylvia atricapilla*). *J. Ornithol.* 131:33–45. (7)

Berthold, P., and S. B. Terrill. 1988. Migratory behaviour and population growth of blackcaps wintering in Britain and Ireland: Some hypotheses. *Ringing & Migr.* 9:153–59. (7)

Bertness, M. D. 1981. Pattern and plasticity in tropical hermit crab growth and reproduction. *Am. Nat.* 117:754–73. (3)

Bhuvanakumar, C. K., C. B. Lynch, R. C. Roberts, and W. G. Hill. 1985. Heterosis among lines of mice selected for body weight. I. Growth. *Theor. Appl. Genet.* 71:44–51. (13)

Blackman, R. L. 1975. Photoperiodic determination of the male and female sexual morphs of *Myzus persicae*. *J. Insect Physiol.* 21:435–53. (7)

Boag, P. T. 1983. The heritability of external morphology in Darwin's ground finches (*Geospiza*) on Isla Daphne Major, Galapagos. *Evolution* 37:877–94. (14)

Boake, C.R.B. 1983. Mating systems and signals in crickets. In *Orthopteran Mating Systems,* ed. D. T. Gwynne and G. K. Morris, 28–44. Boulder: Westview Press. (11)

Boake, C.R.B. 1986. A method for testing adaptive hypotheses of mate choice. *Am. Nat.* 127:654–66. (11)

Boake, C.R.B. 1989a. Correlations between courtship success, aggressive success, and body size in the picture-winged fly, *Drosophila silvestris*. *Ethology* 80:318–29. (9)

Boake, C.R.B. 1989b. Repeatability: Its role in evolutionary studies of mating behavior. *Evol. Ecol.* 3:173–82. (2, 4, 11, 12, 14)

Bock, R. D. 1975. *Multivariate Statistical Methods in Behavioral Research.* New York: McGraw-Hill. (2)

Bogyo, T. P. 1964. Coefficients of variation of heritability estimates obtained from variance analysis. *Biometrics* 20:122–29. (3)

Bohren, B. B., W. G. Hill, and A. Robertson. 1966. Some observations on asymmetrical correlated responses to selection. *Genet. Res.* 7:44–57. (3)

Bolles, R. C. 1975. *Theory of Motivation.* 2d ed. New York: Harper and Row. (12)

Bomze, I. M., P. Schuster, and K. Sigmund. 1983. The role of Mendelian genetics in strategic models of animal behaviour. *J. Theor. Biol.* 101:19–38. (3)

Bondari, K., R. L. Willham, and A. Freeman. 1978. Estimates of direct and maternal genetic correlations for pupa weight and family size of *Tribolium*. *J. Anim. Sci.* 47:358–65. (4)

Bookstein, F., B. Chernoff, R. Elder, J. Humphries, G. Smith, and R. Strauss. 1985. *Morphometrics in Evolutionary Biology, the Geometry of Size and Shape Change, with Examples from Fishes.* Special publication no. 15. Philadelphia: Academy of Natural Sciences of Philadelphia. (8)

Borgia, G. 1979. Sexual selection and the evolution of mating systems. In *Sexual Selection and Reproductive Competition in Insects,* ed. M. S. Blum and N. A. Blum, 19–80. New York: Academic Press. (11)

Borowsky, R. L. 1973. Relative size and the development of fin coloration in *Xiphophorus variatus. Physiol. Zool.* 46:22–28. (8)

Bouchard, P. R., and C. B. Lynch. 1989. Burrowing behavior in wild house mice: Variation within and between populations. *Behav. Genet.* 19:447–56. (13)

Boyd, R., and P. J. Richerson. 1980. Effect of phenotypic variation on kin selection. *Proc. Nat. Acad. Sci. USA.* 77:7506–9. (4)

Bradbury, J. W., and M. B. Andersson, eds. 1987. *Sexual Selection: Testing the Alternatives.* Chichester: Wiley. (1, 6)

Bradford, G. E. 1972. The role of maternal effects in animal breeding. IV. Maternal effects in sheep. *J. Anim. Sci.* 35:1315–25. (4)

Bradley, G. L. 1975. *A Primer of Linear Algebra.* Englewood Cliffs, N.J.: Prentice-Hall. (2)

Bradshaw, W. E. 1986. Pervasive themes in insect life cycle strategies. In *The Evolution of Insect Life Cycles,* ed. F. Taylor and R. Karban, 261–75. New York: Springer-Verlag. (7)

Breden, F. 1990. Partitioning covariance as a method of studying kin selection. *Trends Ecol. Evol.* 5:224–28. (4)

Breden, F. and G. Stoner. 1987. Male predation risk determines female preference in the Trinidad guppy. *Nature* 329:831–33. (5)

Broadhurst, P. L. 1967. An introduction to the diallel cross. In *Behavior-Genetic Analysis,* ed. J. Hirsch, 287–304. New York: McGraw-Hill. (13)

Broadhurst, P. L. 1979. The experimental approach to behavioural evolution. In *Theoretical Advances in Behaviour Genetics,* ed. J. R. Royce and L. P. Mos, 43–95. Alphen aan de Rijn: Sijthoff and Noordhoff. (9)

Broadhurst, P. L., and J. L. Jinks. 1974. What genetical architecture can tell us about the natural selection of behavioral traits. In *The Genetics of Behaviour,* ed. J.H.F. van Abeelen, 43–63. New York: American Elsevier. (1, 12, 14)

Brodie, E. D. III. 1989a. Behavioral modification as a means of reducing the cost of reproduction. *Am. Nat.* 134:225–38. (12)

Brodie, E. D. III. 1989b. Genetic correlations between morphology and antipredator behaviour in natural populations of the garter snake *Thamnophis ordinoides. Nature* 342:542–43. (12)

Brodie, E. D. III. 1991. Functional and genetic integration of color pattern and antipredator behavior in the garter snake *Thamnophis ordinoides.* Ph.D. diss., Univ. of Chicago. (12)

Brodie, E. D. III. 1992. Correlational selection for color pattern and antipredator behavior in the garter snake *Thamnophis ordinoides. Evolution* 46:1284–98. (12)

Brodie, E. D. III. 1993. Homogeneity of the genetic variance-covariance matrix for antipredator traits in two natural populations of the garter snake *Thamnophis ordinoides. Evolution* 44:844–54. (12)

Brodie, E. D. III., and E D. Brodie, Jr. 1991. Evolutionary response of predators to dangerous prey: reduction of toxicity of newts and resistance of garter snakes in island populations. *Evolution* 45:221–224. (12)

Brodie, E. D. III, and T. Garland, Jr. 1993. Quantitative genetics of snake populations. In *Snakes: Ecology and Behavior*, ed. R. A. Siegel and J. T. Collins, 315–62. New York: McGraw Hill.

Bronson, F. H. 1979. The reproductive ecology of the house mouse. *Q. Rev. Biol.* 54:265–99. (13)

Brooks, D. R., and D. A. McClennan. 1991. *Phylogeny, Ecology, and Behavior: A Research Program in Comparative Biology.* Chicago: University of Chicago Press. (12)

Brooks, G. A., and T. D. Fahey, 1984. *Exercise Physiology: Human Bioenergetics and its Applications.* New York: John Wiley and Sons. (12)

Bruell, J. H. 1967. Behavioral heterosis. In *Behavior-Genetic Analysis*, ed. J. Hirsch, 270–304. New York: McGraw-Hill. (13)

Bull, J. J. 1980. Sex determination in reptiles. *Q. Rev. Biol.* 55:3–21. (4)

Bull, J. J. 1985. Models of parent-offspring conflict: Effect of environmental variance. *Heredity* 55:1–8. (4)

Bulmer, M. G. 1972. The genetic variability of polygenic characters under optimizing selection, mutation and drift. *Genet. Res.* 19:17–25. (5)

Bulmer, M. G. 1980. *The Mathematical Theory of Quantitative Genetics.* Oxford: Clarendon Press. (8)

Bulmer, M. G. 1985. *The Mathematical Theory of Quantitative Genetics.* Reprint. Oxford: Oxford University Press. (2, 3)

Bulmer, M. G. 1989. Structural instability in models of sexual selection. *Theor. Popul. Biol.* 35:195–206. (5)

Bult, A., and C. B. Lynch. 1990. Heterosis in mice selected for nesting behavior and a new potential to break selection limits. *Behav. Genet.* 20:707–8. (13)

Bult, A., E. A. van der Zee, J. C. Compaan, and C. B. Lynch. 1992. Differences in the number of arginine-vasopressin-immunoreactive neurons exist in the suprachiasmatic nuclei of house mice selected for differences in nest-building behavior. *Brain Res.* 578:335–38. (13, 14)

Burfening, P. J., D. D. Kress, and R. L. Friedrich. 1981. Calving ease and growth rate of Simmental-sired calves. III. Direct and maternal effects. *J. Anim. Sci.* 53:1210–16. (4)

Bürger, R. 1986. Constraints for the evolution of functionally coupled characters: A nonlinear analysis of a phenotypic model. *Evolution* 40:182–93. (4)

Burghardt, G. M. 1970. Intraspecific geographical variation in chemical food cue preferences of newborn garter snakes (*Thamnophis sirtalis*). *Behaviour* 36:246–57. (12)

Burk, T. 1983. Male aggression and female choice in a field cricket. In *Orthopteran Mating Systems*, ed. D. T. Gwynne and G. K. Morris, 97–119. Boulder: Westview Press. (11)

Butlin, R. K. 1993. A comment on the evidence for a genetic correlation between the sexes in *Drosophila melanogaster. Anim. Behav.* 45:403–4. (6)

Cabanac, M. 1975. Temperature regulation. *Annu. Rev. Physiol.* 37:415–39. (13)

Cade, W. H. 1979. The evolution of alternative male reproductive strategies in field crickets. In *Sexual Selection and Reproductive Competition in Insects*, ed. M. Blum and N. A. Blum, 343–79. London: Academic Press. (6)

Cade, W. H. 1981. Alternative male strategies: Genetic differences in crickets. *Science* 212:563–64. (8, 11)

Caldwell, R. L., and J. P. Hegmann. 1969. Heritability of flight duration in the milkweed bug, *Lygaeus kalmii. Nature* 223:91–92. (7)

Campbell, H. G. 1965. *An Introduction to Matrices, Vectors and Linear Programming.* New York: Appleton-Century-Crofts. (2)

Carey, G. 1988. Inference about genetic correlations. *Behav. Genet.* 18:329–38. (8)

Carson, H. L. 1985. Genetic variation in a courtship-related male character in *Drosophila silvestris* from a single Hawaiian locality. *Evolution* 39:678–86. (11)

Caspari, E. 1958. Genetic basis of behavior. In *Behavior and Evolution*, ed. A. Roe and G. G. Simpson, 103–27. New Haven: Yale University Press. (13)

Castle, W. E. 1921. An improved method of estimating the number of genetic factors concerned in the case of blending inheritance. *Science* 54:223. (10)

Cavalli, L. L. 1952. An analysis of linkage in quantitative inheritance. In *Quantitative Inheritance*, ed. E.C.R. Reeve and C. H. Waddington, 135–44. London: Her Majesty's Stationery Office. (10)

Cawthon, D. A., and D. B. Mertz. 1975. Reproductive failure of females of genetic strain bI of *Tribolium confusum* in crosses with other strains. Tribolium *Info. Bull.* 18:82–83. (10)

Chaffee, R.R.J., and J. C. Roberts. 1971. Temperature acclimation in birds and mammals. *Annu. Rev. Physiol.* 33:155–97. (13)

Charlesworth, B. 1973. Selection in populations with overlapping generations. V. Natural selection and life histories. *Am. Nat.* 107:303–11. (3)

Charlesworth, B. 1980. *Evolution in Age Structured Populations.* Cambridge: Cambridge University Press. (3)

Charlesworth, B. 1987. The heritability of fitness. In *Sexual Selection: Testing the Alternatives*, ed. J. W. Bradbury and M. B. Andersson, 21–40. Chichester: Wiley. (1, 11)

Charlesworth, B. 1990. Optimization models, quantitative genetics, and mutation. *Evolution* 44:520–38. (1, 3, 4)

Charlesworth, B. 1993. Natural selection on multivariate traits in age-structured populations. *Proc. R. Soc. Lond.*, B 251:47–52. (4)

Charlesworth, B., and J. A. León. 1976. The relation of reproductive effort to age. *Am. Nat.* 110:449–59. (3)

Charlesworth, B., and J. A. Williamson. 1975. The probability of survival of a mutant gene in an age-structured population and implications for the evolution of life-histories. *Genet. Res.* 26:1–10. (3)

Charnov, E. L. 1989. Phenotypic evolution under Fisher's Fundamental Theorem of Natural Selection. *Heredity* 62:113–16. (3)

Cheng, K. M., and P. B. Siegel. 1990. Quantitative genetics of multiple mating. *Anim. Behav.* 40:406–7. (6)

Cheung, T. K., and R. J. Parker. 1974. Effect of selection on heritability and genetic correlation of two quantitative traits in mice. *Can. J. Genet. Cytol.* 16:599–609. (3)

Cheverud, J. M. 1984a. Evolution by kin selection: A quantitative genetic model illustrated by maternal performance in mice. *Evolution* 38:766–77. (4)

Cheverud, J. M. 1984b. Quantitative genetics and developmental constraints on evolution by selection. *J. Theor. Biol.* 110:155–171. (12)

Cheverud, J. M. 1985. A quantitative genetic model of altruistic selection. *Behav. Ecol. Sociobiol.* 16:239–43. (4)

Cheverud, J. M. 1988a. A comparison of genetic and phenotypic correlations. *Evolution* 42:958–68. (3, 14)

Cheverud, J. M. 1988b. The evolution of genetic correlation and developmental constraint. In *Population Genetics and Evolution*, ed. D. de Jong, 94–101. Berlin: Springer-Verlag. (12)

Cheverud, J. M., and L. J. Leamy. 1985. Quantitative genetics and the evolution of ontogeny. III. Ontogenetic changes in correlation structure among live body traits in randombred mice. *Genet. Res.* 46:325–35. (4)

Cheverud, J. M., L. J. Leamy, W. R. Atchley, and J. J. Rutledge. 1983. Quantitative genetics and the evolution of ontogeny. I. Ontogenetic changes in quantitative genetic variance components in randombred mice. *Genet. Res.* 42:65–75. (2, 4)

Christman, S. P. 1980. Patterns of geographic variation in Florida snakes. *Bull. Fla. State Mus. Biol. Sci.* 25:157–256. (12)

Clark, A. B. 1978. Sex ratio and local resource competition in a prosimian primate. *Science* 201:163–65. (4)

Clark, A. G. 1987. Senescence and the genetic-correlation hang-up. *Am. Nat.* 129:932–40. (7, 12)

Clutton-Brock, T. H., ed. 1988. *Reproductive Success.* Chicago: University of Chicago Press. (8)

Clutton-Brock, T. H. 1991. *The Evolution of Parental Care.* Princeton, N.J.: Princeton University Press. (2)

Clutton-Brock, T. H., S. D. Albon, and F. E. Guinness. 1985. Parental investment and sex differences in juvenile mortality in birds and mammals. *Nature* 313:131–33. (6)

Clutton-Brock, T. H., F. E. Guinness, and S. D. Albon. 1982. *Red Deer: Behavior and Ecology of Two Sexes.* Chicago: University of Chicago Press. (8)

Coates, D. 1988. Length-dependent changes in egg size and fecundity in females, and brooded embryo size in males, of fork-tailed catfishes (Pisces: Ariidae) from the Sepik River, Papua New Guinea with some implications for stock assessments. *J. Fish Biol.* 33:455–64. (3)

Cock, A. G. 1966. Genetical aspects of metrical growth and form in animals. *Q. Rev. Biol.* 41:131–90. (8)

Cockerham, C. C. 1963. Estimation of genetic variances. In *Statistical Genetics and Plant Breeding*, ed. W. D. Hanson and H. F. Robinson, 53–94. Washington, D.C.: National Academy of Sciences Research Council Publications. (2)

Cohan, F. M., and A. A. Hoffmann. 1986. Genetic divergence under uniform selection. II. Different responses to selection for knockdown resistance to ethanol among *Drosophila melanogaster* populations and their replicate lines. *Genetics* 114:145–63. (14)

Cohen, D. 1966. Optimizing reproduction in a randomly varying environment. *J. Theor. Biol.* 12:119–29. (3)

Comstock, R. E., and H. F. Robinson. 1952. Estimation of average dominance of genes. In *Heterosis*, ed. J. W. Gowen, 494–516. Ames, Iowa: Iowa State College Press. (13)

Conner, J. 1988. Field measurements of natural and sexual selection in the fungus beetle, *Bolitotherus cornutus. Evolution* 42:736–49. (4)

Conner, J. 1989. Density-dependent sexual selection in the fungus beetle, *Bolitotherus cornutus. Evolution* 43:1378–86. (14)

Conover, W. J., M. E. Johnson, and M. M. Johnson. 1981. A comparative study of tests for homogeneity of variances, with applications to the outer continental shelf bidding data. *Technometrics* 23:351–61. (12)

Cowley, D. E. 1990. Prenatal effects on mammalian study: Embryo transfer results. In *The Unity of Evolutionary Biology: Proceedings of the Fourth International Congress of Systematic and Evolutionary Biology,* vol. 2, ed. E. C. Dudley, 762–79. Portland, Oreg.: Dioscorides Press. (2)

Cowley, D. E. 1991. Genetic prenatal maternal effects on organ size in mice and their potential contribution to evolution. *J. Evol. Biol.* 3:363–81. (4)

Cowley, D. E., and W. R. Atchley. 1992. Comparison of quantitative genetic parameters. *Evolution* 46:1965–67. (2)

Cowley, D. E., D. Pomp, W. R. Atchley, E. J. Eisen, and D. Hawkins-Brown. 1989. The impact of maternal uterine genotype on postnatal growth and adult body size in mice. *Genetics* 122:193–203. (4)

Coyne, J. A., and E. Beecham. 1987. Heritability of two morphological characters within and among natural populations of *Drosophila melanogaster. Genetics* 117:727–37. (6, 11, 14)

Craig, D. M. 1986. Stimuli governing intraspecific egg predation in the flour beetles, *Tribolium confusum* and *T. castaneum. Res. Popul. Ecol.* 28:173–83. (10)

Crankshaw, O. S. 1979. Female choice in relation to calling and courtship songs in *Acheta domesticus. Anim. Behav.* 27:1274–75. (11)

Creel, S. 1990. How to measure inclusive fitness. *Proc. R. Soc. Lond.*, B 241:229–31. (3)

Crews, D. 1992. Behavioural endocrinology from an evolutionary perspective. In *Oxford Reviews of Reproductive Biology,* ed. S. R. Milligan, 303–70. Oxford: Oxford University Press. (12)

Crow, J. F., and K. Aoki. 1982. Group selection for a polygenic behavioral trait: A differential proliferation model. *Proc. Natl. Acad. Sci. USA* 79:2628–31. (10)

Crow, J. F., W. R. Engels, and C. Denniston. 1990. Phase three of Wright's shifting balance theory. *Evolution* 44:233–47. (10)

Crusio, W. E., J.M.L. Kerbusch, and J.H.F. van Abeelen. 1984. The replicated diallel cross: A generalized method of analysis. *Behav. Genet.* 14:81–104. (13)

Cundiff, L. V. 1972. The role of maternal effects in animal breeding: VIII. Comparative aspects of maternal effects. *J. Anim. Sci.* 35:1335–37. (4)

Curtsinger, J. W., and I. L. Heisler. 1988. A diploid "sexy son" model. *Am. Nat.* 132:437–53. (5)

Curtsinger, J. W., and I. L. Heisler. 1989. On the consistency of sexy-son models: A reply to Kirkpatrick. *Am. Nat.* 134:978–81. (5)

Darroch, J. N., and J. E. Mosimann. 1985. Canonical and principal components of shape. *Biometrika* 72:241–52. (8)

Darwin, C. 1859. *On the Origin of Species by Natural Selection.* London: John Murray. (1, 4)

Darwin, C. 1871. *The Descent of Man and Selection in Relation to Sex.* London: John Murray. (4, 5, 6)

Davies, N. B., and T. R. Halliday. 1978. Deep croaks and fighting assessment in toads *Bufo bufo. Nature* 274:683–85. (6)

Davies, R. W. 1971. The genetic relationship of two quantitative characters of *Drosophila melanogaster.* II. Location of the effects. *Genetics* 69:363–75. (8)

Davies, R. W., and P. L. Workman. 1971. The genetic relationship of two quantitative characters in *Drosophila melanogaster.* I. Responses to selection and whole chromosome analysis. *Genetics* 69:353–61. (8)

DeFries, J. C., and J. P. Hegmann. 1970. Genetic analysis of open-field behavior. In *Contributions of Behavior-genetic Analysis: The Mouse as a Prototype,* ed. G. Lindzey and D. D. Thiessen, 23–56. New York: Appleton-Century-Crofts. (13)

DeFries, J. C., and G. E. McClearn. 1970. Social dominance and Darwinian fitness in the laboratory mouse. *Am. Nat.* 104:408–11. (13)

DeFries, J. C., and R. W. Touchberry. 1961. A "maternal effect" on body weight in *Drosophila*. *Genetics* 46:1261–66. (4)

Denny, M. W., T. L. Daniel, and M.A.R. Koehl. 1985. Mechanical limits to size in wave-swept organisms. *Ecol. Monogr.* 55:69–102. (3)

Desharnais, R. A., and R. F. Costantino. 1983. Natural selection and density-dependent population growth. *Genetics* 105:1029–40. (3)

Dessauer, H. C., J. E. Cadle, and R. Lawson. 1987. Patterns of snake evolution suggested by their proteins. *Fieldiana Zool.* 34:1–34. (12)

Devine, M. C. 1975. Copulatory plugs in snakes: Enforced chastity. *Science* 187:844–45. (14)

Dhondt, A. A. 1982. Heritability of blue tit tarsus length from normal and cross-fostered broods. *Evolution* 36:418–19. (14)

Dickerson, G. E. 1947. Composition of hog carcasses as influenced by heritable differences in rate and economy of gain. *Res. Bull. Iowa Agric. Exp. Stat.* 354:489–524. (2, 4)

Dingle, H. 1980. Ecology and evolution of migration. In *Animal Migration, Orientation, and Navigation*, ed. S. A. Gauthreaux, Jr., 1–101. New York: Academic Press. (7)

Dingle, H. 1984. Behavior, genes and life histories: Complex adaptations in uncertain environments. In *A New Ecology: Novel Approaches to Interactive Systems*, ed. P. Price, C. Slobodchikoff, and W. Gaud, 169–94. New York: John Wiley and Sons. (11)

Dingle, H. 1985. Migration. In *Comprehensive Insect Physiology, Biochemistry, and Pharmacology*, ed. G. A. Kerkut and L. I. Gilbert, 375–415. Oxford: Pergamon. (7)

Dingle, H. 1986. Evolution and genetics of insect migration. In *Insect Flight, Dispersal, and Migration*, ed. W. Danthanarayana, 11–26. Berlin: Springer-Verlag. (7)

Dingle, H. 1988. Quantitative genetics of life history variation in a migrant insect. In *Population Genetics and Evolution*, ed. G. de Jong, 83–93. Berlin: Springer-Verlag. (7)

Dingle, H. 1991. Evolutionary genetics of animal migration. *Am. Zool.* 31:253–64. (7)

Dingle, H. 1992. Food level reaction norms in size-selected milkweed bugs (*Oncopeltus fasciatus*). *Ecol. Entomol.* 17:121–26. (1, 14)

Dingle, H., and K. E. Evans. 1987. Responses in flight to selection on wing length in non-migratory milkweed bugs, *Oncopeltus fasciatus*. *Entomol. Exp. Appl.* 45:289–96. (7)

Dingle, H., K. E. Evans, and J. O. Palmer. 1988. Responses to selection among life-history traits in a nonmigratory population of milkweed bugs (*Oncopeltus fasciatus*). *Evolution* 42:79–92. (7)

Dingle, H., and J. P. Hegmann, eds. 1982. *Genetics and Evolution of Life Histories.* New York: Springer-Verlag. (11)

Djawdan, M. 1993. Locomotor performance of bipedal and quadrupedal heteromyid rodents. *Funct. Ecol.*, in press. (12)

Dobzhansky, T. 1952. Nature and origin of heterosis. In *Heterosis*, ed. J. W. Gowen, 218–23. Ames, Iowa: Iowa State College Press. (13)

Dohm, M. R. and T. Garland, Jr. 1993. Quantitative genetics of scale counts in the garter snake *Thamnophis sirtalis*. *Copeia* 1993:987–1002. (12)

Dominey, W. J. 1980. Female mimicry in bluegill sunfish—a genetic polymorphism? *Nature* 284:546–48. (8)

Dominey, W. J. 1983. Sexual selection, additive genetic variance and the "phenotypic handicap". *J. Theor. Biol.* 101:495–502. (11)

Dominey, W. J. 1984. Alternative mating tactics and evolutionarily stable strategies. *Am. Zool.* 24:385–96. (8)

Dow, M. A. and F. von Schilcher. 1975. Aggression and mating success in *Drosophila melanogaster. Nature* 254:511–12. (6, 9)

Dowling, H. G., R. Highton, G. C. Maha, and L. R. Maxson. 1983. Biochemical evaluation of colubrid snake phylogeny. *J. Zool.* (Lond.) 201:309–29. (12)

Drummond, H., and G. M. Burghardt. 1983. Geographic variation in the foraging behavior of the garter snake, *Thamnophis elegans. Behav. Ecol. Sociobiol.* 12:43–48. (12)

Edwards, M. D., C. W. Stuber, and J. F. Wendel. 1987. Molecular-marker facilitated investigations of quantitative-trait loci in maize. I. Numbers, genomic distribution and types of gene action. *Genetics* 116:113–25. (10)

Ehrman, L. 1960. Genetics of hybrid sterility in *Drosophila paulistorum. Evolution* 14:212–23. (1)

Ehrman, L. 1961. The genetics of sexual isolation in *Drosophila paulistorum. Genetics* 46:1025–38. (1)

Ehrman, L., and P. A. Parsons. 1976. *The Genetics of Behavior.* Sunderland, Mass.: Sinauer Associates. (1)

Ehrman, L., and P. A. Parsons. 1981. *Behavior Genetics and Evolution.* New York: McGraw-Hill. (1)

Eisen, E. J. 1967. Mating designs for estimating direct and maternal genetic variances and direct-maternal genetic covariances. *Can. J. Genet. Cytol.* 9:13–22. (2, 4)

Elder, B. J. 1989. Quantitative genetic aspects of thermoregulatory adaptation in five geographic populations of *Mus domesticus.* Ph.D. diss., Wesleyan University. (13)

El Oksh, H. A., T. M. Sutherland, and J. S. Williams. 1967. Prenatal and postnatal influence on growth in mice. *Genetics* 57:79–94. (4)

Emlen, S. T. 1984. Cooperative breeding in birds and mammals. In *Behavioural Ecology: An Evolutionary Approach,* 2d ed., ed. J. R. Krebs and N. B. Davies, 305–39. Oxford: Blackwell Scientific Publications. (4)

Endler, J. A. 1983. Natural and sexual selection on color patterns in poeciliid fishes. *Environ. Biol. Fish.* 9:173–90. (6, 11)

Endler, J. A. 1986. *Natural Selection in the Wild.* Princeton, N.J.: Princeton University Press. (4)

Endler, J. A. 1987. Predation, light intensity, and courtship behaviour in *Poecilia reticulata* (Pisces: Poeciliidae). *Anim. Behav.* 35:1376–85. (8)

Endler, J. A. 1988. Sexual selection and predation risk in guppies. *Nature* 332:593–94. (6)

Engelhard, G., S. P. Foster, and T. H. Day. 1989. Genetic differences in mating success and female choice in seaweed flies (*Coelopa frigida*). *Heredity* 62:123–31. (5)

Engels, W. R. 1983. Evolution of altruistic behavior by kin selection: An alternative approach. *Proc. Natl. Acad. Sci. USA* 80:515–18. (4)

Engen, S., and B. Sæther. 1985. The evolutionary significance of sexual selection. *J. Theor. Biol.* 117:277–89. (5)

Eshel, I. 1982. Evolutionarily stable strategies and viability selection in Mendelian populations. *Theor. Popul. Biol.* 22:204–17. (3)

Eshel, I., and M. W. Feldman. 1991. The handicap principle in parent-offspring conflict: Comparison of optimality and population-genetic analyses. *Am. Nat.* 137:167–85. (4)

Ewing, A. W. 1961. Body size and courtship behaviour of *Drosophila melanogaster. Anim. Behav.* 9:93–99. (1, 6)

Ewing, A. W. 1964. The influence of wing area on the courtship behaviour of *Drosophila melanogaster. Anim. Behav.* 12:316–20. (6)

Ewing, A. W. 1989. *Arthropod Bioacoustics: Neurobiology and Behaviour.* Ithaca, N.Y.: Cornell University Press. (11)

Ewing, L. S. 1973. Territoriality and the influence of females on the spacing of males in the cockroach, *Nauphoeta cinerea. Behaviour* 45:281–304. (8)

Fain, P. R. 1978. Characteristics of simple sibship variance tests for the detection of major loci and application to height, weight and spatial performance. *Ann. Hum. Genet.* 42:109–20. (10)

Fairbairn, D. J. 1977. Why breed early? A study of reproductive tactics in *Peromyscus. Can. J. Zool.* 55:862–71. (3)

Fairbairn, D. J. 1984. Microgeographic variation in body size and development time in the waterstrider, *Limnoporus notabilis. Oecologia* 61:126–33. (3)

Fairbairn, D. J., and D. A. Roff. 1990. Genetic correlations among traits determining migratory tendency in the sand cricket, *Gryllus firmus. Evolution* 44:1787–95. (7)

Falconer, D. S. 1960. Selection of mice for growth on high and low planes of nutrition. *Genet. Res.* 1:91–113. (2)

Falconer, D. S. 1964. *Introduction to Quantitative Genetics.* New York: Ronald Press. (8)

Falconer, D. S. 1965. Maternal effects and selection response. In *Genetics Today: Proceedings of 11th International Congress of Genetics,* ed. S. J. Geerts, 763–74. Oxford: Pergamon Press. (4)

Falconer, D. S. 1973. Replicated selection for body weight in mice. *Genet. Res.* 22:291–321. (13)

Falconer, D. S. 1977. Why are mice the size they are? In *Proceedings of the International Conference on Quantitative Genetics,* ed. E. Pollak, O. Kempthorne, and T. B. Bailey, Jr., 19–21. Ames, Iowa: Iowa State University Press. (13)

Falconer, D. S. 1989. *Introduction to Quantitative Genetics.* 3d ed. New York: Wiley. (1, 2, 3, 4, 5, 7, 9, 10, 11, 12, 13, 14)

Falconer, D. S. 1990. Selection in different environments: Effects on environmental sensitivity (reaction norm) and on mean performance. *Genet. Res.* 56:57–70. (13)

Farr, J. A. 1976. Social facilitation of male sexual behavior, intrasexual competition, and sexual selection in the guppy, *Poecilia reticulata* (Pisces: Poeciliidae). *Evolution* 30:707–17. (8)

Farr, J. A. 1980. The effects of sexual experience and female receptivity on courtship-rape decisions in male guppies, *Poecilia reticulata* (Pisces: Poeciliidae). *Anim. Behav.* 28:1195–1201. (8)

Farr, J. A. 1983. The inheritance of quantitative fitness traits in guppies, *Poecilia reticulata* (Pisces: Poeciliidae). *Evolution* 37:1193–1209. (9)

Farr, J. A. 1989. Sexual selection and secondary sexual differentiation in poeciliids: Determinants of male mating success and the evolution of female choice. In *Ecology and Evolution of Livebearing Fishes (Poeciliidae),* ed. G. K. Meffe and F. F. Snelson, Jr., 91–123. Englewood Cliffs, N.J.: Prentice Hall. (8)

Farr, J. A., and W. F. Herrnkind. 1974. A quantitative analysis of social interaction of the guppy, *Poecilia reticulata* (Pisces: Poeciliidae), as a function of population density. *Anim. Behav.* 22:582–91. (8)

Farr, J. A., and J. T. Travis. 1986. Fertility advertisement by female sailfin mollies, *Poecilia latipinna* (Pisces: Poeciliidae). *Copeia* 1986:467–72. (8)

Farr, J. A., and J. Travis. 1989. The effect of ontogenetic experience on variation in growth, maturation, and sexual behavior in the sailfin molly *Poecilia latipinna* (Pisces: Poeciliidae). *Environ. Biol. Fish.* 26:39–48. (8)

Farr, J. A., J. Travis, and J. C. Trexler. 1986. Behavioural allometry and interdemic variation in sexual behaviour of the sailfin molly *Poecilia latipinna* (Pisces: Poeciliidae). *Anim. Behav.* 34:497–509. (8)

Feller, W. 1940. On the logistic law of growth and its empirical verification. *Acta Biotheor.* (Leiden) 5:51–65. (3)

Felsenstein, J. 1976. The theoretical population genetics of variable selection and migration. *Annu. Rev. Genet.* 10:253–80. (11)

Findlay, C. S., and F. Cooke. 1983. Genetic and environmental components of clutch size variance in a wild population of lesser snow geese (*Anser caerulescens caerulescens*). *Evolution* 37:724–34. (14)

Finkbeiner, D. T. II. 1966. *Introduction to Matrices and Linear Transformations.* San Francisco: W. H. Freeman & Co. (2)

Finley, L. M., and L. E. Haley, 1983. The genetics of aggression in the juvenile American lobster, *Homarus americanus. Aquaculture* 33:135–39. (9)

Fisher, R. A. 1915. The evolution of sexual preference. *Eugen. Rev.* 7:184–92. (4, 5)

Fisher, R. A. 1918. The correlations between relatives on the supposition of Mendelian inheritance. *Trans. R. Soc. Edinb.* 52:399–433. (2, 4)

Fisher, R. A. 1928. The possible modification of the response of the wild type to recurrent mutations. *Am. Nat.* 62:115–26. (13)

Fisher, R. A. 1930. *The Genetical Theory of Natural Selection.* Oxford: Clarendon Press. (5, 10, 13)

Fisher, R. A. 1958. *The Genetical Theory of Natural Selection,* 2d rev. ed. New York: Dover. (1, 4, 11)

Fitch, H. S. 1965. An ecological study of the garter snake, *Thamnophis sirtalis. Univ. Kans. Mus. Nat. Hist. Publ.* 15:493–564. (12)

Fitch, H. S., and T. P. Maslin. 1961. Occurrence of the garter snake, *Thamnophis sirtalis,* in the Great Plains and Rocky Mountains. *Univ. Kans. Mus. Nat. Hist. Publ.* 13:289–308. (12)

Flury, B. 1988. *Common Principal Components and Related Multivariate Models.* New York: Wiley. (2)

Ford, N. B., and G. A. Shuttlesworth. 1986. Effects of variation in food intake on locomotory performance of juvenile garter snakes. *Copeia* 1986:999–1001. (12)

Forester, D. C. 1978. Laboratory encounters between attending *Desmognathus ochrophaeus* (Amphibia, Urodela, Plethodontidae) females and potential predators. *J. Herpetol.* 12:537–41. (2)

Forrest, T. G. 1983. Calling songs and mate choice in mole crickets. In *Orthopteran Mating Systems,* ed. D. T. Gwynne and G. K. Morris, 185–204. Boulder: Westview Press. (11)

Fowler, K., and L. Partridge. 1989. A cost of mating in female fruitflies. *Nature* 388:760–61. (6)

Francis, R. C. 1984. The effects of bidirectional selection for social dominance on agonistic behaviour and sex ratios in the paradise fish (*Macropodus opercularis*). *Behaviour* 90:25–45. (9)

Fritz, R. S., and D. H. Morse. 1985. Reproductive success and foraging of the crab spider *Misumena vatia. Oecologia* 65:194–200. (3)

Fulker, D. W. 1966. Mating speed in male *Drosophila melanogaster:* A psychogenetic analysis. *Science* 153:203–5. (9)

Fuller, J. L., and W. R. Thompson. 1960. *Behavior Genetics.* New York: Wiley. (1)

Gadgil, M. 1972. Male dimorphism as a consequence of sexual selection. *Am. Nat.* 106:574–80. (9)

Galton, F. 1889. *Natural Inheritance.* London: Macmillan. (2)

Gans, C. 1989. On phylogenetic constraints. *Acta Morphologica Neerlando-Scandinavica* 27:133–38. (3)

Garland, T., Jr. 1984. Physiological correlates of locomotory performance in a lizard: An allometric approach. *Am. J. Physiol.* (*Regul. Integrat. & Comp. Physiol.* 16) 247:R806–15. (12)

Garland, T., Jr. 1985. Ontogenetic and individual variation in size, shape, and speed in the Australian agamid lizard *Amphibolurus nuchalis. J. Zool.* (Lond.) A 207:425–39. (12)

Garland, T., Jr. 1988. Genetic basis of activity metabolism. I. Inheritance of speed, stamina, and antipredator displays in the garter snake *Thamnophis sirtalis. Evolution* 42:335–50. (12)

Garland, T., Jr. 1993. Locomotor performance and activity metabolism of *Cnemidophorus tigris* in relation to natural behaviors. In *Biology of Whiptail Lizards (Genus* Cnemidophorus*)*, ed. J. W. Wright and L. J. Vitt, 163–210. Norman, OK: Oklahoma Museum of Natural History. (12)

Garland, T., Jr., and S. C. Adolph. 1991. Physiological differentiation of vertebrate populations. *Annu. Rev. Ecol. Syst.* 22:193–228. (12)

Garland, T., Jr., and S. J. Arnold. 1983. Effects of a full stomach on locomotory performance of juvenile garter snakes (*Thamnophis elegans*). *Copeia* 1983:1092–96. (12)

Garland, T., Jr., and A. F. Bennett. 1990. Quantitative genetics of maximal oxygen consumption in a garter snake. *Am. J. Physiol.* (*Regul. Integrat. & Comp. Physiol.* 28) 259:R986–92. (12)

Garland, T., Jr., A. F. Bennett, and C. B. Daniels. 1990. Heritability of locomotor performance and its correlates in a natural population. *Experientia* 46:530–33. (12)

Garland, T., Jr., and P. A. Carter. 1994. Evolutionary physiology. *Annu. Rev. Physiol.* 56:579–621. (12)

Garland, T., Jr., A. W. Dickerman, C. M. Janis, and J. A. Jones. 1993. Phylogenetic analysis of covariance by computer simulation. *Syst. Biol.* 42:265–92. (12)

Garland, T., Jr., F. Gelser, and R. V. Baudinette. 1988. Comparative locomotor performance of marsupial and placental mammals. *J. Zool.* (Lond.) 215:505–22. (12)

Garland, T., Jr., E. Hankins, and R. B. Huey. 1990. Locomotor capacity and social dominance in male lizards. *Funct. Ecol.* 4:243–50. (12)

Garland, T., Jr., and R. B. Huey. 1987. Testing symmorphosis: Does structure match functional requirements? *Evolution* 41:1404–09. (12)

Garland, T., Jr., and J. B. Losos. 1994. Ecological morphology of locomotor performance in squamate reptiles. In *Ecological Morphology: Integrative Organismal Biology*, ed. P. C. Wainwright and S. M. Reilly, 240–302. Chicago: University of Chicago Press. (12)

Gatehouse, A. G. 1986. Migration in the African armyworm, *Spodoptera exempta:* Genetic determination of migratory capacity and a new synthesis. In *Insect*

Flight, Dispersal, and Migration, ed. W. Danthanarayana, 128–44. Berlin: Springer-Verlag. (7)

Gatehouse, A. G., and D. S. Hackett. 1980. A technique for studying flight behaviour of tethered *Spodoptera exempta* moths. *Physiol. Entomol.* 5:215–22. (7)

Gerhardt, H. C. 1991. Female mate choice in treefrogs: Static and dynamic acoustic criteria. *Anim. Behav.* 42:615–35. (5, 14)

Gerhardt, H. C., and G. M. Klump. 1988. Masking of acoustic signals by the chorus background noise in the green treefrog: A limitation on mate choice. *Anim. Behav.* 36:1247–49. (11)

Gibson, R. M., and J. W. Bradbury. 1985. Sexual selection in lekking sage grouse: Phenotypic correlates of male mating success. *Behav. Ecol. Sociobiol.* 18:117–23. (11)

Giesel, J. T. 1988. Effects of parental photoperiod on development time and density sensitivity of progeny of *Drosophila melanogaster. Evolution* 42:1348–50. (7)

Giesel, J. T., C. A. Lanciani, and J. F. Anderson. 1989. Effects of parental photoperiod on metabolic rate in *Drosophila melanogaster. Fla. Entomol.* 71:499–503. (7)

Gillespie, J. H. 1977. Natural selection for variance in offspring numbers: A new evolutionary principle. *Am. Nat.* 111:1010–14. (3)

Gillespie, J. H., and M. Turelli. 1989. Genotype-environment interactions and the maintenance of polygenic variation. *Genetics* 121:129–38. (3, 9)

Gomulkiewicz, R., and M. Kirkpatrick. 1992. Quantitative genetics and the evolution of reaction norms. *Evolution* 46:390–411. (13)

Gomulkiewicz, R. S., and A. Hastings. 1990. Ploidy and evolution by sexual selection: A comparison of haploid and diploid female choice models near fixation equilibria. *Evolution* 44:757–70. (5)

Goodman, D. 1982. Optimal life histories, optimal notation, and the value of reproductive value. *Am. Nat.* 119:803–23. (3)

Goodnight, C. J., J. M. Schwartz, and L. Stevens. 1992. Contextual analysis of models of group selection, soft selection, hard selection and the evolution of altruism. *Am. Nat.* 140:743–61. (10)

Gould, S. J. 1966. Allometry and size in ontogeny and phylogeny. *Biol. Rev.* 41:587–640. (8)

Gould, S. J., and R. C. Lewontin. 1979. The spandrels of San Marco and the Panglossian paradigm—A critique of the adaptionist program. *Proc. R. Soc. Lond.,* B 205:581–98. (1, 3)

Grafen, A. 1982. How not to measure inclusive fitness. *Nature* 298:425–26. (3, 4)

Grafen, A. 1984. Natural selection, kin selection and group selection. In *Behavioural Ecology,* 2d ed., ed. J. R. Krebs and N. B. Davies, 62–84. Sunderland, Mass.: Sinauer Associates. (3)

Grafen, A. 1988. On the uses of data on lifetime reproductive success. In *Reproductive Success,* ed. T. H. Clutton-Brock, 454–71. Chicago: University of Chicago Press. (4, 6)

Grant, B. R., and P. R. Grant. 1989. *Evolutionary Dynamics of a Natural Population: The Large Cactus Finch of the Galápagos.* Chicago: University of Chicago Press. (2)

Grant, P. R. 1986. *Ecology and Evolution of Darwin's Finches.* Princeton, N.J.: Princeton University Press. (14)

Greene, H. W. 1988. Antipredator mechanisms in reptiles. In *Biology of the Reptilia,* vol. 16, Ecology B. *Defense and Life Histories,* ed. C. Gans and R. B. Huey, 1–152. New York: Alan R. Liss. (12)

Griffing, B. 1956a. Concept of general and specific combining ability in relation to diallel crossing systems. *Aust. J. Biol. Sci.* 9:463–93. (13)

Griffing, B. 1956b. A generalized treatment of the use of diallel crosses in quantitative inheritance. *Heredity* 10:31–50. (13)

Griffing, B. 1981. A theory of natural selection incorporating interaction among individuals. I. The modeling process. *J. Theor. Biol.* 89:635–58. (10)

Groeters, F. R. 1989. Geographic and clonal variation in the milkweed-oleander aphid, *Aphis nerii* (Homoptera: Aphididae), for winged morph production, life history, and morphology in relation to host plant permanence. *Evol. Ecol.* 3:327–41. (7)

Groeters, F. R., and H. Dingle. 1987. Genetic and maternal influences on life history plasticity in response to photoperiod by milkweed bugs (*Oncopeltus fasciatus*). *Am. Nat.* 129:332–46. (7)

Groeters, F. R., and H. Dingle. 1988. Genetic and maternal influences on life history plasticity in milkweed bug (*Oncopeltus fasciatus*): Response to temperature. *J. Evol. Biol.* 1:317–33. (7)

Gromko, M. H., and M.E.A. Newport. 1988. Genetic basis for remating in *Drosophila melanogaster.* II. Response to selection based on the behavior of one sex. *Behav. Genet.* 18:621–32. (6, 9)

Gross, M. R. 1982. Sneakers, satellites and parentals: Polymorphic mating strategies in North American sunfishes. *Z. Tierpsychol.* 60:1–26. (3, 8)

Gross, M. R. 1985. Disruptive selection for alternative life histories in salmon. *Nature* 313:47–48. (8)

Gross, M. R., and E. L. Charnov. 1980. Alternative male life histories in bluegill sunfish. *Proc. Natl. Acad. Sci. USA* 77:6937–40. (3)

Gross, M. R., and R. C. Sargent. 1985. The evolution of male and female parental care in fishes. *Am. Zool.* 25:807–22. (4)

Gross, M. R., and R. Shine. 1981. Parental care and mode of fertilization in ectothermic vertebrates. *Evolution* 35:775–93. (4)

Gupta, A. P., and R. C. Lewontin. 1982. A study of reaction norms in natural populations of *Drosophila pseudoobscura. Evolution* 36:934–48. (2, 13)

Gwynne, D. T. 1983. Male nutritional investment and the evolution of sexual differences in Tettigoniidae and other Orthoptera. In *Orthopteran Mating Systems,* ed. D. T. Gwynne and G. K. Morris, 337–66. Boulder: Westview Press. (11)

Hahn, M., J. Hewitt, N. Henderson, and R. Benno, eds. 1990. *Developmental Behavior Genetics: Neural, Biometrical, and Evolutionary Approaches.* New York: Oxford University Press. (2)

Hairston, N. G., Jr., W. E. Walton, and K. T. Li. 1983. The causes and consequences of sex-specific mortality in a freshwater copepod. *Limnol. Oceanogr.* 28:935–47. (3)

Haldane, J.B.S. 1932. *The Causes of Evolution.* New York: Longmans Green. (13)

Halliday, T. 1983. The study of mate choice. In *Mate Choice,* ed. P. Bateson, 3–32. Cambridge: Cambridge University Press. (4, 6, 11)

Halliday, T. 1987. Physiological constraints on sexual selection. In *Sexual Selection: Testing the Alternatives,* ed. J. W. Bradbury and M. B. Andersson, 247–64. Chichester: Wiley. (12)

Halliday, T., and S. J. Arnold. 1987. Multiple mating by females: A perspective from quantitative genetics. *Anim. Behav.* 35:939–41. (6)

Hamilton, W. D. 1964a. The genetical evolution of social behaviour. I. *J. Theor. Biol.* 7:1–16. (3, 4, 10)

Hamilton, W. D. 1964b. The genetical evolution of social behaviour. II. *J. Theor. Biol.* 7:17–52 (10)
Hamilton, W. D. 1967. Extraordinary sex ratios. *Science* 156:477–88. (4)
Hamilton, W. D. 1970. Selection of selfish and altruistic behavior in some extreme models. In *Man and Beast: Comparative Social Behavior,* ed. J. F. Eisenberg and W. S. Dillon, 59–91. Washington, D.C.: Smithsonian Institution. (10)
Hamilton, W. D., and M. Zuk. 1982. Heritable true fitness and bright birds: A role for parasites? *Science* 218:384–87. (5, 11)
Hamilton, W. J. 1962. Reproductive adaptations in the red tree mouse. *J. Mammal.* 43:486–504. (3)
Hanrahan, J. P. 1976. Maternal effects and selection response with an application to sheep data. *Anim. Prod.* 22:359–69. (4)
Hanrahan, J. P., and E. J. Eisen. 1973. Sexual dimorphism and direct and maternal genetic effects on body weight in mice. *Theor. Appl. Genet.* 43:39–45. (2, 4)
Hanrahan, J. P., and E. J. Eisen. 1974. Genetic variation in litter size and 12-day weight in mice and their relationships with post-weaning growth. *Anim. Prod.* 19:13–23. (4)
Harper, A. B. 1989. Evolutionary stability for interactions among kin under quantitative inheritance. *Genetics* 121:877–89. (10)
Harris, R. J. 1975. *A Primer of Multivariate Statistics.* New York: Academic Press. (2)
Hart, J. S. 1971. Rodents. In *Comparative Physiology of Thermoregulation.* vol. 2. *Mammals,* ed. G. C. Whittow, 1–151. New York: Academic Press. (13)
Hart, R. C., and I. A. McLaren. 1978. Temperature acclimation and other influences on embryonic duration in the copepod, *Pseudocalanus* sp. *Mar. Biol.* 45:23–30. (3)
Harvey, P. H., and M. D. Pagel. 1991. *The Comparative Method in Evolutionary Biology.* Oxford: Oxford University Press. (12)
Hasselquist, D., and S. Bensch. 1991. Trade-off between mate guarding and mate attraction in the polygynous great reed warbler. *Behav. Ecol. Sociobiol.* 28:187–93. (6)
Hayman, B. I. 1954a. The analysis of variance of diallel tables. *Biometrics* 10:235–44. (13)
Hayman, B. I. 1954b. The theory and analysis of diallel crosses. *Genetics* 39:789–809. (13)
Hedrick, A. V. 1986. Female preferences for male calling bout duration in a field cricket. *Behav. Ecol. Sociobiol.* 19:73–77. (11)
Hedrick, A. V. 1988. Female choice and the heritability of attractive male traits: An empirical study. *Am. Nat.* 132:267–76. (11)
Hedrick, A. V., and L. M. Dill. 1993. Mate choice by female crickets is influenced by predation risk. *Anim. Behav.* 46:193–96. (11)
Hedrick, P. W. 1986. Genetic polymorphism in heterogeneous environments: A decade later. *Annu. Rev. Ecol. Syst.* 17:535–66. (9)
Hegmann, J. P., and H. Dingle. 1982. Phenotypic and genetic covariance structure in milkweed bug life history traits. In *Evolution and Genetics of Life Histories,* ed. H. Dingle and J. P. Hegmann, 177–86. New York: Springer-Verlag. (7)
Heisler, I. L. 1984a. Inheritance of female mating propensities for *yellow* locus genotypes in *Drosophila melanogaster. Genet. Res.* 44:133–49. (5)
Heisler, I. L. 1984b. A quantitative genetic model for the origin of mating preferences. *Evolution* 38:1283–95. (4, 5, 11)

Heisler, I. L. 1985. Quantitative genetic models of female choice based on "arbitrary" male characters. *Heredity* 55:187–98. (4, 5)

Heisler, I. L., M. Andersson, S. J. Arnold, C. R. Boake, G. Borgia, G. Hausfater, M. Kirkpatrick, R. Lande, J. Maynard Smith, P. O'Donald. A. R. Thornhill, and F. J. Weissing. 1987. Evolution of mating preferences and sexually selected traits. In *Sexual Selection: Testing the Alternatives*, ed. J. W. Bradbury and M. B. Andersson, 97–118. Chichester: Wiley. (5)

Heisler, I. L., and J. W. Curtsinger. 1990. Dynamics of sexual selection in diploid populations. *Evolution* 44:1164–76. (5)

Heisler, I. L., and J. D. Damuth. 1987. A method for analyzing selection in hierarchically-structured populations. *Am. Nat.* 130:582–602. (5, 10)

Henderson, N. D. 1989. Interpreting studies that compare high- and low-selected lines on new characters. *Behav. Genet.* 19:473–502. (9, 12)

Hendrix, S. D. 1984. Variation in seed weight and its effects on germination in *Pastinaca sativa* L. (Umbelliferae). *Am. J. Bot.* 71:795–802. (3)

Herbert, J., K. Kidwell, and H. Chase. 1979. The inheritance of growth and form in the mouse. IV. Changes in the variance components of weight, tail length and tail width during growth. *Growth* 43:36–46. (4)

Hertz, P. E., R. B. Huey, and T. Garland, Jr. 1988. Time budgets, thermoregulation, and maximal locomotor performance: Are ectotherms Olympians or Boy Scouts? *Am. Zool.* 28:927–38. (12)

Herzog, H. H., Jr. 1990. Experimental modification of defensive behaviors in garter snakes (*Thamnophis sirtalis*). *J. Comp. Psychol.* 104:334–39. (12)

Herzog, H. H., Jr., and B. D. Bailey. 1987. Development of antipredator responses in snakes: II. Effects of recent feeding on defensive behaviors of juvenile garter snakes (*Thamnophis sirtalis*). *J. Comp. Psychol.* 101:387–89. (12)

Herzog, H. H., Jr., B. B. Bowers, and G. M. Burghardt. 1989a. Development of antipredator responses in snakes. IV. Interspecific and intraspecific differences in habituation of defensive behavior. *Dev. Psychobiol.* 22:489–508. (12)

Herzog, H. H., Jr., B. B. Bowers, and G. M. Burghardt. 1989b. Stimulus control of antipredator behavior in newborn and juvenile garter snakes (*Thamnophis*). *J. Comp. Psychol.* 103:233–42. (12)

Herzog, H. H., Jr., and G. M. Burghardt. 1986. Development of antipredator responses in snakes: I. Defensive and open-field behaviors in newborns and adults of three species of garter snakes (*Thamnophis melanogaster, T. sirtalis, T. butleri*). *J. Comp. Psychol.* 100:372–79. (12)

Herzog, H. H., Jr., and G. M. Burghardt. 1988. Development of antipredator responses in snakes. III. Stability of individual and litter differences over the first year of life. *Ethology* 77:250–58. (12)

Herzog, H. H., Jr., and J. M. Schwartz. 1990. Geographic variation in the antipredator behavior of neonate and adult garter snakes (*Thamnophis s. sirtalis*). *Anim. Behav.* 40:597–98. (12)

Hewitt, J. K., and D. W. Fulker. 1981. Using the triple test cross to investigate the genetics of behavior in wild populations. I. Methodological considerations. *Behav. Genet.* 11:23–35. (13)

Hewitt, J. K., and D. W. Fulker. 1983. Using the triple test cross to investigate the genetics of behavior in wild populations. II. Escape-avoidance conditioning in *Rattus norvegicus*. *Behav. Genet.* 13:1–15. (13)

Hewitt, J. K., and D. W. Fulker. 1984. Using the triple test cross to investigate the

genetics of behavior in wild populations. III. Activity and reactivity. *Behav. Genet.* 14:125–35. (13)
Hews, D. K. 1990. Examining hypotheses generated by field measures of sexual selection on male lizards, *Uta palmeri. Evolution* 44:1956–66. (6)
Heywood, J. S. 1989. Sexual selection by the handicap mechanism. *Evolution* 43:1387–97. (5)
Hill, W. G. 1972. Estimation of realized heritabilities from selection experiments. II. Selection in one direction. *Biometrics* 28:767–80. (13)
Hill, W. G. 1980. Design of quantitative genetic selection experiments. In *Selection Experiments in Laboratory and Domestic Animals,* ed. A. Robertson, 1–3. Slough, U.K.: Commonwealth Agricultural Bureaux. (2)
Hiraizumi, Y. 1961. Negative genetic correlation between rate of development and female fertility in *Drosophila melanogaster. Genetics* 46:615–24. (12)
Hirsch, J., ed. 1967. *Behavior-Genetic Analysis.* New York: McGraw-Hill. (1)
Hirsch, J., and L. Erlenmeyer-Kimling. 1962. Studies in experimental behavior genetics. IV. Chromosome analyses for geotaxis. *J. Comp. Physiol. Psychol.* 55:732–39. (1, 9)
Hoelzer, G. A. 1989. The good parent process of sexual selection. *Anim. Behav.* 38:1067–78. (4)
Hoffmann, A. A. 1987a. A laboratory study of male territoriality in the sibling species *Drosophila melanogaster* and *D. simulans. Anim. Behav.* 35:807–18. (9)
Hoffmann, A. A. 1987b. Territorial encounters between *Drosophila* males of different sizes. *Anim. Behav.* 35:1899–1901. (9)
Hoffmann, A. A. 1988. Heritable variation for territorial success in two *Drosophila melanogaster* populations. *Anim. Behav.* 36:1180–89. (9)
Hoffmann, A. A. 1989. Geographic variation in the territorial success of *Drosophila melanogaster* males. *Behav. Genet.* 19:241–55. (9)
Hoffmann, A. A. 1990. The influence of age and experience with conspecifics on territorial behavior in *Drosophila melanogaster. J. Insect Behav.* 3:1–12. (9)
Hoffmann, A. A. 1991. Heritable variation for territorial success in field-collected *Drosophila melanogaster. Am. Nat.* 138:668–79. (9)
Hoffmann, A. A., and Z. Cacoyianni. 1989. Selection for territoriality in *Drosophila melanogaster:* Correlated responses in mating success and other fitness components. *Anim. Behav.* 38:23–34. (9)
Hoffmann, A. A., and Z. Cacoyianni. 1990. Territoriality in *Drosophila melanogaster* as a conditional strategy. *Anim. Behav.* 40:526–37. (8, 9)
Hohenboken, W. D., and J. S. Brinks. 1971. Relationships between direct and maternal effects on growth in Herefords. II. Partitioning of covariance between relatives. *J. Anim. Sci.* 32:26–34. (4)
Hopper, J. L., and J. D. Matthews. 1982. Extensions to multivariate normal models for pedigree analysis. *Ann. Hum. Genet.* 46:373–83. (2, 14)
Houck, L. D., S. J. Arnold, and R. A. Thisted. 1985. A statistical study of mate choice: Sexual selection in a plethodontid salamander (*Desmognathus ochrophaeus*). *Evolution* 39:370–86. (2)
Houde, A. E. 1987. Mate choice based upon naturally occurring color-pattern variation in a guppy population. *Evolution* 41:1–10. (11)
Houle, D. 1992. Comparing evolvability and variability of quantitative traits. *Genetics* 130:195–204. (14)
Hoy, R. R., G. S. Pollack, and A. Moiseff. 1982. Species-recognition in the field

cricket, *Teleogryllus oceanicus:* Behavioral and neural mechanisms. *Am. Zool.* 22:597–607. (11)

Hudson, R. R. 1990. Gene genealogies and the coalescent process. In *Oxford Surveys in Evolutionary Biology,* ed. D. J. Futuyma and J. Antonovics, 1–44. Oxford: Oxford University Press. (8)

Huey, R. B., and A. F. Bennett. 1987. Phylogenetic studies of coadaptation: Preferred temperatures versus optimal performance temperatures of lizards. *Evolution* 41:1098–1115. (12)

Huey, R. B., A. F. Bennett, H. B. John-Alder, and K. A. Nagy. 1984. Locomotor capacity and foraging behavior of Kalahari lacertid lizards. *Anim. Behav.* 32:41–50. (12)

Huey, R. B., and R. D. Stevenson. 1979. Integrating thermal physiology and ecology of ectotherms: A discussion of approaches. *Am. Zool.* 19:357–66. (12)

Humphries, J. M., F. L. Bookstein, B. Chernoff, G. R. Smith, R. L. Elder, and S. G. Poss. 1981. Multivariate discrimination by shape in relation to size. *Syst. Zool.* 30:291–308. (8)

Huntingford, F. A. 1993. Behavioural mechanisms in evolutionary perspective. *Trends Ecol. Evol.* 8:81–84. (1)

Istock, C. 1983. The extent and consequences of heritable variation for fitness characters. In *Population Biology: Retrospect and Prospect,* ed. C. King and P. Dawson, 61–96. New York: Columbia University Press. (11)

Iwasa, Y., and E. Teramoto. 1980. A criterion of life history evolution based on density dependent selection. *J. Theor. Biol.* 84:545–66. (3)

Jacobs, M. E. 1960. Influence of light on mating of *Drosophila melanogaster. Ecology* 41:182–88. (9)

Jaenike, J. 1989. Genetic population structure of *Drosophila tripunctata:* Patterns of variation and covariation of traits affecting resource use. *Evolution* 43:1467–82. (9)

Janssen, G. M., G. de Jong, E.N.G. Joose, and W. Scharloo. 1988. A negative maternal effect in springtails. *Evolution* 34:292–305. (4)

Jayne, B. C., and A. F. Bennett. 1989. The effect of tail morphology on locomotor performance of snakes: A comparison of experimental and correlative methods. *J. Exp. Zool.* 252:126–33. (12)

Jayne, B. C., and A. F. Bennett. 1990a. Scaling of speed and endurance in garter snakes: A comparison of cross-sectional and longitudinal allometries. *J. Zool* (Lond.) 220:257–77. (12)

Jayne, B. C., and A. F. Bennett. 1990b. Selection on locomotor performance capacity in a natural population of garter snakes. *Evolution* 44:1204–29. (12)

Jinks, J. L., and P. L. Broadhurst. 1974. How to analyse the inheritance of behaviour in animals—the biometrical approach. In *The Genetics of Behaviour,* ed. J.H.F. van Abeelen, 1–41. New York: American Elsevier. (1, 14)

Jonsson, B., and K. Hindar. 1982. Reproductive strategy of dwarf and normal Arctic charr (*Salvelinus alpinus*) from Vangsvatnet Lake, western Norway. *Can. J. Fish. Aquat. Sci.* 39:1404–13. (3)

Kacser, H., and J. A. Burns. 1981. The molecular basis of dominance. *Genetics* 97:639–66. (9, 13)

Kalisz, S. 1986. Variable selection on the timing of germination in *Collinsia verna* (Scrophulariaceae). *Evolution* 40:479–91. (14)

Kallman, K. D. 1984. A new look at sex determination in poeciliid fishes. In *Evolutionary Genetics of Fishes,* ed. B. J. Turner, 95–171. New York: Plenum Press. (8)

Kallman, K. D. 1989. Genetic control of size at maturity in *Xiphophorus*. In *Ecology and Evolution of Livebearing Fishes (Poeciliidae)*, ed. G. K. Meffe and F. F. Snelson, Jr., 163–84. Englewood Cliffs, N.J.: Prentice Hall. (8)
Kearsey, M. J., and J. L. Jinks. 1968. A general method of detecting additive, dominance and epistatic variation for metrical traits: I. Theory. *Heredity* 23:403–9. (13)
Kearsey, M. J., and K. I. Kojima. 1967. The genetic architecture of body weight and egg hatchability in *Drosophila melanogaster. Genetics* 56:23–37. (9)
Kempthorne, O. 1957. *An Introduction to Genetic Statistics*. New York: Wiley. (10)
Kempthorne, O. 1977. Status of quantitative genetic theory. In *Proceedings of the International Conference on Quantitative Genetics*, ed. E. Pollak, O. Kempthorne, and T. B. Bailey, Jr., 719–67. Ames, Iowa: Iowa State University Press. (3)
Kempthorne, O. 1983. Evolution of current population genetics theory. *Am. Zool.* 23:111–21. (3)
Kempthorne, O., and O. B. Tandon. 1953. The estimation of heritability by regression of offspring on parent. *Biometrics* 9:90–100. (11)
Kephart, D. G. 1982. Microgeographic variation in the diets of garter snakes. *Oecologia* 52:287–91. (12)
Kephart, D. G., and S. J. Arnold. 1982. Garter snake diets in a fluctuating environment: A seven-year study. *Ecology* 63:1232–36. (12)
King, R. B. 1987. Color pattern polymorphism in the Lake Erie water snake, *Nerodia sipedon insularum. Evolution* 41:241–55. (12)
Kirkpatrick, M. 1982. Sexual selection and the evolution of female choice. *Evolution* 36:1–12. (5, 11)
Kirkpatrick, M. 1985. Evolution of female choice and male parental investment in polygynous species: The demise of the "sexy son." *Am. Nat.* 125:788–810. (5, 14)
Kirkpatrick, M. 1986a. The handicap mechanism of sexual selection does not work. *Am. Nat.* 127:222–40. (5)
Kirkpatrick, M. 1986b. Sexual selection and cycling parasites: A simulation study of Hamilton's hypothesis. *J. Theor. Biol.* 119:263–71. (5)
Kirkpatrick, M. 1987a. The evolutionary forces acting on female mating preferences in polygynous animals. In *Sexual Selection: Testing the Alternatives*, ed. J. W. Bradbury and M. B. Andersson, 67–82. Chichester: Wiley (5, 11, 14)
Kirkpatrick, M. 1987b. Sexual selection by female choice in polygynous animals. *Annu. Rev. Ecol. Syst.* 18:43–70. (5, 11)
Kirkpatrick, M. 1988. Consistency in genetic models of the sexy son: Reply to Curtsinger and Heisler. *Am. Nat.* 132:609–10. (5)
Kirkpatrick, M., and R. Lande. 1989. The evolution of maternal characters. *Evolution* 43:485–503. (2, 4, 7)
Kirkpatrick M., and R. Lande. 1992. The evolution of maternal characters: Errata. *Evolution* 46:284. (2)
Kirkpatrick, M., T. Price, and S. J. Arnold. 1990. The Darwin-Fisher theory of sexual selection in monogamous birds. *Evolution* 44:180–93. (5)
Kirkpatrick, M., and M. J. Ryan. 1991. The evolution of mating preferences and the paradox of the lek. *Nature* 350:33–38. (6, 11)
Klein, T. W. 1974. Heritability and genetic correlation: Statistical power, population comparisons, and sample size. *Behav. Genet.* 4:171–89. (2, 4, 14)
Klein, T. W., J. C. DeFries, and C. T. Finkbeiner. 1973. Heritability and genetic correlation: Standard errors of estimates and sample size. *Behav. Genet.* 3:355–64. (2, 3, 4, 14)

Klomp, H., and B. J. Teerink. 1967. The significance of oviposition rates in the egg parasite, *Trichogramma embryophagum. Archives Neerlandaises de Zoologie* 17:350–75. (3)

Klump, G. M., and H. C. Gerhardt. 1987. Use of non-arbitrary acoustic criteria in mate choice by female grey frogs. *Nature* 326:286–88. (11)

Kodric-Brown, A., and J. H. Brown. 1984. Truth in advertising: The kinds of traits favored by sexual selection. *Am. Nat.* 124:309–23. (4, 8, 11)

Kohler, W. 1977. Investigations on the phototactic behavior of *Drosophila melanogaster.* I. Selection response in the presence of multiply marked X-chromosome. *Genetica* 47:93–100. (3)

Kohn, L.A.P., and W. R. Atchley. 1988. How similar are genetic correlation structures? Data from mice and rats. *Evolution* 42:467–81. (2)

Koufopanou, V., and G. Bell. 1984. Measuring the cost of reproduction. IV. Predation experiments with *Daphnia pulex. Oecologia* 64:81–86. (3)

Krebs, J. R., and N. B. Davies, eds. 1987. *An Introduction to Behavioural Ecology.* Oxford: Blackwell Scientific Publications. (6)

Krebs, J. R., and N. B. Davies, eds. 1991. *Behavioural Ecology: An Evolutionary Approach.* 3d ed. London: Blackwell Scientific Publications. (4, 10)

Krebs, J. R., and A. Kacelnik. 1991. Decision-making. In *Behavioural Ecology: An Evolutionary Approach,* 3d ed., ed. J. R. Krebs and N. B. Davies, 105–36. London: Blackwell Scientific Publications. (4)

Kroodsma, D. E. 1989. Suggested experimental designs for song playbacks. *Anim. Behav.* 37:600–609. (11)

Kuhlers, D. L., A. B. Chapman, and N. L. Furst. 1977. Estimates of maternal and grandmaternal influences on weights and gains of pigs. *J. Anim. Sci.* 44:181–88. (4)

Kurtén, B. 1959. Rates of evolution in fossil mammals. *Cold Spring Harb. Symp. Quant. Biol.* 24:205–15. (2)

Kusano, T. 1982. Postmetamorphic growth, survival, and age at first reproduction of the salamander, *Hynobius nebulosus tokyoensis* Tago in relation to a consideration on the optimal timing of first reproduction. *Res. Popul. Ecol.* 24:329–44. (3)

Lacey, E. P., L. Real, J. Antonovics, and D. G. Heckel. 1983. Variance models in the study of life histories. *Am. Nat.* 122:114–31. (3)

Lacy, R. C., and C. B. Lynch. 1979. Quantitative genetic analysis of temperature regulation in *Mus musculus.* I. Partitioning of variance. *Genetics* 91:743–53. (13)

Lacy, R. C., C. B. Lynch, and G. R. Lynch. 1978. Developmental and acclimation effects of ambient temperature regulation of mice selected for high and low levels of nest-building. *J. Comp. Physiol.* 123:185–92. (13)

Laffan, E. A. 1989. Artificial selection for thermoregulatory nest-building in the House Mouse: Analysis of the lines at their limits. Ph.D. diss., Wesleyan University. (13)

Lande, R. 1976a. The maintenance of genetic variability by mutation in a polygenic character with linked loci. *Genet. Res.* 26:221–35. (1, 2, 5, 11)

Lande, R. 1976b. Natural selection and random genetic drift in phenotypic evolution. *Evolution* 30:314–34. (2)

Lande, R. 1977. Statistical tests for natural selection on quantitative characters. *Evolution* 31:442–44. (10)

Lande, R. 1979. Quantitative genetic analysis of multivariate evolution, applied to brain: body size allometry. *Evolution* 33:402–16. (2, 3, 5)

Lande, R. 1980a. The genetic covariance between characters maintained by pleiotropic mutations. *Genetics* 94:203–15. (2, 3, 12)
Lande, R. 1980b. Sexual dimorphism, sexual selection and adaptation in polygenic characters. *Evolution* 34:292–305. (2, 5, 6)
Lande, R. 1981a. The minimum number of genes contributing to quantitative variation between and within populations. *Genetics* 99:541–53. (10)
Lande, R. 1981b. Models of speciation by sexual selection on polygenic traits. *Proc. Natl. Acad. Sci. USA* 78:3721–25. (1, 4, 5, 6, 11, 14)
Lande, R. 1982a. A quantitative genetic theory of life history evolution. *Ecology* 63:607–15. (1, 2, 3, 11)
Lande, R. 1982b. Rapid origin of sexual isolation and character divergence in a cline. *Evolution* 36:213–23. (5)
Lande, R. 1984. The genetic correlation between characters maintained by selection, linkage and inbreeding. *Genet. Res.* 44:309–20. (2)
Lande, R. 1988. Quantitative genetics and evolutionary theory. In *Proceedings of the Second International Conference on Quantitative Genetics,* ed B. Weir, E. Eisen, M. Goodman, and G. Namkoong, 71–84. Sunderland, Mass.: Sinauer Associates. (2, 12)
Lande, R., and S. J. Arnold. 1983. The measurement of selection on correlated characters. *Evolution* 37:1210–26. (2, 4, 5, 14)
Lande, R., and S. J. Arnold. 1985. Evolution of mating preference and sexual dimorphism. *J. Theor. Biol.* 117:651–64. (5)
Lande, R., and M. Kirkpatrick. 1988. Ecological speciation by sexual selection. *J. Theor. Biol.* 133:85–98. (5)
Lande, R., and M. Kirkpatrick. 1990. Selection response in traits with maternal inheritance. *Genet. Res.* 55:189–97. (4)
Lande, R., and T. Price. 1989. Genetic correlations and maternal effects coefficients obtained from offspring-parent regression. *Genetics* 122:915–22. (2, 4)
Lange, K., D. Weeks, and M. Boehnke. 1988. Programs for pedigree analysis: MENDEL, FISHER and dGENE. *Genet. Epidemiol.* 5:471–72. (2, 14)
Lange, K., J. Westlake, and M. A. Spence. 1976. Extensions to pedigree analysis. III. Variance components by the scoring method. *Ann. Hum. Genet.* 39:485–91. (2, 14)
Laurie, E.M.O. 1946. The reproduction of the house mouse (*Mus musculus*) living in different environments. *Proc. R. Soc. Lond* B 133:248–81. (13)
Lawson, R. 1985. Molecular studies of thamnophine snakes. Ph.D. diss., Louisiana State University. (12)
Lee, B.T.O., and P. A. Parsons. 1968. Selection, prediction and response. *Biol. Rev.* 43:139–74. (3)
Legates, J. E. 1972. The role of maternal effects in animal breeding. IV. Maternal effects in laboratory species. *J. Anim. Sci.* 35:1294–1302. (4)
Leggett, W. C. 1985. The role of migrations in the life history evolution of fish. In *Migration: Mechanisms and Adaptive Significance,* ed. M. A. Rankin. *Contrib. Mar. Sci.* 27 (Suppl.):277–95. (7)
Lerner, I. M. 1954. *Genetic Homeostasis.* New York: John Wiley. (13)
Leslie, J. F. 1990. Geographical and genetic structure of life-history variation in milkweed bugs (Hemiptera: Lygaeidae: *Oncopeltus*). *Evolution* 44:295–304. (7)
Lessells, C. M., and P. T. Boag. 1987. Unrepeatable repeatabilities: A common mistake. *Auk* 104: 116–21. (11, 14)

Levins, R. 1966. The strategy of model building in population biology. *Am. Sci.* 54:421–31. (5)

Lewis, W. M. 1987. The cost of sex. In *The Evolution of Sex and Its Consequences,* ed. S. C. Stearns, 33–57. Basel: Birkhauser. (6)

Lewontin, R. C. 1965. Selection for colonizing ability. In *The Genetics of Colonizing Species,* ed. H. G. Baker and G. L. Stebbins, 77–94. New York: Academic Press. (3)

Lewontin, R. C. 1974. *The Genetic Basis of Evolutionary Change.* New York: Columbia University Press. (14)

Li, C. C. 1975. *Path Analysis: A Primer.* Pacific Grove, Calif.: Boxwood Press. (4)

Liley, N. R. 1966. Ethological isolating mechanisms in four sympatric species of poeciliid fishes. *Behaviour Suppl.* 13:1–197. (8)

Lofsvold, D. 1986. Quantitative genetics of morphological differentiation in *Peromyscus.* I. Tests of homogeneity of genetic covariance structure among species and subspecies. *Evolution* 40:559–73. (2, 3)

Lorenz, K. Z. 1958. The evolution of behavior. *Sci. Am.* 199:67–78. (1)

Luckinbill, L. S., M. J. Clare, W. L. Krell, W. C. Cirocco, and P. A. Richards. 1987. Estimating the number of genetic elements that defer senescence in *Drosophila. Evol. Ecol.* 1:37–46. (3)

Luckner, C. L. 1979. Morphological and behavioral polymorphism in *Poecilia latipinna* males (Pisces: Poeciliidae). Ph.D. diss., Louisiana State University. (8)

Lush, J. L. 1945. *Animal Breeding Plans.* 3d ed. Ames, Iowa: Iowa State University Press. (2, 5)

Lynch, C. B. 1977. Inbreeding effects upon animals derived from a wild population of *Mus musculus. Evolution* 31:526–37. (13)

Lynch, C. B. 1980. Response to divergent selection for nesting behavior in *Mus musculus. Genetics* 96:757–65. (13)

Lynch, C. B. 1992. Clinal variation in cold adaptation in *Mus domesticus:* Verification of predictions from laboratory populations. *Am. Nat.* 139:1219–36. (13)

Lynch, C. B., and B. P. Possidente, Jr. 1978. Relationship of maternal nesting to thermoregulatory nesting in House Mice (*Mus musculus*) at warm and cold temperatures. *Anim. Behav.* 26:1136–43. (13)

Lynch, C. B., and R. C. Roberts. 1984. Aspects of temperature regulation in mice selected for large and small size. *Genet. Res.* 43:299–306. (13)

Lynch, C. B., R. C. Roberts, and W. G. Hill. 1986. Heterosis among lines of mice selected for body weight. 3. Thermoregulation. *Genet. Res.* 48:95–100. (13)

Lynch, C. B., and D. S. Sulzbach. 1984. Quantitative genetic analysis of temperature regulation in *Mus musculus.* II. Diallel analysis of individual traits. *Evolution* 38:527–40. (13)

Lynch, C. B., D. S. Sulzbach, and M. S. Connolly. 1988. Quantitative genetic analysis of temperature regulation in *Mus domesticus.* IV. Pleiotropy and genotype-by-environment interaction. *Am. Nat.* 132:521–37. (13)

Lynch, G. R., and C. B. Lynch. 1992. Using genetic approaches to understand the physiological basis of mammalian photoperiodism. In *Techniques for the Genetic Analysis of Brain and Behavior: Focus on the Mouse,* ed. D. Goldowitz, D. Wahlston, and R. E. Wimer, 251–68. Amsterdam; New York: Elsevier. (13)

Lynch, G. R., C. B. Lynch, M. Dube, and C. Allen. 1976. Early cold exposure: Effects on behavioral and physiological thermoregulation in the house mouse, *Mus musculus. Physiol. Zool.* 49:191–99. (13)

Lynch, G. R., C. B. Lynch, and R. M. Kliman. 1989. Genetic analysis of photoresponsiveness in the Djungarian hamster, *Phodopus sungorus. J. Comp. Physiol.* A 164:475–81. (13)

Lynch, M. 1987. Evolution of intrafamilial interactions. *Proc. Natl. Acad. Sci. USA.* 84:8507–11. (4)

Lynch, M. 1988. Path analysis of ontogenetic data. In *Size-structured Populations,* ed. B. Ebenman and L. Persson, 29–46. Berlin: Springer-Verlag. (4)

MacArthur, R. H., and E. R. Pianka. 1966. On optimal use of a patchy environment. *Am. Nat.* 100:603–9. (3)

McCauley, D. E. 1981. Application of the Kence-Bryant model of mating behavior to a natural population of soldier beetles. *Am. Nat.* 117:400–402. (6)

McClearn, G. E., J. R. Wilson, and W. Meredith. 1970. The use of isogenic and heterogenic mouse stocks in behavioral research. In *Contributions to Behavior-Genetic Analysis: The Mouse as a Prototype,* ed. G. Lindsey and D. D. Thiessen, 3–22. New York: Appleton-Century-Crofts. (13)

McComb, K. E. 1991. Female choice for high roaring rates in red deer *Cervus elaphus. Anim. Behav.* 41:79–88. (6)

McGinley, M. A. 1989. The influence of a positive correlation between clutch size and offspring fitness on the optimal offspring size. *Evol. Ecol.* 3:150–56. (3)

MacKay, P. A., and W. G. Wellington. 1977. Maternal age as a source of variation in the ability of an aphid to produce dispersing forms. *Res. Popul. Ecol.* 18:195–209. (7)

Magurran, A. E., and M. A. Nowak. 1991. Another battle of the sexes: The consequences of sexual asymmetry in mating costs and predation risk in the guppy, *Poecilia reticulata. Proc. R. Soc. Lond.* B 246:31–38. (8)

Magurran, A. E., and B. H. Seghers. 1990. Risk sensitive courtship in the guppy (*Poecilia reticulata*). *Behaviour* 112:194–201. (8)

Majerus, M.E.N., P. O'Donald, and J. Weir. 1982. Female mating preference is genetic. *Nature* 300:521–23. (5)

Mann, R.H.K., and C. A. Mills. 1979. Demographic aspects of fish fecundity. *Symp. Zool. Soc. Lond.* 44:161–77. (3)

Manning, A. 1961. The effects of artificial selection for mating speed in *Drosophila melanogaster. Anim. Behav.* 9:82–92. (1)

Manning, A. 1963. Selection for mating speed in *Drosophila melanogaster* based on the behaviour of one sex. *Anim. Behav.* 11:116–20. (1, 6)

Markow, T. A., and A. G. Clark. 1984. Correlated response to phototactic selection. *Behav. Genet.* 14:279–93. (3)

Marsteller, F. A., and C. B. Lynch. 1983. Reproductive consequences of food restriction at low temperature in lines of mice divergently selected for themoregulatory nesting. *Behav. Genet.* 13:397–410. (13)

Marsteller, F. A., and C. B. Lynch. 1987a. Energetic aspects of mammalian reproduction. II. Mating and pregnancy. *Biol. Reprod.* 37:838–43. (13)

Marsteller, F. A., and C. B. Lynch. 1987b. Energetic aspects of mammalian reproduction. III. Lactation in house mice. *Biol. Reprod.* 37:844–50. (13)

Martins, E. P., and T. Garland, Jr. 1991. Phylogenetic analyses of the correlated evolution of continuous traits: A simulation study. *Evolution* 45:534–57. (12)

Mather, K. and B. J. Harrison. 1949. The manifold effects of selection. Part 1. *Heredity* 3:1–52. (3)

Mather, K., and J. L. Jinks. 1971. *Biometrical Genetics.* London: Chapman and Hall. (13)

Mather, K., and J. L. Jinks. 1977. *Introduction to Biometrical Genetics.* Ithaca, N.Y.: Cornell University Press. (10)

Mather, K., and J. L. Jinks, 1982. *Biometrical Genetics: The Study of Continuous Variation.* 3d ed. London: Chapman and Hall. (1, 12)

Maynard Smith, J. 1978. *The Evolution of Sex.* Cambridge: Cambridge University Press. (11, 14)

Maynard Smith, J. 1981. Will a sexual population evolve to an ESS? *Am. Nat.* 117:1015–18. (3)

Maynard Smith, J. 1982. *Evolution and the Theory of Games.* Cambridge: Cambridge University Press. (3, 4)

Maynard Smith, J. 1985. Sexual selection, handicaps and true fitness. *J. Theor. Biol.* 115:1–8. (4, 11, 14)

Maynard Smith, J. 1987. Sexual selection: a classification of models. In *Sexual Selection: Testing the Alternatives,* ed. J. W. Bradbury and M. B. Andersson, 9–20. Chichester: Wiley. (6)

Maynard Smith, J., R. Burian, S. Kauffman, P. Alberch, J. Campbell, B. Goodwin, R. Lande, D. Raup, and L. Wolpert. 1985. Developmental constraints and evolution. *Q. Rev. Biol.* 60:265–87. (3, 12)

Maynard Smith, J., and G. A. Parker. 1976. The logic of asymmetric contests. *Anim. Behav.* 24:159–75. (5)

Maynard Smith, J., and S. E. Riechert. 1984. A conflicting-tendency model of spider agonistic behaviour: Hybrid-pure population line comparisons. *Anim. Behav.* 32:564–78. (1)

Mayo, O., T. W. Hancock, and P. A. Baghurst. 1980. Influence of major genes on variance within sibships for a quantitative trait. *Ann. Hum. Genet.* 43:419–21. (10)

Mayr, E. 1963. *Animal Species and Evolution.* Cambridge, Mass.: Belknap Press of Harvard University Press. (13)

Mayr, E. 1974. Behavior programs and evolutionary strategies. *Am. Sci.* 62:650–59. (7)

Mayr, E. 1983. How to carry out the adaptationist program? *Am. Nat.* 121:324–34. (3)

Medawar, P. B. 1979. *Advice to a Young Scientist.* New York: Basic Books. (14)

Mertz, D. B., and D. A. Cawthon. 1973. Sex differences in the cannibalistic roles of adult flour beetles. *Ecology* 54:1400–1402. (10)

Mertz, D. B., T. Park, and W. J. Youden. 1965. Mortality patterns in eight strains of flour beetle. *Biometrics* 21:99–114. (10)

Messina, F. J. 1987. Genetic contribution to the dispersal polymorphism of the cowpea weevil (Coleoptera: Bruchidae). *Ann. Entomol. Soc. Am.* 80:12–16. (7)

Mitchell, E. R. 1979. Migration by *Spodoptera exigua* and *S. frugiperda* North American style. In *Movement of Highly Mobile Insects: Concepts and Methodology in Research,* ed. R. L. Rabb and G. G. Kennedy, 386–93. Raleigh: North Carolina State University. (7)

Mitchell-Olds, T. 1986. Quantitative genetics of survival and growth in *Impatiens capensis. Evolution* 40:107–16. (14)

Mitchell-Olds, T., and J. Bergelson. 1990. Statistical genetics of an annual plant, *Impatiens capensis.* I. Genetic basis of quantitative inheritance. *Genetics* 124:407–16. (10)

Mitchell-Olds, T., and J. J. Rutledge. 1986. Quantitative genetics in natural plant populations: A review of the theory. *Am. Nat.* 127:379–402. (11, 10)

Mitchell-Olds, T., and R. G. Shaw. 1987. Regression analysis of natural selection: Statistical inference and biological interpretation. *Evolution* 41:1149–61. (5)
Møller, A. P. 1988. Female choice selects for male sexual tail ornaments in the monogamous swallow. *Nature* 332:640–42. (6)
Møller, A. P. 1989. Viability costs of male tail ornaments in a swallow. *Nature* 339:32–35. (6)
Moore, A. J . 1990a. The evolution of sexual dimorphism by sexual selection: The separate effects of intrasexual and intersexual selection. *Evolution* 44:315–34. (4)
Moore, A. J. 1990b. The inheritance of social dominance, mating behaviour and attractiveness to mates in male *Nauphoeta cinerea*. *Anim. Behav.* 39:388–97. (11)
Moore, F. R., and R. E. Gatten, Jr. 1989. Locomotor performance of hydrated, dehydrated, and osmotically stressed anuran amphibians. *Herpetologica* 45:101–10. (12)
Morris, R. F. 1971. Observed and simulated changes in genetic quality in natural populations of *Hyphantria cunea*. *Can. Entomol.* 102:893–906. (3)
Morton, N. E. 1982. *Outline of Genetic Epidemiology.* New York: Karger. (10)
Mosimann, J. E. 1975. Statistical problems of size and shape. I. Biological applications and basic theorems. In *Statistical Distributions in Scientific Work*, ed. G. P. Patil, S. Kotz, and K. Ord, 187–217. Dordrecht: D. Reidel. (8)
Mosimann, J. E., and F. C. James. 1979. New statistical methods for allometry with application to Florida red-winged blackbirds. *Evolution* 33:444–59. (8)
Mousseau, T. A. 1991. Geographic variation in maternal-age effects on diapause in a cricket. *Evolution* 45:1053–59. (4)
Mousseau, T. A., and H. Dingle. 1990. Maternal effects in insects: Examples, constraints, and geographic variation. In *The Unity of Evolutionary Biology: Proceedings of the Fourth International Congress of Systematic and Evolutionary Biology*, vol. II, ed. E. C. Dudley, 745–76. Portland, Oreg.: Dioscorides Press. (2)
Mousseau, T. A., and H. Dingle. 1991. Maternal effects in insect life histories. *Annu. Rev. Entomol.* 36:511–34. (4, 7)
Mousseau, T. A., and D. A. Roff. 1987. Natural selection and the heritability of fitness components. *Heredity* 59:181–97. (1, 2, 3, 4, 7, 10, 11, 12, 13)
Murdoch, W. W., and A. A. Sih. 1978. Age-dependent interference in a predatory insect. *J. Anim. Ecol.* 47:581–92. (3)
Nagai, J., E. J. Eisen, J.A.B. Emsley, and N. L. Furst. 1978. Selection for increased nursing ability and adult weight in mice. *Genetics* 88:761–80. (4)
Nee, S. 1989. Does Hamilton's rule describe the evolution of reciprocal altruism? *J. Theor. Biol.* 141:81–91. (4)
Newman, R. A. 1988. Adaptive plasticity in development of *Scaphiopus couchii* tadpoles in desert ponds. *Evolution* 42:744–83. (2)
Nichols, R. A., and R. K. Butlin. 1989. Does runaway sexual selection work in finite populations? *J. Evol. Biol.* 2:299–313. (5)
Nur, N., and O. Hasson. 1984. Phenotypic plasticity and the handicap principle. *J. Theor. Biol.* 110:275–97. (5, 8)
O'Donald, P. 1967. A general model of sexual and natural selection. *Heredity* 22:499–518. (5)
O'Donald, P. 1980. *Genetic Models of Sexual Selection.* Cambridge: Cambridge University Press. (5, 11)
Olsson, G. 1960. Some relations between number of seeds per pod, seed size and oil

content and the effects of selection for these characters in *Brassica* and *Sinapis. Hereditas* 46:29–70. (3)

O'Neill, K. M., and S. W. Skinner. 1990. Ovarian egg size in five species of parasitoid wasps. *J. Zool.* (Lond) 220:115–22. (3)

Otronen, M. 1984. Male contests for territories and females in the fly *Dryomyza anilis. Anim. Behav.* 32:891–98. (8)

Palmer, J. O., and H. Dingle. 1986. Direct and correlated responses to selection among life-history traits in milkweed bugs (*Oncopeltus fasciatus*). *Evolution* 40:767–77. (7)

Palmer, J. O., and H. Dingle. 1989. Responses to selection on flight behavior in a migratory population of milkweed bug (*Oncopeltus fasciatus*). *Evolution* 43:1805–8. (7)

Park, T., P. H. Leslie, and D. B. Mertz. 1964. Genetic strains and competition in populations of *Tribolium. Physiol. Zool.* 37:97–162. (10)

Park, T., D. B. Mertz, W. Grodzinski, and T. Prus. 1965. Cannibalistic predation in populations of flour beetles. *Physiol. Zool.* 38:289–321. (10)

Park, T., D. B. Mertz, and K. Petrusewicz. 1961. Genetic strains of *Tribolium:* Their primary characteristics. *Physiol. Zool.* 34:62–80. (10)

Parker, G. A., and M. Begon. 1986. Optimal egg size and clutch size: Effects of environmental and maternal phenotype. *Am. Nat.* 128:573–92. (3)

Parker, W. E., and A. G. Gatehouse. 1985a. The effect of larval rearing conditions on flight performance of the African armyworm *Spodoptera exempta* (Walker) (Lepidoptera: Noctuidae). *Bull. Entomol. Res.* 75:35–47. (7)

Parker, W. E., and A. G. Gatehouse. 1985b. Genetic factors in the regulation of migration in the African armyworm moth, *Spodoptera exempta* (Walker) (Lepidoptera: Noctuidae). *Bull. Entomol. Res.* 75:49–63. (7)

Parsons, P. A. 1973. *Behavioural and Ecological Genetics: A Study in* Drosophila. London: Oxford University Press. (1, 9)

Partridge, L. 1980. Mate choice increases a component of fitness in fruit flies. *Nature* 283:290–91. (9)

Partridge, L., and R. Andrews. 1985. The effect of reproductive activity on the longevity of male *Drosophila melanogaster* is not caused by an acceleration of ageing. *J. Insect Physiol.* 31:393–95. (6)

Partridge, L., and N. H. Barton. 1993. Optimality, mutation and the evolution of ageing. *Nature* 362:305–11. (6)

Partridge, L., and J. A. Endler. 1987. Life history constraints on sexual selection. In *Sexual Selection: Testing the Alternatives,* ed. J. W. Bradbury and M. B. Andersson, 265–277. Chichester: Wiley. (6)

Partridge, L., E. Ewing, and A. Chandler. 1987. Male size and mating success in *Drosophila melanogaster:* The role of male and female behaviour. *Anim. Behav.* 35:555–62. (6)

Partridge, L., and M. Farquhar. 1981. Sexual activity reduces lifespan of male fruitflies. *Nature* 294:580–82. (3, 6)

Partridge, L., and M. Farquhar. 1983. Lifetime mating success of male fruitflies (*Drosophila melanogaster*) is related to their size. *Anim. Behav.* 31:871–77. (6)

Partridge, L., and K. Fowler. 1993. Responses and correlated responses to selection on thorax length in *Drosophila melanogaster. Evolution* 47:213–26. (6)

Partridge, L., and P. H. Harvey. 1985. Costs of reproduction. *Nature* 316:20. (3)

Partridge, L., A. Hoffmann, and J. S. Jones. 1987. Male size and mating success in

Drosophila melanogaster and *D. pseudoobscura* under field conditions. *Anim. Behav.* 35:468–76. (6, 9)

Partridge, L., and R. Sibly. 1991. Constraints in the evolution of life histories. *Phil. Trans. R. Soc. Lond.* B. 332:3–13. (6)

Pearson, K. 1903. Mathematical contributions to the theory of evolution. XI. On the influence of natural selection on the variability and correlation of organs. *Phil. Trans. R. Soc. Lond.* A 200:1–66. (2)

Pearson, K. 1904. On a criterion which may serve to test various theories of inheritance. *Proc. R. Soc. Lond.* 73:262–80. (10)

Pease, C. M., and J. J. Bull. 1988. A critique of methods for measuring life history trade-offs. *J. Evol. Biol.* 1:293–303. (3, 6, 12)

Petersson, E. 1991. Effects of remating on the fecundity and fertility of female caddis flies, *Mystacides azurea. Anim. Behav.* 41:813–17. (6)

Phillips, P. C., and S. J. Arnold. 1989. Visualizing multivariate selection. *Evolution* 43:1209–22. (2)

Pitafi, K. D., R. Simpson, J. J. Stephen, and T. H. Day. 1990. Adult size and mate choice in seaweed flies (*Coelopa frigida*). *Heredity* 65:91–97. (6)

Pitnick, S. 1991. Male size influences mate fecundity and remating interval in *Drosophila melanogaster. Anim. Behav.* 41:735–45. (6)

Plomin, R., J. C. DeFries, and G. E. McClearn. 1980. *Behavioral Genetics, a Primer.* San Francisco: W. H. Freeman. (1)

Pomiankowski, A. 1987a. The costs of choice in sexual selection. *J. Theor. Biol.* 128:195–218. (11)

Pomiankowski, A. 1987b. Sexual selection: The handicap principle does work—sometimes. *Proc. R. Soc. Lond.* B. 231:123–45. (5)

Pomiankowski, A. 1988. The evolution of female mate preferences for male genetic quality. *Oxf. Surv. Evol. Biol.* 5:136–84. (2, 5)

Poramarcom, R., and C.R.B. Boake. 1991. Behavioural influences on male mating success in the Oriental fruit fly, *Dacus dorsalis* Hendel. *Anim. Behav.* 42:453–60. (9)

Pough, F. H. 1989. Organismal performance and Darwinian fitness: Approaches and interpretations. *Physiol. Zool.* 62:199–236. (12)

Price, T. D., and P. R. Grant. 1985. The evolution of ontogeny in Darwin's finches: A quantitative genetic approach. *Am. Nat.* 125:169–88. (4)

Price, T. D., M. Kirkpatrick, and S. J. Arnold. 1988. Directional selection and evolution of breeding date in birds. *Science* 240:789–99. (12)

Price, T. D., and D. Schluter. 1991. On the low heritability of life-history traits. *Evolution* 45:853–61. (1, 11, 12, 14)

Prout, T. 1968. Sufficient conditions for multiple niche polymorphism. *Am. Nat.* 102:493–96. (5)

Prout, T., and J.S.F. Barker. 1989. Ecological aspects of the heritability of body size in *Drosophila buzzatii. Genetics* 123:803–13. (9, 11, 14)

Provine, W. B. 1971. *The Origins of Theoretical Population Genetics.* Chicago: University of Chicago Press. (1)

Queller, D. C. 1989. Inclusive fitness in a nutshell. *Oxf. Surv. Evol. Biol.* 6:71–109. (4)

Queller, D. C. 1992. A general model for kin selection. *Evolution* 46:376–80. (4)

Rankin, M. A., ed. 1985. *Migration: Mechanisms and Adaptive Significance. Contrib. Mar. Sci.* 27 (Suppl.). (7)

Rankin, M. A. 1991. Endocrine effects on migration. *Am. Zool.* 31:217–30. (7)

Rankin, M. A., M. L. McAnelly, and J. E. Bodenhamer. 1986. The oogenesis-flight syndrome revisited. In *Insect Flight: Dispersal and Migration,* ed. W. Danthanarayana, 27–48. Berlin: Springer-Verlag. (7)

Rasmuson, B., M. Rasmuson, and J. Nygren. 1977. Genetically controlled differences in behavior between cycling and noncycling populations of field voles (*Microtus agrestis*). *Hereditas* 87:33–42. (7)

Real, L. 1990. Search theory and mate choice. I. Models of single-sex discrimination. *Am. Nat.* 136:376–405. (11)

Reynolds, J. D. N.d. Condition-dependent courtship: Theory and a test with Trinidadian guppies. *Am. Nat.* in press. (8)

Reznick, D. 1981. "Grandfather effects": the genetics of interpopulation differences in offspring size in the mosquito fish. *Evolution* 35:941–53. (4, 7)

Reznick, D. 1982. Genetic determination of offspring size in the guppy (*Poecilia reticulata*). *Am. Nat.* 120:181–88. (4)

Reznick, D. 1985. Costs of reproduction: An evaluation of the empirical evidence. *Oikos* 44:257–67. (3, 6)

Reznick, D. N. 1990. Maternal effects in fish life histories. In *The Unity of Evolutionary Biology: Proceedings of the Fourth International Congress of Systematic and Evolutionary Biology,* vol. II, ed. E. C. Dudley, 780–93. Portland, Oreg.: Dioscorides Press. (2)

Reznick, D. N., E. Perry, and J. Travis. 1986. Measuring the cost of reproduction: A comment on papers by Bell. *Evolution* 40:1338–44. (3)

Rich, E. R. 1956. Egg cannibalism and fecundity in *Tribolium. Ecology* 37:109–20. (10)

Ricker, J. P., and J. Hirsch. 1988. Genetic changes occurring over 500 generations in lines of *Drosophila melanogaster* selected divergently for geotaxis. *Behav. Genet.* 18:13–25. (1)

Ridley, M. 1978. Parental care. *Anim. Behav.* 26:904–32. (4)

Riechert, S. E. 1993. The evolution of behavioral phenotypes: Lessons learned from divergent spider populations. *Adv. Stud. Behav.* 22:103–34. (4)

Riechert, S. E., and J. Maynard Smith. 1989. Genetic analyses of two behavioural traits linked to individual fitness in the desert spider *Agelenopsis aperta. Anim. Behav.* 37:624–37. (1, 14)

Riska, B. 1989. Composite traits, selection response, and evolution. *Evolution* 43:1172–91. (12)

Riska, B., T. Prout, and M. Turelli. 1989. Laboratory estimates of heritabilities and genetic correlations in nature. *Genetics* 123:865–71. (6, 9, 11, 12, 14)

Riska, B. J., J. J. Rutledge, and W. R. Atchley. 1985. Covariance between direct and maternal genetic effects in mice, with a model of persistent environmental influences. *Genet. Res.* 45:287–97. (2, 4)

Roach, D. A., and R. D. Wulff. 1987. Maternal effects in plants. *Annu. Rev. Ecol. Syst.* 18:209–36. (4)

Roberts, R. C. 1967a. Some concepts and methods in quantitative genetics. In *Behavior-Genetic Analysis,* ed. J. Hirsch, 214–57. New York: McGraw-Hill. (13)

Roberts, R. C. 1967b. Some evolutionary implications of behavior. *Can. J. Genet. Cytol.* 9:419–35. (13)

Robertson, A. 1959a. Experimental design in the evaluation of genetic parameters. *Biometrics* 15:219–26. (2, 3, 14)

Robertson, A. 1959b. The sampling variance of the genetic correlation coefficient. *Biometrics* 15:469–85. (2, 14)
Robertson, A. 1966. A mathematical model of the culling process in dairy cattle. *Anim. Prod.* 8:93–108. (2)
Robertson, A. 1977. The effect of selection on the estimation of genetic parameters. *Zeitschrift für Tierzüechtung Züechtungsbiologie* 94:131–35. (3)
Robertson, F. W. 1960. The ecological genetics of growth in *Drosophila.* 1. Body size and development time on different diets. *Genet. Res.* 1:288–304. (6)
Robertson, F. W. 1963. The ecological genetics of growth in *Drosophila.* 6. The genetic correlation between the duration of the larval period and body size in relation to larval diet. *Genet. Res.* 4:74–92. (6)
Robertson, F. W., and E. Reeve. 1952. Studies in quantitative inheritance. 1. The effects of selection on wing and thorax length in *Drosophila melanogaster. J. Genet.* 50:414–48. (6)
Robison, O. W. 1972. The role of maternal effects in animal breeding. V. Maternal effects in swine. *J. Anim. Sci.* 35:1303–15. (4)
Roff, D. A. 1981. On being the right size. *Am. Nat.* 118:405–22. (3)
Roff, D. A. 1984. The evolution of life history parameters in teleosts. *Can. J. Fish. Aquat. Sci.* 41:984–1000. (3)
Roff, D. A. 1990a. The evolution of flightlessness in insects. *Ecol. Monogr.* 60:389–421. (7)
Roff, D. A. 1990b. Understanding the evolution of insect life cycles: The role of genetical analysis. In *Genetics, Evolution and Coordination of Insect Life Cycles,* ed. F. Gilbert, 5–27. New York: Springer-Verlag. (3)
Roff, D. A. 1991. Life history consequences of bioenergetic and biomechanical constraints on migration. *Am. Zool.* 31:205–15. (3)
Roff, D. A. 1992. *The Evolution of Life Histories.* New York: Chapman and Hall. (3)
Roff, D. A., and T. A. Mousseau. 1987. Quantitative genetics and fitness: Lessons from *Drosophila. Heredity* 58:103–18. (1, 3, 7, 10, 12)
Rohlf, F. J., and R. R. Sokal. 1969. *Statistical Tables.* San Francisco: W. H. Freeman. (2)
Rose, M. R. 1982. Antagonistic pleiotropy, dominance, and genetic variation. *Heredity* 48:63–78. (11, 14)
Rose, M. R., P. M. Service, and E. W. Hutchinson. 1987. Three approaches to trade-offs in life-history evolution. In *Genetic Constraints on Adaptive Evolution,* ed. V. Loeschcke, 91–105. Berlin: Springer-Verlag. (3, 4, 12)
Rubenstein, D. I. 1981. Population density, resource patterning, and territoriality in the Everglades pygmy sunfish. *Anim. Behav.* 29:155–72. (8)
Rubenstein, D. I., and R. W. Wrangham, eds. 1986. *Ecological Aspects of Social Evolution: Birds and Mammals.* Princeton, N.J.: Princeton University Press. (4)
Ruse, M. 1979. *The Darwinian Revolution: Science Red in Tooth and Claw.* Chicago: University of Chicago Press. (4)
Ruthven, A. G. 1908. *Variations and genetic relationships of the garter-snakes.* Smithsonian Institution U.S. National Museum Bulletin 61. (12)
Rutledge, J. J., O. W. Robison, E. J. Eisen, and J. E. Legates. 1972. Dynamics of genetic and maternal effects in mice. *J. Anim. Sci.* 35:911–18. (4)
Ryan, M. J. 1985. *The Tungara Frog: A Study in Sexual Selection and Communication.* Chicago: University of Chicago Press. (11)
Ryan, M. J., and B. J. Causey. 1989. "Alternative" mating behavior in the swordtails

Xiphophorus nigrensis and *Xiphophorus pygmaeus. Behav. Ecol. Sociobiol.* 24:341–48. (8)

Ryan, M. J., C. M. Pease, and M. R. Morris. 1992. A genetic polymorphism in the swordtail *Xiphophorus nigrensis:* Testing the prediction of equal fitnesses. *Am. Nat.* 139:21–31. (8)

Sakaluk, S. K. 1990. Sexual selection and predation: Balancing reproductive and survival needs. In *Insect Defenses: Adaptive Mechanisms and Strategies of Prey and Predators,* ed. D. L. Evans and J. O. Schmidt, 63–90. Albany: State University of New York Press. (11)

Sakaluk, S. K., and J. J. Belwood. 1984. Gecko phonotaxis to cricket calling song: A case of satellite predation. *Anim. Behav.* 32:659–62. (11)

Sakaluk, S. K., and R. L. Smith. 1988. Inheritance of male parental investment in an insect. *Am. Nat.* 132:594–601. (4)

Schaal, B. A. 1984. Life history variation, natural selection, and maternal effects in plant populations. In *Perspectives in Plant Population Ecology,* ed. R. Dirzo and J. Sarukhan, 188–206. Sunderland, Mass.: Sinauer Associates. (4)

Schaffer, W. M. 1979. Equivalence of maximizing reproductive value and fitness in the case of reproductive strategies. *Proc. Natl. Acad. Sci. USA* 76:3567–69. (3)

Schaffer, W. M. 1981. On reproductive value and fitness. *Ecology* 62:1683–85. (3)

Scharloo, W. 1987. Constraints on selection response. In *Genetic Constraints on Adaptive Evolution,* ed. V. Loeschke, 125–49. Berlin: Springer. (7)

Schiefflen, C. D., and A. de Queiroz. 1991. Temperature and defense in the common garter snake: Warm snakes are more aggressive than cold snakes. *Herpetologica* 47:230–37. (12)

Schluter, D. 1988. Estimating the form of natural selection on a quantitative trait. *Evolution* 42:849–61. (5)

Schmalhausen, I. I. 1949. *Factors of Evolution.* Philadelphia: Blakeston. (13)

Schnee, F. B., and J. N. Thompson. 1984. Conditional polygenic effects in the sternopleural bristle system of *Drosophila melanogaster. Genetics* 108:409–24. (9)

Scholander, P. F., R. Hock, V. Walters, and L. Irving. 1950. Adaptation to cold in arctic and tropical mammals and birds in relation to body temperature, insulation, and basal metabolic rate. *Biol. Bull.* 99:259–71. (13)

Schwartz, J. M. 1989. Multiple paternity and offspring variability in wild populations of the garter snake *Thamnophis sirtalis* (Colubridae). Ph.D. diss., University of Tennessee. (12)

Schwartz, J. M., and C.R.B. Boake. 1992. Sexual dimorphism in remating in Hawaiian *Drosophila* species. *Anim. Behav.* 44:231–38. (6)

Schwartz, J. M., and H. A. Herzog, Jr. 1993. Estimates of heritability of antipredator behavior in three garter snake species (*Thamnophis butleri, T. melanogaster,* and *T. sirtalis*) in nature. *Behav. Genet.* In press. (12)

Schwartz, J. M., G. F. McCracken, and G. M. Burghardt. 1989. Multiple paternity in wild populations of the garter snake, *Thamnophis sirtalis. Behav. Ecol. Sociobiol.* 25:269–73. (12, 14)

Scott, J. P., and J. L. Fuller. 1965. *Genetics and the Social Behavior of the Dog.* Chicago: University of Chicago Press. (1)

Searle, S. R. 1982. *Matrix Algebra Useful for Statistics.* New York: Wiley. (2)

Seger, J. 1985. Unifying genetic models for the evolution of female choice. *Evolution* 39:1185–93. (5)

Seger, J., and H. J. Brockmann. 1987. What is bet-hedging? *Oxf. Surv. Evol. Biol.* 4:182–211. (3)

Seger, J., and R. Trivers. 1986. Asymmetry in the evolution of female mating preferences. *Nature* 319:771–73. (5)
Seigel, R. A., M. M. Huggins, and N. B. Ford. 1987. Reduction in locomotor ability as a cost of reproduction in gravid snakes. *Oecologia* 73:481–85. (12)
Shaffer, H. B., G. C. Austin, and R. B. Huey. 1991. The consequences of metamorphosis on salamander (*Ambystoma*) locomotor performance. *Physiol. Zool.* 64:212–31. (12)
Sharpe, R. S., and P. A. Johnsgard. 1966. Inheritance of behavioral characters in F_2 mallard and pintail (*Anas platyrhynchos* L. × *Anas acuta* L.) hybrids. *Behaviour* 27:259–72. (1)
Shaw, R. G. 1987. Maximum-likelihood approaches applied to quantitative genetics of natural populations. *Evolution* 41:812–26. (2, 4, 12, 14)
Shaw, R. G. 1991. The comparison of quantitative genetic parameters between populations. *Evolution* 45:143–51. (2)
Shaw, R. G. 1992. Comparison of quantitative genetic parameters: Reply to Cowley and Atchley. *Evolution* 46:1967–69. (2)
Shelly, T. E. 1988. Lek behavior in *Drosophila cnecopleura* in Hawaii. *Ecol. Entomol.* 13:51–55. (9)
Sheridan, A. K., and J.S.F. Barker. 1974. Two-trait selection and the genetic correlation. II. Changes in the genetic correlation during two-trait selection. *Aust. J. Biol. Sci.* 27:89–101. (3)
Sherman, P. W., and D. F. Westneat. 1988. Multiple mating and quantitative genetics. *Anim. Behav.* 36:1545–47. (6)
Sih, A. 1980. Optimal behavior: Can foragers balance two conflicting demands. *Science* 210:1041–43. (3)
Sih, A. 1982. Foraging strategies and the avoidance of predation by an aquatic insect *Notonecta hoffmanni*. *Ecology* 63:786–96. (3)
Simmons, L. W. 1986. Female choice in the field cricket, *Gryllus bimaculatus* (de Geer). *Anim. Behav.* 34:1463–70. (11)
Simmons, L. W. 1990. Post-copulatory guarding, female choice and the levels of gregarine infections in the field cricket, *Gryllus bimaculatus*. *Behav. Ecol. Sociobiol.* 26:403–07. (11)
Simms, E. L., and D. S. Burdick. 1988. Profile analysis of variance as a tool for analyzing correlated responses in experimental ecology. *Biometrical J.* 1:1–14. (2)
Simms, E. L., and M. D. Rausher. 1989. The evolution of resistance to herbivory in *Ipomoea purpurea*. II. Natural selection by insects and costs of resistance. *Evolution* 43:573–85. (2)
Simonson, E., and P. C. Weiser, eds. 1976. *Psychological Aspects and Physiological Correlates of Work and Fatigue.* Springfield, Ill.: Charles C. Thomas. (12)
Skrzipek, K. H., B. Kroner, and H. Hager. 1979. Aggression bei *Drosophila melanogaster*-Laboruntersuchungen. *Z. Tierpsychol.* 49:87–103. (9)
Slatkin, M. 1974. Hedging one's evolutionary bets. *Nature* 250:704–5. (3)
Slatkin, M. 1984. Ecological causes of sexual dimorphism. *Evolution* 38:622–30. (6)
Smith, J.N.M., and A. A. Dhondt. 1980. Experimental confirmation of heritable morphological variation in a natural population of song sparrows. *Evolution* 34:1155–58. (14)
Smith, R. H. 1991. Genetic and phenotypic aspects of life-history evolution in animals. *Adv. Ecol. Res.* 21:63–120. (4)
Snelson, F. F., Jr. 1982. Indeterminate growth in males of the sailfin molly, *Poecilia latipinna*. *Copeia* 1982:296–304. (8)

Snelson, F. F., Jr. 1984. Seasonal maturation and growth of males in a natural population of *Poecilia latipinna. Copeia* 1984:252–55. (8)

Snelson, F. F., Jr. 1985. Size and morphological variation in males of the sailfin molly, *Poecilia latipinna. Environ. Biol. Fish.* 13:35–47. (8)

Snyder, R. 1988. Migration and evolution of life histories of the threespine stickleback (*Gasterosteus aculeatus* L.). Ph.D. diss., University of California, Davis. (7)

Snyder, R., and H. Dingle. 1989. Adaptive, genetically based differences in life history between estuary and freshwater threespine sticklebacks (*Gasterosteus aculeatus* L.). *Can. J. Zool.* 67:2448–54. (7)

Sokal, R. R., and F. J. Rohlf. 1981. *Biometry, the Principles and Practice of Statistics in Biological Research.* 2d ed. San Francisco: W. H. Freeman. (1, 2, 11, 14)

Sokolowski, M. G., S. J. Bauer, V. Wai-Ping, L. Rodriguez, J. Wong, and C. Kent. 1986. Ecological genetics and behaviour of *Drosophila melanogaster* larvae in nature. *Anim. Behav.* 32:403–8. (9)

Southwood, O. I., and B. W. Kennedy. 1990. Estimation of direct and maternal genetic variance for litter size in Canadian Yorkshire and Landrace swine using an animal model. *J. Anim. Sci.* 68:1841–47. (4)

Southwood, O. I., B. W. Kennedy, K. Meyer, and J. P. Gibson. 1989. Estimation of additive maternal and cytoplasmic genetic variance under animal models. *J. Dairy Sci.* 72:3006–12. (4)

Spieth, H. T. 1968. The evolutionary implications of sexual behavior in *Drosophila. Evol. Biol.* 2:157–91. (9)

Sprent, D. 1972. The mathematics of size and shape. *Biometrics* 28:23–37. (8)

Stamenkovic-Radak, M., L. Partridge, and M. Andjelkovic. 1992. A genetic correlation between the sexes for mating speed in *Drosophila melanogaster. Anim. Behav.* 43:389–96. (6)

Stamenkovic-Radak, M., L. Partridge, and M. Andjelkovic. 1993. Genetic correlation between the sexes in *Drosophila melanogaster:* a reply to Butlin. *Anim. Behav.* 45:405. (6)

Stamps, J. A. 1991. Why evolutionary issues are reviving interest in proximate behavioral mechanisms. *Am. Zool.* 31:338–48. (4)

Stearns, S. C. 1989. The evolutionary significance of phenotypic plasticity. *Bioscience* 59:436–45. (7)

Stearns, S. C. 1992. *The Evolution of Life Histories.* Oxford: Oxford University Press. (14)

Stearns, S. C., and R. E. Crandall. 1981. Quantitative predictions of delayed maturity. *Evolution* 35:455–63. (3)

Stearns, S. C., and J. C. Koella. 1986. The evolution of phenotypic plasticity in life-history traits: Predictions of reaction norms for age and size at maturity. *Evolution* 40:893–913. (3)

Stebbins, R. C. 1985. *A Field Guide to Western Reptiles and Amphibians.* 2d ed. Boston: Houghton Mifflin. (12)

Stevens, L. 1986. The genetics and evolution of cannibalism in flour beetles, genus *Tribolium.* Ph.D. diss., University of Illinois at Chicago. (10)

Stevens, L. 1989. The genetics and evolution of cannibalism in flour beetles, genus *Tribolium. Evolution* 43:169–79. (10)

Stevens, L. 1992. Cannibalism in beetles. In *Cannibalism: Ecology and evolution among diverse taxa,* ed. M. A. Elgar and B. J. Crespi, 157–76. Oxford: Oxford University Press. (10)

Stevens, L., and D. B. Mertz. 1985. Genetic stability of cannibalism in *Tribolium confusum*. *Behav. Genet.* 15:549–59. (10)

Stratton, G. E., and G. W. Uetz. 1986. The inheritance of courtship behavior and its role as a reproductive isolating mechanism in two species of *Schizocosa* wolf spiders (Araneae: Lycosidae). *Evolution* 40:129–41. (1)

Strauss, R. E. 1985. Evolutionary allometry and variation in body form in the South American catfish genus *Corydoras* (Callichthyidae). *Syst. Zool.* 34:381–96. (8)

Streams, F. A., and T. P. Shubeck. 1982. Spatial structure and intraspecific interactions in *Notonecta* populations. *Environ. Entomol.* 11:652–59. (3)

Strong, D. 1972. Life history variation among the populations of an amphipod. *Ecology* 53:1103–11. (3)

Sumner, I. T., J. Travis, and C. D. Wilson. N.d. Methods of female fertility advertisement and variation among males in responsiveness in the sailfin molly (*Poecilia latipinna*). *Copeia*, in press. (8)

Taylor, C. E., and V. Kekic. 1988. Sexual selection in a natural population of *Drosophila melanogaster*. *Evolution* 42:197–99. (6, 9)

Taylor, C. E., A. D. Pereda, and J. A. Ferrari. 1987. On the correlation between mating success and offspring quality in *Drosophila melanogaster*. *Am. Nat.* 129:721–29. (9)

Taylor, E. B. 1988. Adaptive variation in rheotactic and agonistic behavior in newly emerged fry of chinook salmon, *Oncorhynchus tshawytscha*, from ocean- and stream-type populations. *Can. J. Fish. Aquat. Sci.* 45:237–43. (9)

Taylor, H. M., R. S. Gourby, C. E. Lawrence, and R. S. Kaplan. 1974. Natural selection of life history attributes: An analytical approach. *Theor. Popul Biol.* 5:104–22. (3)

Taylor, P. T., and G. C. Williams. 1981. On the modelling of sexual selection. *Q. Rev. Biol.* 56:305–13. (5)

Thoday, J. M., and J.J.N. Thompson. 1976. The number of segregating genes implied by continuous variation. *Genetica* 46:335–44. (3)

Thompson, R. 1976. The estimation of maternal genetic variances. *Biometrics* 32:903–17. (4)

Thompson, R. 1977. Estimation of genetic parameters. In *Proceedings of the International Conference on Quantitative Genetics*, ed. E. Pollak, O. Kempthorne, and T. B. Bailey, Jr., 639–57. Ames, Iowa: Iowa State University Press. (2)

Thornhill, R. 1976. Mate choice in *Hylobittacus apicalis* (Insecta: Mecoptera) and its relation to some models of mate choice. *Evolution* 34:519–38. (4)

Thornhill, R. 1981. *Panorpa* (Mecoptera: Panorpidae) scorpionflies: Systems for understanding resource-defense polygyny and alternative male reproductive efforts. *Annu. Rev. Ecol. Syst.* 12:355–86. (8)

Thornhill, R. 1983. Cryptic female choice and its implication in the scorpionfly *Harpobittacus nigriceps*. *Am. Nat.* 122:765–88. (4)

Thornhill, R. 1986. Relative parental contribution of the sexes to their offspring and the operation of sexual selection. In *Evolution of Animal Behavior: Paleontological and Field Approaches*, ed. M. H. Nitecki and J. A. Kitchell, 113–36. Oxford: Oxford University Press. (4)

Thornhill, R., and J. Alcock. 1983. *The Evolution of Insect Mating Systems*. Cambridge, Mass.: Harvard University Press. (4, 9)

Thornhill, R., and D. T. Gwynne. 1986. The evolution of sexual differences in insects. *Am. Sci.* 74:382–89. (4)

Thornhill, R., and P. Sauer. 1992. Genetic sire effects on the fighting ability of sons and daughters and mating success of sons in a scorpionfly. *Anim. Behav.* 43:255–64. (9)

Tolman, E. C. 1924. The inheritance of maze-learning ability in rats. *J. Comp. Psychol.* 4:1–18. (1)

Travis, J. 1989a. Ecological genetics of life-history traits in poeciliid fishes. In *Ecology and Evolution of Livebearing Fishes (Poeciliidae)*, ed. G. K. Meffe and F. F. Snelson, Jr., 185–200. New York: Prentice-Hall. (8)

Travis, J. 1989b. The role of optimizing selection in natural populations. *Annu. Rev. Ecol. Syst.* 20:279–96. (12)

Travis, J., J. A. Farr, and J. C. Trexler. N.d. Body-size variation in the sailfin molly, *Poecilia latipinna* (Pisces: Poeciliidae): II. Genetics of male behaviors and secondary sex characters. *J. Evol. Biol.*, in press. (8)

Travis, J., and J. C. Trexler. 1987. Regional variation in habitat requirements of the sailfin molly, with special reference to the Florida Keys. Florida Game and Fresh-Water Fish Commission, Tallahassee. (8)

Travis, J., J. C. Trexler, and J. A. Farr. N.d. Body-size variation in the sailfin molly, *Poecilia latipinna* (Pisces: Poeciliidae): I. Sex-limited genetic variation for size and age of maturation. *J. Evol. Biol.*, in press. (8)

Travis, J., and B. D. Woodward. 1989. Social context and courtship flexibility in male sailfin mollies, *Poecilia latipinna* (Pisces: Poeciliidae). *Anim. Behav.* 38:1001–11. (8)

Trexler, J. C., and J. Travis. 1990. Phenotypic plasticity in the sailfin molly, *Poecilia latipinna* (Pisces: Poeciliidae). I. Field experiments. *Evolution* 44:143–56. (2)

Trexler, J. C., J. Travis, and M. Trexler. 1990. Phenotypic plasticity in the sailfin molly, *Poecilia latipinna* (Pisces: Poeciliidae). II. Laboratory experiment. *Evolution* 44:157–67. (8)

Trivers, R. L. 1972. Parental investment and sexual selection. In *Sexual Selection and the Descent of Man, 1871–1971*, ed. B. Campbell, 136–79. Chicago: Aldine. (4, 11)

Trivers, R. L. 1974. Parent-offspring conflict. *Am. Zool.* 14:249–64. (4)

Tryon, R. C. 1934. Individual differences. In *Comparative Psychology*, ed. F. A. Moss, 409–45. Englewood Cliffs, N.J.: Prentice-Hall. (1)

Tsuji, J. S., R. B. Huey, F. H. van Berkum, T. Garland, Jr., and R. G. Shaw. 1989. Locomotor performance of hatchling fence lizards (*Sceloporus occidentalis*): quantitative genetics and morphometric correlates. *Evol. Ecol.* 3:240–52. (12)

Turelli, M. 1984. Heritable genetic variation via mutation-selection balance: Lerch's zeta meets the abdominal bristle. *Theor. Popul. Biol.* 25:138–93. (2, 5, 11)

Turelli, M. 1985. Effects of pleiotropy on predictions concerning mutation-selection balance for polygenic traits. *Genetics* 111:165–95. (2)

Turelli, M. 1986. Gaussian vs. non-Gaussian genetic analyses of polygenic mutation-selection balance. In *Evolutionary Processes and Theory*, ed. S. Karlin and E. Nevo, 607–28. New York: Academic Press. (2)

Turelli, M. 1988a. Phenotypic evolution, constant covariances, and the maintenance of additive variance. *Evolution* 42:1342–47. (3, 12)

Turelli, M. 1988b. Population genetic models for polygenic variation and evolution. In *Proceedings of the Second International Conference on Quantitative Genetics*, ed. B. Weir, E. Eisen, M. Goodman, and G. Namkoong, 601–18. Sunderland, Mass.: Sinauer Associates. (2)

Urarov, B. P. 1921. A revision of the genus *Locusta* L. (*Pachytelus* Fab.) with a new theory as to the periodicity and migration of locusts. *Bull. Entomol. Res.* 12:135–63. (7)

van Abeelen, J.H.F., ed. 1974. *The Genetics of Behaviour.* New York: American Elsevier. (1)

van Berkum, F. H., R. B. Huey, J. S. Tsuji, and T. Garland, Jr. 1989. Repeatability of individual differences in locomotor performance and body size during early ontogeny of the lizard *Sceloporus occidentalis* (Baird & Girard). *Funct. Ecol.* 3:97–105. (12)

van Noordwijk, A. J. 1984. Quantitative genetics in natural populations of birds illustrated with examples from the great tit, *Parus major.* In *Population Biology and Evolution,* ed. K. Wöhman and V. Loeschke, 67–79. Heidelberg: Springer-Verlag. (14)

van Noordwijk, A. J. 1990. The methods of genetical ecology applied to the study of evolutionary change. In *Population Biology: Ecological and Evolutionary Viewpoints,* ed. K. Wohrmann and S. Jain, 291–319. Berlin: Springer. (7)

van Noordwijk, A. J., and G. de Jong. 1986. Acquisition and allocation of resources: Their influence on variation in life history tactics. *Am. Nat.* 128:137–42. (3)

van Noordwijk, A. J., and M. Gebhardt. 1987. Reflections on the genetics of quantitative traits with continuous environmental variation. In *Genetic Constraints on Adaptive Evolution,* ed. V. Loeschke, 73–90. Berlin: Springer-Verlag. (7)

van Oortmerssen, G. A., and T.C.M. Bakker. 1981. Artificial selection for short and long attack latencies in wild *Mus musculus domesticus. Behav. Genet.* 11:115–26. (9)

Van Vleck, L. D., and C. R. Henderson. 1961. Empirical sampling estimates of genetic correlations. *Biometrics* 17:359–71. (3)

Vesely, J. A., and O. W. Robison. 1971. Growth and preweaning growth and type score in beef calves. *J. Anim. Sci.* 32:825–31. (4)

Via, S. 1984. The quantitative genetics of polyphagy in an insect herbivore. I. Genetic correlations in larval performance within and among host plants. *Evolution* 38:896–905. (14)

Via, S., and R. Lande. 1985. Genotype-environment interaction and the evolution of phenotypic plasticity. *Evolution* 39:505–22. (2, 3, 7, 13, 14)

Wade, M. J. 1984. Changes in group selected traits that occur when group selection is relaxed. *Evolution* 38:1039–46. (10)

Wade, M. J. 1985. Soft selection, hard selection, kin selection, and group selection. *Am. Nat.* 125:61–73. (4)

Wagner, G. 1988. The influence of variation and of developmental constraints on the rate of multivariate phenotypic evolution. *J. Evol. Biol.* 1:45–66. (4, 8)

Walker, T. J. 1957. Specificity in the response of female tree crickets to calling songs of males. *Ann. Entomol. Soc. Am.* 50:626–36. (11)

Wallace, B. 1991. The manly art of self-defense: On the neutrality of fitness components. *Q. Rev. Biol.* 66:455–65. (12, 14)

Walton, B. M. 1988. Relationships among metabolic, locomotor, and field measures of organismal performance in the Fowler's toad (*Bufo woodhousei fowleri*). *Physiol. Zool.* 61:107–18. (12)

Walton, M., B. C. Jayne, and A. F. Bennett. 1990. The energetic cost of limbless locomotion. *Science* 249:524–27. (12)

Warner, R. R., D. R. Robertson, and E. G. Leigh. 1975. Sex change and sexual selection. *Science* 190:633–38. (8)

Weatherhead, P. J. 1990. Secondary sexual traits, parasites, and polygyny in red-winged blackbirds, *Agelaius phoeniceus. Behav. Ecol.* 1:125–30. (6)

Weatherhead, P. J., and R. H. Robertson. 1979. Offspring quality and the polygyny threshold: the "sexy son" hypothesis. *Am. Nat.* 113:201–8. (5)

Weinberg, W. 1908. Über den Nachweis der Vererbung beim Menschen. *Jh. Ver. vaterl. Naturk. Wurttemb.* 4:369–82. (2)

Weisberg, S. 1980. *Applied Linear Regression.* New York: Wiley. (11)

Weissburg, M. 1986. Risky business: On the ecological relevance of risk-sensitive foraging. *Oikos* 46:261–62. (3)

Weller, J. I., M. Soller, and T. Brody. 1988. Linkage analysis of quantitative traits in an interspecific cross of tomato (*Lycopersicon esculentum* × *Lycopersicon pimpinellifolium*) by means of genetic markers. *Genetics* 118:329–39. (10)

Werner, E. E., J. F. Gilliam, D. J. Hall, and G. G. Mittelbach. 1983. An experimental test of the effects of predation on habitat use in fish. *Ecology* 64:1540–48. (3)

Westneat, D. F., P. W. Sherman, and M. L. Morton. 1990. The ecology and evolution of extra-pair copulations in birds. In *Current Ornithology,* vol. 7, ed. D. M. Power, 331–69. New York: Plenum Press. (14)

Wilcox, R. S., and T. Ruckdeschel 1982. Food threshold territoriality in a water strider (*Gerris remiges*). *Behav. Ecol. Sociobiol.* 11:85–90. (8)

Wilkinson, G. S. 1987. Equilibrium analysis of sexual selection in *Drosophila melanogaster. Evolution* 41:11–21. (6, 9)

Wilkinson, J., K. Fowler, and L. Partridge. 1990. Resistance of genetic correlation structure to directional selection in *Drosophila melanogaster. Evolution* 44:1990–2003. (6, 2)

Willham, R. L. 1963. The covariance between relatives for characters composed of components contributed by related individuals. *Biometrics* 19:8–27. (2, 4)

Willham, R. L. 1972. The role of maternal effects in animal breeding. III. Biometrical aspects of maternal effects in animals. *J. Anim. Sci.* 35:1288–93. (2, 4)

Williams, B. K. 1983. Some observations on the use of discriminant analysis in ecology. *Ecology* 64:1283–91. (7)

Williams, G. C. 1966. *Adaptation and Natural Selection.* Princeton, N.J.: Princeton University Press. (4)

Willis, J. H., J. A. Coyne, and M. Kirkpatrick. 1991. Can one predict the evolution of quantitative characters without genetics? *Evolution* 45:441–44. (3, 6, 12, 14)

Wilson, D. S., and A. V. Hedrick. 1982. Speciation and the economics of mate choice. *Evol. Theory* 6:15–24. (11)

Wilson, D. S., M. Leighton, and D. R. Leighton. 1978. Interference competition in a tropical ripple bug (Hemiptera: Veliidae). *Biotropica* 10:302–6. (3)

Wilson, E. O. 1971. *The Insect Societies.* Cambridge, Mass.: Belknap Press of Harvard University Press. (4)

Wilson, E. O. 1975. *Sociobiology.* Cambridge, Mass.: Belknap Press of Harvard University Press. (4)

Winemiller, K. O., M. Leslie, and R. Roche. 1990. Phenotypic variation in male guppies from natural inland populations: An additional test of Haskins' sexual selection/predation hypothesis. *Environ. Biol. Fish.* 29:179–91. (8)

Winer, B. J. 1971. *Statistical Principles in Experimental Design.* New York: McGraw-Hill. (2)

Wingfield, J. C., R. E. Hegner, A. M. Dufty, Jr., and G. F. Ball. 1990. The "challenge hypothesis": Theoretical implications for patterns of testosterone secretion, mating systems, and breeding strategies. *Am. Nat.* 136:829–46. (12)

Woolfenden, G. E., and J. W. Fitzpatrick. 1984. *The Florida Scrub Jay: Demography of a Cooperatively Breeding Bird.* Princeton, N.J.: Princeton University Press. (7)

Wrangham, R. W., and D. I. Rubenstein. 1986. Social evolution in birds and mammals. In *Ecological Aspects of Social Evolution: Birds and Mammals,* ed. D. I. Rubenstein and R. W. Wrangham, 452–70. Princeton, N.J.: Princeton University Press. (4)

Wright, S. 1921. Systems of mating. *Genetics* 6:111–78. (2)

Wright, S. 1929. Fisher's theory of dominance. *Am. Nat.* 63:274–79. (13)

Wright, S. 1930. Review of *The Genetical Theory of Natural Selection. J. Hered.* 21:349–56. (14)

Wright, S. 1952. The genetics of quantitative variability. In *Quantitative Inheritance,* ed. E.C.R. Reeve and C. H. Waddington, 5–41. London: Her Majesty's Stationery Office. (10)

Wright, S. 1960. Physiological genetics, ecology of populations and natural selection. In *Evolution After Darwin,* ed. S. Tax, 429–75. Chicago: University of Chicago Press. (10)

Wright, S. 1968. *Evolution and the Genetics of Populations.* Vol. 1, *Genetic and Biometric Foundations.* Chicago: University of Chicago Press. (4, 5, 10, 14)

Wright, S. 1978. *Evolution and the Genetics of Populations.* Vol. 4, *Variability Within and Among Natural Populations.* Chicago: University of Chicago Press. (10)

Wright, S. 1978. *Evolution and the Genetics of Populations.* Vol.4, *Variability Within and Among Natural Populations.* Chicago: University of Chicago Press. (10)

Wright, S. 1982. Character change, speciation, and the higher taxa. *Evolution* 36:427–43. (4)

Wright, S. 1988. Surfaces of selective value revisited. *Am. Nat.* 131:115–23. (4)

Yokoyama, S., and J. Felsenstein. 1978. A model of kin selection for an altruistic trait considered as a quantitative character. *Proc. Natl. Acad. Sci. USA.* 75:420–22. (4)

Young, C. W., and J. E. Legates. 1965. Genetic, phenotypic, and maternal interrelationships of growth in mice. *Genetics* 52:563–76. (4)

Young, C. W., J. E. Legates, and B. R. Farthing. 1965. Prenatal and postnatal influences on growth, prolificity, and maternal performance in mice. *Genetics* 52:553–61. (4)

Zahavi, A. 1975. Mate selection—a selection for a handicap. *J. Theor. Biol.* 53:205–14. (5)

Zeng, Z.-B. 1992. Correcting the bias of Wright's estimates of the number of genes affecting a quantitative character: A further improved method. *Genetics* 131:987–1001. (10)

Zeng, Z.-B., D. Houle, and C. C. Cockerham. 1990. How informative is Wright's estimator of the number of genes affecting a quantitative character? *Genetics* 126:235–47. (10)

Zimmerer, E. J., and K. D. Kallman. 1989. Genetic basis for alternative reproductive tactics in the pygmy swordtail, *Xiphophorus nigrensis. Evolution* 43:1298–1307. (8)

Contributors

Stevan J. Arnold, Department of Ecology and Evolution, University of Chicago, 940 E. 57th Street, Chicago, Illinois 60637

Christine R. B. Boake, Department of Zoology, University of Tennessee, Knoxville, Tennessee 37996-0810

James M. Cheverud, Department of Anatomy and Neurobiology, Washington University School of Medicine, 660 S. Euclid Avenue, St. Louis, Missouri 63110

Hugh Dingle, Department of Entomology, University of California, Davis, California 95616

Theodore Garland, Jr., Department of Zoology, University of Wisconsin, Madison, Wisconsin 53706

Ann V. Hedrick, Department of Biology, Reed College, Portland, Oregon 97202-8199

I. Lorraine Heisler, Department of Biology, University of Oregon, Eugene, Oregon 97403-1210

Ary A. Hoffmann, Department of Genetics and Human Variation, La Trobe University, Bundoora, Victoria 3083, Australia

Carol B. Lynch, Department of Environmental, Population, and Organismic Biology, University of Colorado, Boulder, Colorado 80309-0334

Allen J. Moore, Department of Entomology, University of Kentucky, Lexington, Kentucky 40546-0091

Linda Partridge, Institute of Cell, Animal, and Population Biology, University of Edinburgh, West Mains Rd., Edinburgh EH9 3JT, United Kingdom

Derek A. Roff, Department of Biology, McGill University, 1205 Dr. Penfield Avenue, Montreal, Quebec H3A 1B1, Canada

Lori Stevens, Department of Zoology, University of Vermont, Burlington, Vermont 05405-0086

Joseph Travis, Department of Biological Science, Florida State University, Tallahassee, Florida 32306-2043

Index